ADVANCED SAFETY MANAGEMENT

ADVANCED SAFETY MANAGEMENT

Focusing on Z10.0, 45001, and Serious Injury Prevention

Third Edition

FRED A. MANUELE, CSP, PE
President
Hazards Limited

This edition first published 2020
© 2020 John Wiley & Sons, Inc.

Edition History
"John Wiley & Sons Inc. (1e, 2008)".
"John Wiley & Sons Inc. (2e, 2014)".

All rights reserved. No part of this publication may be reproduced, stored in a retrieval system, or transmitted, in any form or by any means, electronic, mechanical, photocopying, recording or otherwise, except as permitted by law. Advice on how to obtain permission to reuse material from this title is available at http://www.wiley.com/go/permissions.

The right of Fred A. Manuele to be identified as the author of this work has been asserted in accordance with law.

Registered Office
John Wiley & Sons, Inc., 111 River Street, Hoboken, NJ 07030, USA

Editorial Office
111 River Street, Hoboken, NJ 07030, USA

For details of our global editorial offices, customer services, and more information about Wiley products visit us at www.wiley.com.

Wiley also publishes its books in a variety of electronic formats and by print-on-demand. Some content that appears in standard print versions of this book may not be available in other formats.

Limit of Liability/Disclaimer of Warranty
In view of ongoing research, equipment modifications, changes in governmental regulations, and the constant flow of information relating to the use of experimental reagents, equipment, and devices, the reader is urged to review and evaluate the information provided in the package insert or instructions for each chemical, piece of equipment, reagent, or device for, among other things, any changes in the instructions or indication of usage and for added warnings and precautions. While the publisher and authors have used their best efforts in preparing this work, they make no representations or warranties with respect to the accuracy or completeness of the contents of this work and specifically disclaim all warranties, including without limitation any implied warranties of merchantability or fitness for a particular purpose. No warranty may be created or extended by sales representatives, written sales materials or promotional statements for this work. The fact that an organization, website, or product is referred to in this work as a citation and/or potential source of further information does not mean that the publisher and authors endorse the information or services the organization, website, or product may provide or recommendations it may make. This work is sold with the understanding that the publisher is not engaged in rendering professional services. The advice and strategies contained herein may not be suitable for your situation. You should consult with a specialist where appropriate. Further, readers should be aware that websites listed in this work may have changed or disappeared between when this work was written and when it is read. Neither the publisher nor authors shall be liable for any loss of profit or any other commercial damages, including but not limited to special, incidental, consequential, or other damages.

Library of Congress Cataloging-in-Publication Data

Names: Manuele, Fred A., author. | American National Standards Institute.
Title: Advanced safety management focusing on Z10.0, 45001 and serious
   injury prevention / Fred A Manuele.
Description: Third edition. | Hoboken, NJ : Wiley, 2020. | Includes
   bibliographical references and index.
Identifiers: LCCN 2019052191 (print) | LCCN 2019052192 (ebook) | ISBN
   9781119605416 (hardback) | ISBN 9781119605447 (adobe pdf) | ISBN
   9781119605409 (epub)
Subjects: MESH: Safety Management–standards | Wounds and
   Injuries–prevention & control | Occupational Health | Risk
   Management–standards | United States
Classification: LCC RA645.T73 (print) | LCC RA645.T73 (ebook) | NLM WA
   485 | DDC 363.11–dc23
LC record available at https://lccn.loc.gov/2019052191
LC ebook record available at https://lccn.loc.gov/2019052192

Cover Design: Wiley
Cover Image: © MisterStock/Shutterstock

Set in 10/12pt, TimesLTStd by SPi Global, Chennai, India.

Printed in the United States of America

10 9 8 7 6 5 4 3 2 1

To my family.

To my family

CONTENTS

PREFACE TO THE THIRD EDITION — ix
PREFACE TO THE SECOND EDITION — xi
PREFACE TO THE FIRST EDITION — xiii
ACKNOWLEDGMENTS — xv
INTRODUCTION — 1

1. AN OVERVIEW OF ANSI/ASSP Z10.0-2019 AND ANSI/ASSP/ISO 45001-2018 — 15
2. ORGANIZATIONAL CULTURE, MANAGEMENT LEADERSHIP, AND WORKER PARTICIPATION — 49
3. SAFETY PROFESSIONALS AS CULTURE CHANGE AGENTS — 69
4. ESSENTIALS FOR THE PRACTICE OF SAFETY — 85
5. A PRIMER ON SYSTEMS/MACRO THINKING — 103
6. A SOCIO-TECHNICAL MODEL FOR AN OPERATIONAL RISK MANAGEMENT SYSTEM — 115
7. INNOVATIONS IN SERIOUS INJURY, ILLNESS, AND FATALITY PREVENTION — 133
8. HUMAN ERROR AVOIDANCE — 157

9	ON HAZARDS ANALYSES AND RISK ASSESSMENTS	169
10	THREE- AND FOUR-DIMENSIONAL RISK SCORING SYSTEMS	201
11	HIERARCHIES OF CONTROL	237
12	SAFETY DESIGN REVIEWS	251
13	PREVENTION THROUGH DESIGN	279
14	MANAGEMENT OF CHANGE	307
15	THE PROCUREMENT PROCESS	349
16	EVALUATION AND CORRECTIVE ACTION	387
17	MANAGEMENT REVIEW/IMPROVEMENT	395
18	AUDIT REQUIREMENTS	401
19	INCIDENT INVESTIGATION	419
20	INCIDENT CAUSATION MODELS	437
21	THE FIVE WHY PROBLEM-SOLVING TECHNIQUE	453
22	MORT—THE MANAGEMENT OVERSIGHT AND RISK TREE	465
23	JAMES REASON'S SWISS CHEESE MODEL	481
24	ON SYSTEM SAFETY	493
25	ACHIEVING ACCEPTABLE RISK LEVELS: THE OPERATIONAL GOAL	511
INDEX		531

PREFACE TO THE THIRD EDITION

Significant developments concerning the practice of safety compelled the writing of this third edition. In 2018, an international standard for occupational health and safety management systems was approved by the International Organization for Standardization. It is known as 45001.

A third version of the American standard for occupational risk management was approved by the American National Standards Institute in 2019. It is known as Z10.0.

Updating of the second edition of this book was necessary to include comments on both the new international standard and the substantial extensions made in Z10.0. Chapter 1 provides an overview of 45001 and Z10.0. There are many similarities and differences in the two standards. When it was feasible, composites are given of the requirements of the standards.

But it was necessary to comment on individual Clauses in the standards when their contents were not similar. Also, statements are made on the subjects pertaining to good risk management that are not included in either standard.

Interest in further reducing serious injuries, illnesses, and fatalities is now more prominent. A good number of safety professionals have recognized that while occupational injury frequency has been notably reduced, the reductions have been greater for incidents that result in lesser severe injuries than in more serious injuries. So, serious injuries and illnesses loom larger in the remaining total.

This author continues to propose that major innovations in the content and focus of occupational risk management systems will be necessary to achieve additional reductions of incidents that result in serious consequences. That idea has achieved more acceptance in recent years. Speeches, seminars, and workshops on the subject are more frequently given.

Major extensions have been made in the chapter titled "Innovations in Serious Injury, Illness, and Fatality Prevention" and in other relative chapters.

For example, immediately following the chapter giving an overview of 45001 and Z10, there is a chapter titled "Organizational Culture, Management Leadership, and Worker Participation". This chapter commences with these statements:

> Safety is culture driven. Everything that occurs or doesn't occur that relates to safety is a reflection of an organization's culture. Over time, every organization develops a culture, positive or negative, although it may not realize that it has one.

That is the most important chapter in the book. Throughout this third edition, more is made of the significance of an organization's culture concerning the quality of a safety management system. A risk-based approach is taken.

Revisions to update and extend content have been made in all the chapters in the second edition that are a part of this third edition. These are the new chapters.

- Essentials for the Practice of Safety
- A Primer on Systems/Macro Thinking
- Incident Causation Models
- The Five Why Problem Solving Technique
- MORT—The Management Oversight and Risk Tree
- James Reason's Swiss Cheese Model

As was the case for previous editions, the principle purposes of this book continue to be providing guidance to managements, safety professionals, educators, and students on the:

- Content of effective operational risk management systems.
- Requirements of Z10.0, and now 45001, and
- Reduction of incidents that result in serious consequences.

This book is used in several university safety degree programs. Input was sought from professors and experienced safety professionals on subjects for which expansion would be appropriate or which should also be addressed.

FRED A. MANUELE, CSP, PE
President, Hazards Limited
2019

PREFACE TO THE SECOND EDITION

This book focuses on Z10, which is the national standard for Occupational Health and Safety Management Systems, and on serious injury and fatality prevention. Impetus for an updated version of the first edition derived from two developments.

- A revised version of Z10 was approved by the American National Standards Institute in June 2012. And comment and guidance for the revisions and extensions would be beneficial.
- While the rates for serious occupational injuries and fatalities continued to drop significantly over past decades, it is now recognized that the rates have plateaued in recent years. This author proposes that major and somewhat shocking innovations in the content and focus of occupational risk management systems will be necessary to achieve additional progress.

The principle purpose of this book continues to be providing guidance to managements, safety professionals, educators, and students on having operational risk management systems that meet the requirements of Z10 and to be informative on reducing the occurrence of accidents that result in serious injuries and fatalities.

This book is used in several university level safety degree programs. Input was sought from professors, and experienced safety professionals who use the book as a reference, on subjects that should be expanded or also addressed.

- Additional emphasis is given to the most important section in Z10— Management Leadership and Employee Involvement, with particular reference to contributions that employees can make.

- A new provision was added to Z10 on Risk Assessment. Its importance is stressed.

This author proposes that risk assessments be established as the core of an Operational Risk Management System as a separately identified element following soon after the first element. Comments are made in several chapters on hazard identification and analysis and risk assessment techniques.

- A significant departure from typical safety management systems is presented in the chapter on "Innovations in Serious Injury and Fatality Prevention" in the form of a Socio-Technical Model for an Operational Risk Management System. This model stresses: The significance of the organizational culture established at the board of directors and senior management levels with respect to attaining and maintaining acceptable risk levels; Providing adequate resources; Risk assessments, prioritization, and management; Prevention through Design; Maintenance for system integrity; and Management of change. It is made clear that there is a notable connection between improving management systems to meet the provisions of Z10 and avoiding serious injuries and fatalities.
- Avoidance of human error is given expanded attention.
- New chapters have been researched and written. Their titles are:
- Macro Thinking—The Socio-Technical Model
- Safety Professionals as Culture Change Agents
- Prevention through Design
- A Primer on System Safety
- Chapters on "Management of Change" and "The Procurement Process" have been revised and expanded considerably.

Soon after approval was given by the American National Standards Institute for the revision of Z10, the Secretariat was transferred to the American Society of Safety Engineer (ASSE). Additional promotion has been given to Z10 by ASSE as the state-of-art occupational safety management system. Interest shown in the standard by safety professionals is impressive. Z10 has achieved recognition as a sound base from which to develop innovations in existing safety management systems.

For a huge percent of organizations, adopting the provision in Z10 will achieve major improvements in their occupational health and safety management systems and serve to reduce the potential for accidents that may result in serious injury or fatality.

FRED A. MANUELE, CSP, PE
2014

PREFACE TO THE FIRST EDITION

The principle purpose of this book is to provide guidance to managements, safety professionals, educators, and students concerning two major, interrelated developments impacting on the occupational safety and health discipline. They are the:

- Issuance, for the first time in the United States, of a national consensus standard for occupational safety and health management systems.
- Emerging awareness that traditional systems to manage safety do not adequately address serious injury prevention.

On July 25, 2005, the American National Standards Institute approved a new standard titled Occupational Health and Safety Management Systems. Its designation is ANSI/AIHA Z10-2005. This standard is a state-of-the-art, best practices guide. Over time, Z10 will revolutionize the practice of safety.

The chapter titled "An Overview of ANSI/AIHA Z10-2005" comments on all the provisions in the standard. The chapter on "Serious Injury Prevention" gives substance to the position that adopting a different mindset is necessary to reduce serious injury potential. Other chapters give implementation guidance with respect to the standard's principal provisions and to serious injury prevention.

Recognition of the significance of Z10 has been demonstrated. Its provisions are frequently cited as representing highly effective safety and health management practices. The sales record for Z10 is impressive. Safety professionals are quietly making gap analyses, comparing existing safety and health management systems to the provisions of Z10.

Even though the standard sets forth minimum requirements, very few organizations have safety and health management systems in place that meet all the provisions of the standard. The provisions for which shortcomings will often exist, and for which emphasis is given in this book, pertain to:

- Risk assessment and prioritization
- Applying a prescribed hierarchy of controls to achieve acceptable risk levels
- Safety design reviews
- Including safety requirements in procurement and contracting papers
- Management of change systems.

As ANSI standards are applied, they acquire a "quasi-official" status as the minimum requirements for the subjects to which they pertain. As Z10 attains that stature, it will become the benchmark, the minimum, against which the adequacy of safety and health management systems will be measured.

The chapter on "Serious Injury Prevention" clearly demonstrates that while occupational injury and illness incident frequency is down considerably, incidents resulting in serious injuries are not down proportionally. The case is made that typical safety and health management systems do not adequately address serious injury prevention. Thus, major conceptual changes are necessary in the practice of safety to reduce serious injury potential. That premise permeates every chapter in this book.

Safety and health professionals are advised to examine and reorient the principles on which their practices are based to achieve the significant changes necessary in the advice they give. Guidance to achieve those changes is provided.

Why use the word "Advanced" in the title of this book? If managements adopt the provisions in Z10 and give proper emphasis to the prevention of serious injuries, they will have occupational health and safety management systems as they should be, rather than as they are. There is a strong relationship between improving management systems to meet the provisions of Z10, a state-of-the-art standard, and minimizing serious injuries.

FRED A. MANUELE, CSP, PE
2008

ACKNOWLEDGMENTS

To recognize by name all who have contributed to my education, and thence to this book, would require a lengthy list. But, particularly, I must express my sincere thanks and gratitude to the people who gave of their time in past years and critiqued individual essays, and to the people who made written contributions to individual chapters.

INTRODUCTION

This book comments on the provisions of ANSI/ASSP Z10.0-2019, the American standard for *Occupational Health and Safety Management Systems,* and ANSI/ASSP/ISO 45001-2018—the international standard titled *Occupational health and safety management systems—Requirements with guidance for use.* It also includes several chapters that relate to the standards and also to serious injury, illness, and fatality prevention. An abstract is provided in this Introduction for each chapter.

In accord with a suggestion made by a professor who uses my books in his classes, an attempt was made to have each chapter be a stand-alone essay. Partial success with respect to that suggestion has been achieved. Although doing so requires some repetition, the reader benefits by not having to refer to other chapters while perusing the subject at hand.

CHAPTERS

1. AN OVERVIEW OF ANSI/ASSP Z10.0-2019 AND ANSI/ASSP/ISO 45001-2018

ANSI/ASSP Z10.0-2019 is titled *Occupational Health and Safety Management Systems* and ANSI/ASSP/ISO 45001-2018 is titled *Occupational health and safety management systems—Requirements with guidance for use.*

Advanced Safety Management: Focusing on Z10.0, 45001, and Serious Injury Prevention,
Third Edition. Fred A. Manuele.
© 2020 John Wiley & Sons, Inc. Published 2020 by John Wiley & Sons, Inc.

2 INTRODUCTION

A third edition of Z10.0 was approved in August 2019. Major changes were made in relation to the content of previous editions. The 45001 standard is a new international standard that was approved in 2018.

While the titles are similar and some of the names for Clauses in the standards are close to the same, the standards have significant differences. They are highlighted. Initial discussions in this chapter pertain to the history and development of the standards; compatibility and ease of harmonization with other ISO standards; continual improvement processes—PDCA; relationship to serious injury prevention; and the standards being management systems-oriented and not specification-oriented.

Then, comments are made on all of the Clauses and sections in these standards. Also, a section in this chapter comments on the subjects for which provisions are not included.

2. ORGANIZATIONAL CULTURE, MANAGEMENT LEADERSHIP, AND EMPLOYEE PARTICIPATION

It is highly probable that this chapter is the most important in the book. Much more is made of the significance of an organization's culture than previously. Safety is culture-driven, and management creates the culture. As top management makes decisions directing the organization, the outcomes of those decisions establish its safety culture.

While numerous articles have appeared in safety-related literature about an organization's culture, it seems that almost all of them have been directed to what senior management is to do. This chapter assists safety professionals in understanding that almost everything that occurs or doesn't occur with respect to the practice of safety is a reflection of an organization's culture—positive or negative.

Safety professionals will surely agree that top management leadership and effective employee participation are crucial for success. An organization's continual improvement processes cannot be achieved without the provision of adequate resources and sincere top management direction.

As management provides direction and leadership, assumes responsibility for the OH&S management system, ensures effective employee participation, and creates the organization's culture, the purpose of the standards must be kept in mind—to reduce the risk of occupational injuries, illnesses, and fatalities.

In both 45001 and Z10.0, the requirements for management leadership and worker participation are extensive. A verbatim copy of the management requirements as in Z10.0 is included in this chapter. Workers are to

- have meaningful involvement in the structure, operation, and pursuit of the objectives of the occupational risk management system;
- identify tasks, hazards and risks, and possible control measures; and
- participate in planning, evaluation, implementation of control measures for risk reduction.

3. SAFETY PROFESSIONALS AS CULTURE CHANGE AGENTS

This chapter promotes the idea that the overarching role of a safety professional is that of a culture change agent. A case is made that actions taken to eliminate or control a hazard/risk situation are not complete if they do not address the relative deficiencies in management systems or processes.

The deficiency can be corrected only if there is a modification in an organization's culture—a modification in the way things get done; and a modification in the *system of expected performance*.

Thus, the primary role for a safety professional is that of a culture change agent. This is the definition of a change agent: A change agent is a person who serves as a catalyst to bring about organizational change: They assess the present, are controllably dissatisfied with it; contemplate a future that should be; and take action to achieve the culture changes necessary to achieve the desired future.

Examples are given for the involvement of safety professionals as culture change agents in hazards/risks situations.

4. ESSENTIALS FOR THE PRACTICE OF SAFETY

For the practice of safety to be recognized as a profession, it must have a sound theoretical and practical base which, if applied, will be effective in hazard avoidance, elimination, and control, in achieving acceptable risk levels and reducing harmful and damaging incidents. Safety practitioners give advice to management based on a variety of premises. They can't all be right.

This chapter presents logical and sound concepts that this author believes are the bases for the practice of safety. They pertain to his experience. It is not suggested that they are complete. Other safety professionals may have different views.

5. A PRIMER ON SYSTEMS/MACRO THINKING

"Systems thinking" is a term now used more often by some safety professionals. Occasionally, the term is found in safety-related literature. Promoting systems/macro thinking should be encouraged. Doing so is progressive and commendable.

This chapter comments on the various definitions of systems thinking; explores complexity, to which some writers on system thinking refer; describes the status quo in the practice of safety; promotes the use of the terms "macro thinking" and "micro thinking"; encourages recognition of the enormity of the culture change necessary in some organizations to adopt systems/macro thinking concepts; and connects systems/macro thinking to having a socio-technical balance in operations.

Also, encouragement is given to the use of the Five Why Problem-Solving Technique in the early stages of applying systems/macro concepts.

6. A SOCIO-TECHNICAL MODEL FOR AN OPERATIONAL RISK MANAGEMENT SYSTEM

It was said in Chapter 2 that safety is culture-driven, and management creates the culture. As top management makes decisions directing the organization, the outcomes of those decisions establish its safety culture.

This chapter presents a model that emphasizes the significance of an organization's culture and management direction and involvement. Comments are made on how the culture affects what becomes an organization's policies, standards, and processes.

Mention of system thinking and a taking a holistic, socio-technical approach in hazard and risk control appears more often in safety-related literature. This is powerful stuff, and needed. It is proposed that for superior risk management (and productivity) can be achieved only if there is a balance in the social aspects and the technical aspects of operations.

This chapter uses the term "macro thinking" principally rather that system thinking because macro implies taking a much larger view. Descriptions are given for all of the individual aspects of the socio-technical model to validate their existence.

7. INNOVATIONS IN SERIOUS INJURY, ILLNESS, AND FATALITY PREVENTION

Preventing occupational incidents and illnesses that have serious results has now become a subject that speakers address in conferences, authors write about, and for which some safety professionals seek information.

While incident frequency has been substantially reduced, the greater part of the reduction has been for incidents resulting in lesser severity. Also, statistics indicate that the incident rate for fatalities, and possibly serious injuries, has plateaued.

This chapter sets forth the actions that can be taken to achieve additional reductions in severity. First, though, illustrative statistical and supportive data is presented.

This author's recent research requires the conclusion that major innovations in safety management systems will be necessary to further reduce serious injuries, illnesses, and fatalities. These are the actions presented and discussed for consideration by safety professionals.

   a. Study the principles on which their practice is based to determine what is sound and not sound. Ask—is it time to recognize evolving principles and practices?
   b. Develop meaningful and convincing data on the activities in which serious injuries, illnesses, and fatalities have occurred in your operations and for the industry of which it is a part.
   c. Evaluate the culture in place.

d. Recognize that for most of the revisions to be made that a culture change will be necessary.
e. Convince management that having good OSHA type incidence rates may not provide assurance that barriers and controls are adequate to prevent serious injuries and fatalities.
f. Educate management on the inappropriateness of Heinrich's principle indicating that 88% of occupational accidents are caused by employee unsafe acts.
g. Persuade decision-makers that focusing on reducing incident frequency may not result in an equivalent reduction in serious injuries.
h. Be aware of the changes in approach that have taken place in some circles with respect to addressing human errors/unsafe acts and determine how they might affect what you do.
i. Appreciate the importance of barriers in safety management.
j. Encourage all engineers to recognize that they are also safety engineers and that they should be adept in making risk assessments both in original designs and in alterations.
k. Promote defining potential problems through making risk assessments.
l. Recognize the beneficial impact of applying prevention through design principles.
m. Encourage the effective application of a Management of Change system.
n. Analyze the incident investigation system in place and propose improvements as necessary.
o. Understand that *tweaking existing systems will not achieve the substantial improvements desired.*

8. HUMAN ERROR AVOIDANCE

Many injuries, illnesses, and fatalities result from avoidable human errors. Organizational, cultural, technical, and management systems deficiencies often lead to those errors. Although focusing on the management decision making that may be the source of human errors was proposed many years ago, doing so is infrequently within the work of safety professionals.

Fortunately, a renewed interest has emerged on being able to explain why human errors occur in the occupational setting. Safety professionals will be more effective as they give counsel on injury and illness prevention if they are aware of the reality of human error causal facto*rs*.

This chapter brings attention to human errors that derive from: management decisions that result in unacceptable risk levels; extremes of cost reduction; safety management systems deficiencies; design and engineering decisions; overly stressing work methods; and error-provocative operations.

Comments are made on current thinking that proposes that if you want to know why human errors occur, inquiry should be made into the decision-making that resulted in the design of the workplace and the work methods.

9. ON HAZARDS ANALYSES AND RISK ASSESSMENTS

The intent here is to provide sufficient knowledge of hazards analysis and risk assessment methods to serve most of a safety and health professional's needs. This chapter explores what a hazard analysis is; discusses how a hazard analysis is extended into a risk assessment; outlines the steps to be followed in conducting a hazard analysis and a risk assessment; includes descriptions of several commonly used risk assessment techniques; and gives examples of risk assessment matrices.

10. THREE- AND FOUR-DIMENSIONAL RISK SCORING SYSTEMS

For risk assessments, the practice—broadly—is to establish qualitative risk levels by considering only two dimensions. They are probability of event occurrence, and the severity of harm or damage that could result. However, systems now in use may be three or four dimensional. This chapter reviews several such systems.

A three-dimensional numerical risk scoring system developed by this author to serve the needs of those who prefer to have numbers in their risk assessment systems is presented.

11. HIERARCHIES OF CONTROLS

A hierarchy is a system of persons or things ranked one above the other. Hierarchies of control in Z10.0 and 45001 provide a systematic way of thinking, considering steps in a ranked and sequential order; and an effective way for decision-makers to eliminate or reduce hazards and the risks that derive from them.

Acknowledging the premise—that risk reduction measures should be considered and taken in a prescribed order—represents an important step in the evolution of the practice of safety. A major premise in applying a hierarchy of controls is that the outcome of the actions taken is to be an acceptable risk level.

Requirements in the hierarchies of control in the 45001 and Z10.0 standards are close to the same, but not quite. This is the hierarchy shown in Z10.0: Elimination; Substitution of less-hazardous materials, processes, operations, or equipment; Engineering controls; Warnings; Administrative controls, and Personal protective equipment.

In 45001, the hierarchy is the same, with one important difference: It does not include a Warnings element. In this chapter, comments are made on all of the elements in the Z10.0 standard, giving examples.

12. SAFETY DESIGN REVIEWS

Design Review and Management of Change requirements are addressed jointly in Z10.0. Although the subjects are interrelated, each has its own importance and uniqueness. Guidance on the management of change concept is provided in Chapter 14.

There is no specifically designated section in 45001 titled design, or design requirements or design review. But the standard includes comments similar to the following: management is to have in place processes to identify hazards arising from product and service design and from the design of the work areas.

To do as Z10.0 implies, systems must be in place to avoid, eliminate, reduce, or control hazards and the risks that derive from them: as early as possible and as often as needed in every aspect of the design and redesign processes; and in all phases of operations.

This chapter includes a review of safety through design concepts; comments on how some safety professionals are engaged in the design process; a composite of safety through design procedures in place; A Safety Design Review and Operations Requirements Guide; and a general design safety checklist.

13. PREVENTION THROUGH DESIGN

Several provisions in Z10.0 and 45001 relate to prevention through design concepts. They are Risk Assessments; Design Requirements; Hierarchy of Controls; and Procurement.

Fortunately, an American National Standard exists that provides guidance on the subject. Its designation is ANSI/ASSE Z590.3—2011(R2016) and its title is *Prevention through Design: Guidelines for Addressing Occupational Hazards and Risks in Design and Redesign Processes.*

This chapter gives a history of the safety through design/prevention through design movement and comments extensively on the continuing activities at the National Institute of Safety and Health on prevention through design; Presents highlights of Z590.3, emphasizing the applicability of the provisions of the standard to all hazards-based initiatives; and encourages safety professionals to become involved in prevention through design for job satisfaction and to be perceived as providing additional value.

14. MANAGEMENT OF CHANGE

Because of this author's belief that management of change should be an important element in an occupational risk management system, a plea is made that readers give particular attention to the requirements for this subject as in Z10.0 and 45001. They both require that management of change processes be in place for the new and the revised. This chapter

- Makes the case that having an effective Management of Change System (MOC) in place as a distinct element in a safety and health management system will reduce the potential for injuries, environmental damage, and other forms of damage at all levels of severity.
- Cites statistics in support of having effective MOC systems.

- Defines the purpose and methodology of a management of change system.
- Outlines management of change procedures, keeping in mind the staffing limitations at other than large locations and their need to avoid burdensome paper work.
- Provides guidelines on how to initiate and utilize a MOC system.
- Emphasizes the significance of communication and training.
- Includes examples of six management of change systems in place and provides access to four other real-world MOC systems.

In accord with the emphasis given here to management of change, the requirements in 45001 and Z10.0 are duplicated in their entirety.

15. THE PROCUREMENT PROCESS

Both the Z10.0 and 45001 standards require that what is purchased conforms to the organization's occupational health and safety management system. In reality, the purpose of the procurement process is to avoid bringing hazards and their accompanying risks into the workplace.

To assist safety professionals as they give advice on implementing those provisions, this chapter duplicates the procurement requirements of the standards; comments on prevalent purchasing practices; establishes the significance of the procurement processes; discusses the prework necessary to include safety specifications in the procurement process; provides some resources; comments on available occupational health and safety purchasing specifications; and gives examples of design specifications that become purchasing specifications.

16. EVALUATION AND CORRECTIVE ACTION

In Z10.0, Clause 9.0 contains specific provisions to achieve both a performance evaluation and to establish corrective action mechanisms as is said in its first two bulleted items as follows:

This section defines requirements for processes to

- Evaluate the performance of the OHSMS through Monitoring, Measurement, and Assessment, Incident Investigation, and Audits and
- Take corrective action when nonconformances, system deficiencies, hazards, and incidents that are not being controlled to an acceptable of risk are found in the OHSMS.

In 45001, the purpose principally of Clause 9 is for performance evaluation. But as is noted in this chapter, the outputs of a management review are to include opportunities for continual improvement; any needed changes to the OH&S management system; and actions if needed.

Having processes for taking corrective action are a requirement of Clause 10 in 45001, the title of which is Improvement.

After introductory material is provided in both chapters, these captions appear: Monitoring, measurement, and assessment in Z10.0 and Monitoring, measurement, analysis and performance evaluation in 45001. Each of the elements required for those subjects are addressed, such as Audits (see Chapter 18); Incident investigation (see Chapter 19).

17. MANAGEMENT REVIEW/IMPROVEMENT

In Z10.0, the title for Clause 10.0 is Management Review. Comparable provisions for management reviews in 45001 are in its Clause 9—which is titled Performance evaluation.

In 45001, the title for Clause 10 is Improvement. Provisions for improvement appear throughout Z10.0, but specifically in its Clause 9.0, which is titled Evaluation and Corrective Action. First, the entirety of the requirements of Clause 10 as in 45001 is duplicated. There is a peculiarity in Clause 10 in that incidents are in the same category as nonconformities. It could be argued that incidents are nonconformities. Nevertheless, the requirements of organizations in 45001 when incidents or nonconformities occur are very much the same as for the investigation of worker incidents.

Requirements of Clause 10 in Z10.0 are also duplicated. An emphasis is given to the leadership aspects in the management review.

18. AUDIT REQUIREMENTS

As is the case with every aspect of an organization's endeavors, making periodic reviews of progress with respect to stated goals is good business practice. Stated goals, in this instance, would be to have processes in place that meet the requirements of 45001 or Z10.0.

Internal audit requirements in 45001 are at 9.2 within Clause 9—Performance evaluation. In Z10.0, they are at 9.3 within Clause 9.0—Evaluation and Corrective Action. A composite of their requirements follows.

> An organization shall periodically have audits made of its health and safety management system to determine whether the requirements of this standard are implemented and maintained.

To assist safety professionals in crafting or recrafting safety and health audit systems to meet the requirements of Z10.0 or 45001, this chapter establishes the purpose of an audit; discusses the implications of observed hazardous situations; explores management's expectations with respect to audits; establishes that safety auditors are also being audited during the audit process; comments on auditor qualifications; discusses the need to have safety and health management system audit guides relate to the hazards and risks in the operations at the location being audited; provides information and resources for the development of suitable audit guides.

It is made clear that the Principle Purpose of a Safety Audit: To Improve the Safety Culture.

19. INCIDENT INVESTIGATION

Incident investigation in 45001 is addressed in Clause 10 Improvement. Within Clause 10, Section 10.2 is titled Incident, nonconformity, and corrective action. Duplication of the entirety of Clause 10 as in 45001 appears in Chapter 17.

In this chapter, all of the requirements for incident investigation as in Z10.0 are shown, expectations of management on incident investigation in Z10.0 are pretty much the same as the requirements in 45001.

Comments on this author's research on incident investigation are a large part of this chapter. An introduction to those comments was given in Chapter 7, Innovations in Serious Injury, Illness, and Fatality Prevention.

20. INCIDENT CAUSATION MODELS

As safety professionals participate in determining causal and contributing factors for an incident, they are applying their adopted causation model. Their models relate to what they have learned and their beliefs concerning how accidents happen. For effectiveness, an organization must have adopted a causation model that gets to the reality of risk-based causal factors. Stellar performance in incident investigation cannot be achieved otherwise.

Unfortunately, there are too many causation models. This chapter defines the requirements for a causation model; establishes that a safety professional who gives counsel on incident investigations has an adopted causation model; gives causation model categories; comments on several causation models; emphasizes the transition taking place whereby safety professionals are encouraged to explore the management system shortcomings that are the sources of human errors (unsafe acts); comments on the significance of inadequacies in barriers and controls as causal factors; and gives resources related to causal factors.

Recommendations are made for safety professionals to acquire knowledge of The Five Why Problem-Solving Technique; MORT—The Management Oversight and Risk Tree; and James Reason's Swiss Cheese Model.

21. THE FIVE WHY PROBLEM-SOLVING TECHNIQUE

This author has noted that the quality of incident investigations, even in some very large organizations, is significantly less than stellar. Yet incident investigation is a vital element within an operational risk management system. For a first step forward to improve on incident investigation, I promote adoption of the five why problem-solving technique, with emphasis.

This chapter gives the reasons why the five why system is supported; comments on for what the five why technique is to be used and when; gives procedures for use of the five why technique; emphasizes the need for training; suggests that a safety professional be aware that to adopt the technique a culture change may be necessary; gives five examples of use of the technique; and offers my cautions that result from having analyzed incident reports on situations in which the five why system was used.

22. MORT—THE MANAGEMENT OVERSIGHT AND RISK TREE

If I taught at a university in a safety and environmental science degree program, students would be exposed to MORT—The Management Oversight and Risk Tree for the thought base it provides on how accidents happen and what is needed for their prevention.

What MORT is and how it can be used is addressed in the Abstract for the third edition of the MORT User's Manual issued by the U.S. Department of Energy (DOE).

> MORT is a comprehensive analytical procedure that provides a disciplined method for determining the causes and contributing factors of major accidents. Alternatively, it serves as a tool to evaluate the quality of an existing program. (p. iii)

Developing skills for the application of MORT requires study and continuous application to develop and retain knowledge of the technique and the skills necessary in its usage. It is not the intent here to have safety professionals become skilled in the application of MORT. But they would add to their knowledge and capabilities by developing an awareness of its content.

MORT is unusual in that it specifically promotes inquiry into management's upstream decision-making for the identification of sources of causal factors in much greater detail than most other incident investigation models. MORT is also unusual because of its premise that where there are deficiencies in operating controls, there will be related deficiencies in management decision-making.

MORT places great significance on having the necessary barriers and controls in place to prevent accidents and to reduce their impact if they occur. Analyzing for the adequacy or inadequacy of barriers and controls is significant in applying the MORT system. Barriers are not exclusively physical. They may include having a qualified staff, training, supervision, sound procedures, maintenance, and communications.

Throughout a MORT exercise, the purpose is to determine whether barriers and controls were sufficient or "Less than adequate"—which is close to a synonym for "Risk was unacceptable." Discussions in this chapter relate to all of the main branches of MORT

23. JAMES REASONS' SWISS CHEESE MODEL

In recent years, no individual has had more influence on how people think about incident causation than James Reason. His concepts are applied throughout the world. Reason recognized that active errors, what have been called unsafe acts, could be committed by operators and be the initiator for an incident. But he also emphasized that what he labeled latent conditions could also be causal factors.

Reason said that latent conditions are present in all organizations; result from decisions made by manufacturers, designers, and organizational managers; and could be the more significant causal factors for accidents. Latent conditions are represented by "holes" in the system.

Reason used the term "holes" symbolically and as a metaphor. He did not imply that there were actual holes in management systems. What the holes represent are inadequacies in barriers and controls that, if adequate and appropriately used, may have prevented an incident from occurring. Such inadequacies result from deficiencies in the management systems that relate to them. Relationship of holes to each other is not fixed: It is fluid and subject to change as are all operational situations.

It is highly recommended that safety professionals become aware of Reason's thinking about the causal factors for accidents.

24. ON SYSTEM SAFETY

This author believes that generalists in the practice of safety will improve the quality of their performance by acquiring knowledge of applied system safety concepts and practices. This chapter relates to the generalist's practice of safety to applied system safety concepts; gives a history of the origin, development, and application of system safety methods; outlines the system safety idea in terms applicable to the generalist's practice of safety; and encourages safety generalists to acquire knowledge and skills in system safety.

25. ACHIEVING ACCEPTABLE RISK LEVELS: THE OPERATIONAL GOAL

Reference is made to several parts of Z10.0 for the need to achieve and maintain acceptable risk levels. Other recently modified standards do the same. This is not so in 45001.

This chapter provides a primer from which an understanding of risk and the concept of acceptable risk can be attained. For that purpose

- Purposely numerous examples are given for the use of the term "acceptable risk."
- Reasons why some people shy away from the acceptable risk concept are given.
- Several examples are given of the use of the term "acceptable risk" as taken from the applicable literature.
- Discussions address the impossibility of achieving zero-risk levels.
- The inadequacy of using "minimum risk" as a replacement term for acceptable risk is explored.
- A Risk Assessment Matrix for use in determining acceptable risk levels is offered as an example.
- The As Low As Reasonably Practicable (ALARP) concept is discussed with an example of how the concept is applied in achieving an acceptable risk level.
- A definition of acceptable risk is given as well as the logic for arriving at that definition.

An all-encompassing premise follows that is basic to the work of all personnel who give counsel to avoid injury and illness, and property and environmental damage.

The entirety of purpose of those responsible for safety, regardless of their titles, is to manage their endeavors with respect to hazards so that the risks deriving from those hazards are acceptable.

CHAPTER 1

AN OVERVIEW OF ANSI/ASSP Z10.0-2019 AND ANSI/ASSP/ISO 45001-2018

In 2019, the American National Standards Institute (ANSI) approved a third edition of a standard titled *Occupational Health and Safety Management Systems*. Its secretariat is the American Society of Safety Professionals (ASSP), and its designation is ANSI/ASSP Z10.0-2019. This American standard was first approved in 2005, revised in 2012, and reaffirmed in 2017. It is commonly referred to as Z10.

Major improvements and extensions were made in the 2019 version. It is a much-improved standard. Revisions are as follows:

- Make it easier to integrate the standard with quality, environmental, and other management systems standards.
- Provide alignment with the 45001 standard.
- Are more specific concerning the significance of management leadership and the culture that management creates.
- Encourage additional worker involvement in the safety and health management system.
- Replace the term "employee" with "worker" to broaden the coverage and include certain workers that are not employed by the organization.
- Add a new section on the Context of the Organization.
- Create a new section on Support which includes resources, education, training and competence, communication, and document control process.

Advanced Safety Management: Focusing on Z10.0, 45001, and Serious Injury Prevention, Third Edition. Fred A. Manuele.
© 2020 John Wiley & Sons, Inc. Published 2020 by John Wiley & Sons, Inc.

- Extend the importance of the design requirements in which prevention through design components are now more prominent.

In 2018, the first international standard for occupational health and safety management systems was approved by an entity with significant credentials—the International Organization for Standardization (ISO, 2015). Full title of the standard is ISO 45001: *Occupational health and safety management systems—Requirements with guidance for use*. This standard is commonly referred to as 45001.

In the same year, approval was given to this international standard by ANSI to become an American standard. Its designation is ANSI/ASSP/ISO 45001-2018, and the secretariat is the American Society of Safety Professionals.

Although 45001 is based principally on BS OHSAS 18001:2007, the title of which is *Occupational health and safety management systems—Requirements*, it is considered a new international standard since numerous changes were made in the process of 18001 becoming an ISO standard. For OHSAS18001, the British Standards Institute (BSI) was the publisher. A composite of what ISO said about 45001 is as follows:

> The goal of ISO 45001 is the drastic reduction of occupational injuries and diseases. It is a milestone. The standard is based on OHSAS 18001, conventions and guidelines of the International Labour Organization including ILO OSH 2001, and national standards.

It is said in the Foreword of BS OHSAS 18001:2007 that "This OSHAS Standard will be withdrawn on publication of its contents in, or as, an International Standard." BSI has done that.

BSI has been a certifying body for 18001. On the Internet, BSI advises those organizations that are now certified to OSHAS 18001 need to migrate to ISO 45001 by March 1, 2021. As would be expected, BSI says its organization has the expertise and knowledge to help upgrade certification to ISO 45001.

SIMILARITIES AND DIFFERENCES

As readers will have noted, the titles of 45001 and Z10.0 are similar. To determine how alike the content of the standards might be, a review was made that included only the titles of the standards and the major captions in their tables of contents. Results are shown in Figure 1.1.

Figure 1.1 shows that, from a macro view, the contents of the standards are pretty much the same. Alignment of the major captions of the two standards was done intentionally by the Z10 committee to facilitate integration (Introduction of Z10.0). But there are significant differences in the subsets of the two standards as will be seen later in this chapter.

Titles of standards and major captions in tables of contents

ANSI/ASSP Z10.0–2019	ANSI/ASSP/ISO 45001 2018
Occupational health and safety Management systems	Occupational health and safety management systems–requirements with guidance for use
1. Scope, purpose and application	1. Scope
2. References	2. Normative references
3. Definitions	3. Terms and definitions
4. Strategic consideration–context of the organization	4. Context of the organization
5. Management leadership and worker participation	5. Leadership and worker participation
6. Planning	6. Planning
7. Support	7. Support
8. Implementation and operation	8. Operation
9. Evaluation and corrective action	9. Performance evaluation
10. Management review	10. Improvement

FIGURE 1.1 Comparison—Z10.0 and 45001.

Comments are made in this chapter on the following:

- History, Development, and Consensus
- Compatibility and Harmonization
- The Continual Improvement Process: The PDCA Concept
- A Major Theme
- Relating This Major Theme to Serious Injury Prevention
- Z10.0 and 45001 Are Management System Standards
- Forewords and Introductions for Z10.0 and 45001
- Word Usage – Requirements, Recommended Practices, Permissions, and Possibilities
- Composites for Z10.0 and 45001: 1.0 Through 3.0—Scope, Purpose, and Application, References, Definitions and Terms and Conditions
- Composite: 4.0 Strategic Considerations: Context of the Organization in Z10.0: 4. Context of the Organization in 45001
- Composite: 5.0 Management Leadership & Employee Participation in Z10.0: 5. Leadership and Worker Participation in 45001
- Composite: 6.0 Planning in Z10.0: 6. Planning in 45001
- Composite: 7.0 Support In Z10.0:7. Support In 45001
- Composite: 8.0 Implementation and Operation in Z10.0: 8. Operation in 45001
- 9.0 Evaluation and Corrective Action in Z10.0: 9. Performance evaluation in 45001
- 10. Management Review in Z10.0: 10. Improvement in 45001

- Annexes, Bibliographies and Guidance Manuals (Not a part of the standards)
- Observations on the Standards, Including Significant Provisions that are Missing

HISTORY, DEVELOPMENT, AND CONSENSUS

When ANSI first approved Z10 in 2005, the secretariat was the American Industrial Hygiene Association. Z10 was revised in 2012 and reaffirmed in 2017. A major revision of Z10.0 was approved by ANSI in 2019. Its secretariat is now the American Society of Safety Professionals.

For the 2005 version, as many as 80 safety professionals were involved as committee members, alternates, resources, and interested commenters over a six-year period. They represented industry, labor, government, business associations, professional organizations, academe, and persons of general interest.

Thus, broad participation in the development of and acceptance of the standard was achieved, and the breadth of that participation was significant. A similar structure was maintained for the latter versions. In the 2019 version of Z10.0, about 75 names are listed as participants and members of the committee.

Imagine the variations in opinions that so many individuals would have about what should or should not be contained in a health and safety management system standard. It was not easy to obtain concurrence. In some instances, the leaders had to accept the negative vote on provisions that they thought should be contained in Z10.0.

One of the reasons for the Z10.0 committee's success was its strict adherence to the due diligence requirements applicable to the development of an ANSI standard. There was a balance of stakeholders providing input and open discussion, which resulted in their vetting of each issue raised to a conclusion. In crafting Z10.0, the intent was not only to achieve significant safety and health benefits through its application but also to impact favorably on productivity, financial performance, quality, and other business goals.

Similar comments can be made about 45001. Through five years of committee activity, representatives from about 80 countries were involved. Thus, a multiplicity of views was represented. One participant said that she likened the difficulty in arriving at consensus to what she imagined would occur when a treaty was drafted.

It must be recognized that getting ISO to approve an international standard on occupational health and safety management systems is a major accomplishment. Leaders of the activity that resulted in the adoption of 45001 should be congratulated for their accomplishments. Knowing how standard committees work and how difficult it is sometimes to get agreement on a provision, it may be that the standard represents all that could be achieved.

ASSP has recognized the international implications of 45001. It has developed training programs and created a certification mechanism for those organizations that decide that it will be to their advantage to have such certification. That would show that the health and safety management systems in those organizations are at least equivalent to the provisions of 45001.

A journey into the Internet will reveal that offers of training and certification for 45001 are numerous.

COMPATIBILITY AND HARMONIZATION

One of the goals of the drafters of Z10.0 was to assure that the standard could be easily integrated into whatever management system an organization has in place. As to structure, both standards are compatible and harmonized with the ISO quality and environmental management system standards—the ISO 9000 and ISO 14000 series.

Also, Z10.0 and 45001 are written as generic standards and patterned after the style of the ISO standards. In this context, generic means that the standards can be applied to all:

- organizations of any size or type.
- sectors of activity, whether a business enterprise, a nonprofit service provider, or a government entity.

THE CONTINUAL IMPROVEMENT PROCESS: THE PDCA CONCEPT

Writers of Z10.0 and 45001 say that each standard is built on the well-known Plan–Do–Check–Act (PDCA) process for continual improvement. In Z10's Introduction, on page iii, Figure 1.2 is a depiction of the PDCA concept. It is shown here as Figure 1.2 in Z10.0.

On page iv in Z10.0, Figure 1.3 is a chart pertaining largely to the Clauses in Z10.0. It is a considerable extension of the PDCA concept. It does not retain a PDCA identification. A chart similar to Figure 1.3 is presented at the beginning of each of the standard's major sections.

In 45001, on page viii, Figure 1.4 is titled "Relationship between PDCA and the framework in this document." That chart does show the PDCA initials. It is not replicated elsewhere in the standard. It is shown here.

Figure 1.2 Depiction of original Plan–Do–Check–Act system.

20 AN OVERVIEW OF ANSI/ASSP Z10.0-2019 AND ANSI/ASSP/ISO 45001-2018

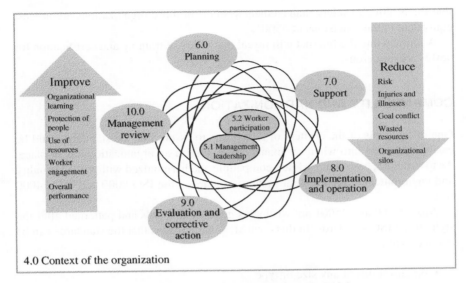

Figure 1.3 Replacement for PDCA as in Z10.0.

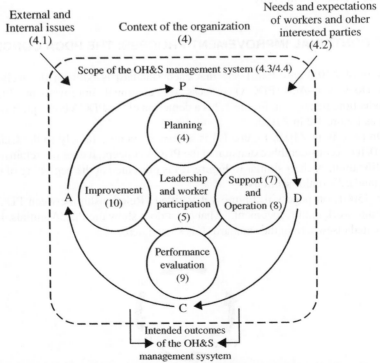

Note: The numbers given in brackets refer to the clause numbers in this document.

Figure 1.4 Relationship between PDCA and the framework in this document.

A MAJOR THEME

Throughout Z10.0 and 45001, the following theme is evident. Organizations are to have processes in place to assure that:

- Management is providing direction and leadership.
- Workers are extensively involved.
- Resources are adequate.
- Hazards are identified and evaluated.
- Risks are assessed and prioritized.
- Management system deficiencies and opportunities for improvement are identified.
- Risk elimination, reduction, or control measures are taken.
- Injuries and illnesses are to be reduced.
- There is continuing communication about the status of compliance or noncompliance with the provisions in an organization's health and safety management system so that continual improvement can be made.
- Management constantly reviews for effectiveness.

RELATING THIS MAJOR THEME TO SERIOUS INJURY PREVENTION

A plea is made in Chapter 7 in this book, Innovations in Serious Injury, Illness, and Fatality Prevention, for organizations to improve their safety cultures so that a focus on the prevention of serious outcomes is embedded into every aspect of their safety and health management systems.

To accomplish the greatest good with the limited resources available, risks presenting the potential for the most serious harm must be given higher priority for management consideration and action. Accomplishments with respect to the major theme previously cited can serve that purpose.

Z10.0 AND 45001 ARE MANAGEMENT SYSTEM STANDARDS

Both Z10.0 and 45001 outline the overall processes to be put in place to have an effective safety and health management system. They are not specification standards. They are management system standards.

What's the difference between a management system standard and a specification standard? In a management system standard, general process and system guidelines are given for a provision without specifying the details on how the provision is to be carried out. In a specification standard, such details are given.

Item B in Section 7.2 Education, Training and Competence in Z10.0 is repeated here to illustrate the difference.

Section 7.2: Education, Training, and Competence.
The organization shall establish processes to:

B. Ensure through appropriate education, training, or other methods that workers understand applicable OHSMS requirements and their importance and are competent to carry out their responsibilities.

That is the extent of the requirements for Section 7.2-B of Z10.0. If the standard was written as a specific standard, requirements comparable to the following might be extensions of 7.2-B.

a. At least 12 hours of training shall be given initially to engineers and safety professionals in prevention through design, to be followed annually with a minimum of six hours spent on refresher materials.
b. All employees shall be given a minimum of six hours training annually in hazard identification.
c. All employees shall be given at least six hours of training on how they can participate in the health and safety management system.
d. All employees shall be given a minimum of four hours training annually in the use of personal protective equipment.
e. All training activities conducted as a part of this provision shall be documented, and the records shall be retained for a minimum of five years.

In the Introduction for Z10.0, they say "ANSI/ASSP Z10.0 focuses primarily on the strategic levels of policy and the processes to ensure the policy is effectively carried out. The standard does not provide detailed procedures, job instructions or documentation mechanisms." (p. iii)

Similarly, 45001 says that "This document does not state specific criteria for OH&S performance, nor is it prescriptive about the design of an OH&S management system." (p. 1)

FOREWORDS AND INTRODUCTIONS FOR Z10.0 AND 45001

Much of the Forward for Z10.0 substantiates the need for and the value of having occupational health and safety management systems and gives a history of the standard. For 45001, the comments pertain mostly to: meeting ISO requirements; that some of the contents of the standard may be subject to patent rights; and that trade names used does not indicate endorsement.

In the Introduction to 45001, the drafters say that the adoption of the standard provides an organization with the means to provide a workplace in which the occupational health and safety risks can be controlled and injuries and illnesses can be prevented.

Success factors are listed, and the list starts with top management leadership, commitment, responsibilities, and accountability and continues with the goals to be reached by applying the provisions of each of the sections of the standard.

In the Introduction to Z10.0, they say that the purpose of the standard is to provide organizations an effective tool for continual improvement of their occupational health and safety performance. Also, an occupational health and safety management system that is implemented in conformance with this standard can help organizations minimize workplace risks and reduce the occurrence and cost of occupational injuries, illnesses, and fatalities.

WORD USAGE—REQUIREMENTS, RECOMMENDED PRACTICES, PERMISSIONS, AND POSSIBILITIES

As is common in ANSI and international standards, requirements are identified by the word "shall." An organization that chooses to conform to the standard is expected to fulfill the "shall" requirements: the word "should" pertains to recommended practices; "may" indicates permission; and "can" indicates a possibility or a capability.

COMPOSITES OF INTRODUCTIONS FOR Z10.0 AND 45001

This statement appears in the Introduction to Z10.0 and establishes the aim of an OS&H management system.

> The purpose of the standard is to provide organizations an effective tool for continual improvement of their occupational health and safety performance." (p. iii)

In the Introduction for 45001, 0.1 is the designation for Background. (p. vi). These are the opening sentences.

> An organization is responsible for the occupational health and safety of workers and others who can be affected by its activities. This responsibility includes promoting and protecting their physical and *mental health.*" (Emphasis added)

Mental health is not defined. It is difficult to fathom how the top management of say a 500-employee organization would interpret the term or what preventive actions would be taken if mental health means more than what have been occupational health exposures.

Still in the Introduction to 45001, comments are made on Success Factors. The following appears (note the word culture):

The level of detail, the complexity, the level of documented information and the resources needed to ensure success of an organization's OH&S will depend on a number of factors, such as: The organization's context (e.g. number of workers, site, geography, culture, legal requirements and other requirements. (p. vii)

At the top of page 27 of the Annex to 45001, there is an explanatory paragraph on culture. An organization's culture can have great influence on the decision making for the operation of an occupational health and safety management system. It would have been beneficial if culture had been defined within the standard and given considerable explanatory and support information.

Chapter 2 in this book is titled "Organizational Culture, Management Leadership, and Worker Participation" and its opening sentences are:

Safety is culture driven. Everything that occurs or doesn't occur that relates to safety is a reflection of an organization's culture.

Nevertheless, the Success factors listed in 45001 cover about a page. They are a good guide for management personnel.

COMPOSITES FOR Z10.0 AND 45001: 1.0 THROUGH 3.0—SCOPE, PURPOSE, AND APPLICATION, REFERENCES, DEFINITIONS, AND TERMS AND CONDITIONS

1.0 Scope, Purpose, & Application in Z10.0: 1. Scope in 45001

Scope This standard defines the requirements for an occupational health and safety (OH&S) management system and gives guidance for its use. Organizations are provided flexibility in how to conform to the requirements of this standard in a manner appropriate to each organization and commensurate with its OH&S risks.

Purpose This standard provides a management tool to achieve the intended purposes of it OH&S management system, improve performance, provide safe workplaces, reduce the risk of occupational injuries, illnesses, and fatalities, achieve continual improvement, and fulfill legal or other requirements.

Application This standard is applicable to organizations of all sizes and types.

2.0 References in Z10.0: 2. Normative References in 45001

Normative references: There are no normative references.

Informative references: There are no informative references.

Composite: 3.0 Definitions in Z10.0: 3. Terms and Definitions in 45001

As is typical in ANSI and ISO standards, definitions are given for some of the terms used in the standard. Safety professionals should become familiar with them. In Z10.0, the definitions are in alphabetical order. In 45001, they are in the order in which they appear in the standard. An alphabetical index of those terms can be found on page 41 of 45001.

In a November 2018 draft of Z10, this was the definition given for **Acceptable Level of Risk**:

A level of risk which can be further lowered by an increase in resource expenditure or decrease in benefit that is disproportionate in relation to the resulting decrease in risk.

Note: This is a typical definition of "as low as reasonably practicable" (ALARP).

This is the definition that the Z10 committee decided upon for the 2019 version of its standard.

Acceptable Level of Risk: The level of risk to workers, resulting from exposure to hazards or system deficiencies, that is tolerated by the organization.

Had the Z10.0 committee decided to accept the November 2018 version, which included reference to ALARP—that would have been progressive.

Although the BSI publication on 18001 includes a definition of acceptable risk, there is no such definition in 45001. That BSI definition is:

Acceptable Risk: risk that has been reduced to a level that can be tolerated by the organization having regard to its legal obligations and its own OH&S policy.

Note the terminology "risk that has been reduced to a level."

COMPOSITE: 4.0 STRATEGIC CONSIDERATIONS: CONTEXT OF THE ORGANIZATION IN Z10.0: 4. CONTEXT OF THE ORGANIZATION IN 45001

This section of Z10.0 and 45001 is a new addition. It seems to represent a European view that has not been incorporated in ANSI standards as yet. A composite of the requirements follows.

An organization aligns the OH&S management system with the context of the organization by understanding the organization and its context; understanding the needs and expectations of workers and other interested parties; determining the scope of the OHSMS; and determining activities under the organization's control.

To understand the organization and its context: An organization determines the external and internal issues that are relevant to its purpose and vision that affect its ability to achieve the intended outcome(s) of its OH&S management system to provide safe workplaces and to prevent injuries, illnesses, and fatalities.

To understand the needs and expectations of workers and other interested parties: An organization determines the relevant needs and expectations (i.e., requirements) of workers and other interested parties; the other interested parties, in addition to workers, that are relevant to the OHS management system; and which of these needs and expectations are or could become legal requirements and other requirements.

To determine the scope of the OHS management system: An organization identifies the boundaries and applicability of the OH&S management system to establish its scope; assures that its OH&S system includes the activities, products, and services within the organization's control or influence that can impact the organization's OH&S performance.

With respect to the OH&S management system: The organization shall establish, implement, maintain, and continually improve an OH&S management system, including the processes needed and their interactions, in accordance with the requirements of this document.

COMPOSITE: 5.0 MANAGEMENT LEADERSHIP & EMPLOYEE PARTICIPATION IN Z10.0: 5. LEADERSHIP AND WORKER PARTICIPATION IN 45001

In both Z10.0 and 45001, Section 5 represents the most important requirement for a successful health and safety management system. Safety professionals will surely agree that top management leadership and effective employee participation are crucial for success.

Top management leadership is vital because it sets the organization's safety culture and because continual improvement processes cannot be achieved without the provision of adequate resources and sincere top management direction. A composite of key statements in the standards is as follows:

- Top management shall take overall responsibility for worker safety.
- Top management shall direct the organization to establish, implement, and maintain an OH&S.
- Top management shall ensure that the resources needed to have an effective occupational health and management system are provided.
- Top management shall provide the leadership to integrate the OH&S system into the general business practice.
- Top management shall ensure that responsibilities of individuals for implementation of an effective occupational health and safety management system are understood.

- An organization's top management shall establish a documented occupational health and safety policy.
- An organization's top management shall establish a process to ensure effective participation in the OH&S system by its employees at all levels.
- Top management must identify and remove, or reduce obstacles or barriers to participation by workers in the safety management system.

As management provides direction and leadership, assumes responsibility for the OH&S management system, ensures effective employee participation, and creates the organization's culture, the purpose of the standard must be kept in mind—to reduce the risk of occupational injuries, illnesses, and fatalities.

That will be done best if personnel in the organization understand that in the application of every safety and health management process, the outcome is to achieve acceptable risk levels, and that a special focus must be given to identifying the reality of causal factors for incidents that may result in injuries.

In both standards, the provisions for worker participation are extensive and accomplish close to the same purposes. A duplication of the opening sentence in 45001, following **5.4, Consultation and participation of workers**, gives an indication of how extensive are the provisions.

The organization shall establish, implement and maintain a process(es) for consultation and participation of workers at all applicable levels and functions, and, where they exist, workers' representatives, in the development, planning, implementation, performance evaluation and actions for improvement of the OH&S management system.

Additional comments on worker participation are made under the Planning requirements, which follow.

COMPOSITE: 6.0 PLANNING IN Z10.0: 6. PLANNING IN 45001

Planning is the first step in the PDCA process. As would be expected, this section sets forth the planning process to implement the standard and to establish plans for improvement. Planning is an ongoing and recurring process to help identify issues, prioritize them, and implement plans to achieve objectives. In the Definitions for Z10.0, OHSMS Issues are defined as:

> Hazards, risks, management system deficiencies, legal requirements and opportunities for improvement (p. 3).

In the continual improvement process, as elements in the standard are applied, information defining opportunities for further improvement in the safety and health management system is to be fed back into the planning process for additional consideration, and for additional risk reduction if feasible.

Reviews are to be made to identify the differences between existing operational safety management systems and the requirements of the standards. Reviews shall include information regarding:

- Relevant business systems and operational processes.
- Operational issues such as, hazards, risks, and controls.
- OH&S opportunities for improvement
- Identification of system deficiencies
- Previously identified OH&S issues.
- Allocation of resources.
- Applicable regulations, standards, and other health and safety requirements.
- Design of the workplace and work methods and how work is organized.
- Hazard identification and analysis.
- Risk assessments and evaluations.
- Process and mechanisms for employee participation.
- Results of audits.
- Legal requirements.
- Other relevant activities.

As was stated previously, both 45001 and Z10.0 require extensive involvement of workers. It is intended that workers be encouraged to

- have meaningful involvement in the structure, operation, and pursuit of the objectives of the occupational risk management system;
- identify tasks, hazards and risks, and possible control measures; and
- participate in planning, evaluation, implementation of control measures for risk reduction.

Chapter 2 in this book is titled Organizational Culture, Management Leadership, and Worker Participation. Worker participation is well covered in that chapter.

In 45001, there is a subsection within the Planning Section titled Hazard identification and assessment of risks and opportunities. To follow the requirements in 45001 under that caption, which leads to risk assessments, an organization shall "Establish, implement and maintain process(es) for hazard identification that is ongoing." A sampling follows of subjects to be considered for hazard identification.

- how the work is organized.
- design of facilities, equipment, and work areas.
- human factors.
- how the work is performed.
- actual and proposed changes in operations.
- changes in knowledge or information about hazards.

For risk assessment, in 45001, an organization shall establish a process(es) to assess, prioritize, and address its OH&S issues on an ongoing basis to:

- identify potential hazards.
- grade the potential severity of hazards.
- assess the risks which derive from recognized hazards.
- give appropriate consideration to exposure data and frequency of exposure.
- ensure that human behavior potentials are a part of the assessment.
- consider the effectiveness of existing controls.
- establish priorities, considering fatal and serious injury and illness potential.
- identify causal factors.
- establish a process to achieve risk reduction.

In addition, the organization, in its Planning, is to assess operations to seek opportunities for improvement of the OH&S system, such as by revising work methods and eliminating or reducing risks.

Priorities and objectives are to be established for amelioration to achieve the objectives to reduce illnesses and injuries. Implementation plans and allocation of resources should be periodically reviewed and updated as necessary to reflect changes in policies, objectives, activities, products, services, or operating conditions of the organization.

For all of the Planning activity, the organization shall maintain and retain documented information on the OH&S objectives and the plans to achieve them.

In Z10.0, it is said that "Assessments made in this section (the Planning section) are only for the purpose of prioritizing OH&S issues and may not be complete or sufficient to determine hazard controls."

COMPOSITE: 7.0 SUPPORT IN Z10.0:7. SUPPORT IN 45001

This section on Support is new for Z10.0. Its purpose is to assure that the resources are adequate to establish and maintain an effective occupational safety and health management system. Support includes the necessary human and financial resources, training and education, having competent personnel, effective communication and documentation.

Contents of the two standards in this section are markedly different. For example, the first subsection for 45001 after "Resources" is titled "Competence". In Z10.0, it is "Education, Training and Competence."

In 45001, for Competence, an organization is required to assure that workers are competent, that they are able to identify hazards through education, training, or experience, and retain the necessary documentation on competence. In a "Note" (not part of the standard) "Applicable actions can include, for example, the provision of training."

Training in 45001 is a "can" option for which the definition is to have a "possibility or a capability." Although the word training appears in several sections of 45001, the standard is short on training.

In Z10.0, under Education, Training and Competence, an organization is to provide for the competence of workers, supervisors, managers, engineers, contractors, etc. and ensure that they are competent through appropriate education, training, and other suitable means.

Training is to be ongoing and timely, and evaluations are to be made to assure that trainers are competent and that training is effective.

In 45001, for Awareness, workers are to be made aware of the occupational health and safety policy, its objectives, their ability to make contributions, incidents and the outcomes of investigations, hazards and risks, and the ability to remove themselves from work situations that present imminent and serious danger to their lives.

Under Communication in 45001, an organization is to determine: what, how, and when there is to be communication; and that information is provided to all who are interested in the organization's OH&S policy and its requirement for continual improvement.

In Z10.0, Awareness and Communication are treated in one subsection. Content includes pretty much of what is said in 45001, but it is more extensive. Certain provisions are of particular interest. One such provision requires that an organization ensures effective access to, and remove barriers to participation in, education and training. And for that subject, a Note is included that says:

> Examples of obstacles or barriers may include illiteracy, language differences reprisals (supervisory and/or peer) or policies, practices or programs that penalize or discourage communication. Incentive programs, drug testing programs, and disciplinary mechanisms, should be carefully designed and implemented to ensure workers are not discouraged from reporting job-related injuries, illnesses, hazards, and risks.

They say in Z10.0 that an organization should communicate on relevant changes to be made or are being made **to all levels** that would be affected. (Emphasis added). They also say that "communication should be tailored to the audience; and clear and timely communications about issues related to the OH&S is a key success factor."

General comments follow reflecting contents of both standards. They pertain to what should be considered when developing communications and the communication process.

- results of OH&S performance such as progress towards established metrics, lessons learned from incident investigations, findings from OH&S assessments, and outcomes from management reviews.
- results of monitoring and measurements to affected parties.
- effective means for obtaining and responding to worker input on OH&S issues.
- what level of understanding has been achieved of the OH&S system, hazards, and risks at various levels of the organization.

- external parties including visitors, neighbors, emergency services, insurers, and regulatory agencies.

COMPOSITE: 8.0 IMPLEMENTATION AND OPERATION IN Z10.0: 8. OPERATION IN 45001

This section defines the operational elements that are required for implementation of an effective OH&S management system. These elements provide the backbone of an OHSMS and the means to pursue the objectives from the planning system.

For both standards, general requirements are that an organization shall plan, implement, control, and maintain the processes needed to meet the requirements of this occupational health and safety management system. Examples of operational control of the processes include the following:

- use of procedures and systems of work.
- ensuring the competence of workers.
- establishing specifications for the procurement of goods and services.
- application of legal and/or manufacturers' instructions for equipment use.
- engineering and administrative controls.
- adapting work to workers.

Whenever practical, the operational elements should be integrated into existing business processes and systems.

Additional comments made here on hazard identification and risk assessment and control pertain to the content of Z10.0. Recall that hazard identification and risk assessment were covered in the Planning section of 45001. Some of the material here may be considered duplicators.

Identification of Hazards

An organization shall establish a process to identify hazards on an ongoing basis. Hazard identification processes should include, but not be limited to

- how work is organized.
- social factors (including workload, work hours, victimization, harassment, bullying, and isolation).
- leadership.
- organizational culture.
- routine and nonroutine activities and situations.
- human factor error potential.
- how work is performed.
- design of the work place.

- changes in knowledge of, and information about, hazards.
- high-risk operations, jobs, activities and tasks.
- potential for severity of injury or illness.
- incident and illness trends.
- occupational health exposure assessments.

Risk Assessment

An organization shall establish a process to assess, prioritize, and address risks on an ongoing basis. During the risk assessment process, routine and nonroutine activities should be considered, including hazards arising from:

- infrastructure, equipment, materials, substances, and the physical conditions of the workplace.
- product and service design, research, development, testing, production, assembly, construction, service delivery, maintenance, and disposal.
- human factors (includes ergonomics).
- how the work is performed.
- past relevant incidents, internal and external to the organization.

Incorporating occupational health and safety into the tools used by the business to operate leads to increased ownership by all members of the organization. OH&S should be a normal, integrated part of the overall operation of the business. Risk assessment processes should include, but not be limited to

- considering OH&S in the procurement process helps to keep hazards from being introduced into the workplace and can be part of the continuity of the supply chain.
- observing OH&S behaviors and condition as part walkthroughs of the operations helps to associate it with the "way business is done."
- including OH&S as an element of the maintenance program allows for rapid mitigation of hazards and opportunities to eliminate risks during preventative maintenance.
- when OH&S uses the same process to identify and track non-conformances as quality, manufacturing, and other parts of the organization, they get increased visibility and access to resources.
- adding OH&S to process improvement activities will further reinforce its importance and serves to better integrate it into the business.

Management system deficiencies may be found during hazards identification and when making risk assessments. What is important when such deficiencies are recognized is that improvement actions are taken. System deficiencies include resources, personnel, policies, practices, poor top management engagement, cultural issues, and condoning unacceptable risks.

8.4 HIERARCHY OF CONTROLS IN Z10.0: 8.1 ELIMINATING HAZARDS AND REDUCING OH&S RISKS IN 45001

In 45001, an organization shall establish, implement, and maintain a process to eliminate or reduce risks using the following hierarchy of controls.

- elimination.
- substitute with less hazardous processes, operations, or equipment.
- use engineering controls and reorganization of work.
- use administrative controls, including training.
- use adequate personal protective equipment.

While the same general direction is given in Z10.0—an organization shall establish a process for achieving an acceptable level of risk reduction based upon the following preferred order of controls—the "preferred order of controls" is duplicated here because it adds "Warnings" to the hierarchy.

- Elimination,
- Substitution of less hazardous materials, processes, operations, or equipment,
- Engineering controls,
- Warnings,
- Administrative controls, and
- Personal protective equipment.

When controlling a hazard, the organization should first consider methods to eliminate the hazard or substitute a less hazardous method or process. This is best accomplished in the concept and design phases for any new project or alteration.

If this is not feasible, engineering controls such as machine guards and ventilation systems should be the next approach. This process continues down the hierarchy until the highest level feasible control is found.

For example, if an equipment modification or noise enclosure (engineering control) is insufficient to reduce noise levels, then limiting exposure through job rotation and use of hearing protection would be an acceptable supplemental means of control.

Warning system effectiveness and the effectiveness of instructions, signs, and warning labels rely considerably on administrative controls, such as training, drills, the quality of maintenance, and the reaction capabilities of people. Further, although vital in many situations, warning systems may be reactionary in that they alert persons only after a hazard's potential is in the process of being realized (e.g., a smoke alarm).

Comment is necessary on this author's preferred use of the term "warning systems" over warnings or warning signs. The terms warnings and warning signs appear in some published hierarchies of control—as is the case in Z10.0. The entirety of the needs of a warning system must be considered, for which warning signs or warning devices alone may be inadequate. For example, the NFPA Life Safety Code 101 may require, among other things:

- detectors for smoke and products of combustion;
- automatic and manual audible and visible alarms;
- lighted exit signs;
- designated, alternate, properly lit exit paths;
- adequate spacing for personnel at the end of the exit path;
- proper hardware for doors; and
- emergency power systems

Obviously, much more verbiage than "Warnings" is needed in a hierarchy of controls.

Administrative controls include training, job planning, rotating and scheduling, changes to work procedures, implementation of work area protection (e.g., temporary barricades), and similar measures.

In 45001, there is a note following the hierarchy of controls which says that in some countries, legal requirements and other requirements say that personal protective equipment be provided at no cost to workers.

In Z10.0, the only mention of personal protective equipment is in a part of a sentence (Note 1) that says "lower order controls, (e.g., warnings, administrative controls, or personal protective equipment) are used to complement engineering controls to reduce risks to an acceptable level."

In Chapter 11—Hierarchies of Control—in this book, it is said that:

> Although the use of personal protective equipment is common and necessary in many occupational situations, it is the least effective method to deal with hazards and risks.
>
> Systems put in place for their use can easily be defeated. In the design process, one of the goals should be to reduce reliance on personal protective equipment to as low as reasonably practicable.

8.5 DESIGN REVIEW AND MANAGEMENT OF CHANGE IN Z10.0: 8.1.3. MANAGEMENT OF CHANGE IN 45001

In 45001, the comparable section is captioned Management of Change, only. There is no particularly identified section in 45001 for design review, although design is mentioned in several sections.

Requirements in Z10.0 are that an organization shall establish a process(es) to identify and take appropriate steps to prevent or otherwise control hazards at the design and redesign stages and when changes are made. The purpose of the process(es) is to avoid bringing new hazards and risks into the work environment, and thereby prevent the occurrence of injuries and illnesses.

8.5 DESIGN REVIEW AND MANAGEMENT OF CHANGE IN Z10.0

Consideration of hazards and risks should continue as the design is detailed and then comes into operation. Consideration must be ongoing during full life cycles to reflect current, changing, and future activities.

Processes instituted by organizations for design and redesign activities and for management of change shall include:

- identification of tasks and related health and safety hazards.
- recognition of hazards associated with human factors including human errors that are provoked by design deficiencies.
- review of applicable regulations, codes, standards, internal and external recognized guidelines.
- application of control measures (hierarchy of controls).
- determination of the appropriate scope and degree of the design review and management of change.
- employee participation.

The following are examples of conditions that should trigger a design review or a management of change process:

- new or modified technology (including software), equipment or facilities.
- new or revised procedures, work practices, and design specifications.
- different types and grades of raw materials.
- significant changes to the site's organizational structure and staffing, including use of contractors.
- modification of health and safety devices.
- new health and safety standards or regulations.

Small improvement events can fall outside the scope of the formal design review and management of change processes. This may result in hazards being brought into the workplace. It is critical that OH&S considerations be included in rapid improvement events and that risk assessments be conducted at the conclusion of small improvement events.

In 45001, the first instruction given for **Management of change** is that an organization shall establish a process(es) for the implementation and control of planned temporary and permanent changes that impact OH&S performance, including:

a. new products, services and processes, or changes to existing products, services and processes, including—work place locations and surroundings; work organization; working conditions; equipment; the work force.
b. changes to legal requirements and other requirements.
c. changes in knowledge or information about hazards and OH&S risks.
d. developments in knowledge and technology.

It also says in 45001 that an organization shall review the consequences of unintended changes, taking action to mitigate any adverse effects, as necessary.

Applicable Life Cycle Phases—Z10.0 Only

During the design review and management of change processes, all applicable life cycle phases shall be taken into consideration.

Note: An effective design review considers all aspects including design, procurement, construction, operation, maintenance, and decommissioning. The life cycle phases should integrate quality, health and safety, production, procurement, and consider potential impacts.

Process Verification in Z10.0 Only

A provision for process verification exists only in Z10.0. This is a composite of the Z10.0 requirement.

> Periodic assessments by internal and external professionals (same organization, different location or true externals) can often identify system deficiencies that could lead to OH&S risks and help to control them. Business reviews can also provide insight into the creation of risks, allowing management to proactively control them and to enhance operational excellence, sustainability and reliability.

Procurement

Although the requirements for procurement in 45001 and Z10.0 are plainly stated and easily understood, they are brief in relation to the enormity of what will be required to implement them. This is a composite of how the two standards read.

> Organizations shall establish, implement and maintain processes to control the procurement of products and services to ensure that they conform with their occupational health and safety management systems.
> It is important that the procurement methods not focus solely on financial or operational considerations, but instead embed occupational health and safety considerations in the same business processes for contractor selection, award and future approvals.

Acquiring procurement specifications pertaining to safety is not easy to do. They are considered proprietary by many companies. In addition to considering quality, financial, manufacturing, and cost, some companies do include OH&S as part of the process of supplier qualification.

Getting these procurement provisions established presents a huge challenge for safety professionals, but the benefits can be immense.

Contractors

This is a composite of what is said in 45001 and Z10.0 under Contractors: The organization shall coordinate its procurement process(es) with its contractors, to identify hazards and to assess and control the OH&S risks arising from:

a. the contractors' activities and operations that impact the organization.
b. the organization's activities and operations that impact the contractor's workers.
c. the contractor's activities and operations that impact other interested parties in the workplace.

Organizations are also to assure that its occupational health and safety management system is adhered to by contractors. It is important to recognize that contractors can create risks not only to themselves but also to the contracting organization.

It is therefore critical that an effective program for managing contractor OH&S be in place prior to beginning the work. Working with Procurement personnel to make OH&S requirements part of the contract agreement establishes a link to an important part of the organization.

Occupational Health

In the Background for 45001, they say that "An organization is responsible for the occupational health and safety of workers and others who can be affected by its activities." (0.1—p. 6). And the term OH&S is used throughout the standard. But there is no separate section for Occupational Health in 45001.

All that follows on occupational health pertains to Z10.0, and it is extensive. A few excerpts follow:

An organization shall establish an occupational health (OH) process(es) to protect the health of workers, including the following:

- anticipation, recognition, evaluation, control, and confirmation (that the control is working and effective) for chemical, physical, and biological agents, and ergonomic and psychosocial stressors that can adversely affect the health of workers.
- prevention, early detection, diagnosis, and treatment of work-related injuries and illnesses, including emergency care.
- recognition and reasonable accommodation for both work-related and non-work-related, medical conditions that may affect workers' abilities to perform their jobs safely and productively.
- integration with other aspects of this management system to ensure occupational health risk issues are addressed.

Outsourcing

This subject appears only in 45001. They say that organizations shall have outsourced functions and processes under control, that they meet legal requirements, and that they achieve the intended outcomes of the OH&S management system.

EMERGENCY PREPAREDNESS IN Z10.0: EMERGENCY PREPAREDNESS AND RESPONSE IN 45001

To meet the requirements of this provision, an organization is to have management systems in place to identify, prevent, prepare for, and/or respond to emergencies. A composite of requirements in both standards follows. Processes are to

- comply with legal and other requirements.
- schedule adequate training and retraining for all members at all levels of the organization.
- have provisions to periodically test the emergency response system through drills and to evaluate its performance.
- assure that sufficient emergency response resources are available, e.g., medical rescue, crisis response, law enforcement, fire departments, etc.
- contain the necessary information, internal communication, and coordination to protect all people, including contractors and visitors, in the event of an emergency.
- have provisions to inform and communicate with all workers, contractors, and visitors, as well as with the relevant authorities, the neighboring community, and emergency response and medical services.

9.0 EVALUATION AND CORRECTIVE ACTION IN Z10.0: 9. PERFORMANCE EVALUATION IN 45001

Generally, this Clause in both standards outlines the requirements for processes to evaluate the performance of a health and safety management system. Z10.0 includes requirements for taking corrective action when shortcomings are found. In 45001, the comparable requirements are in Clause 10—Improvement.

Monitoring, Measurement, Analysis and Performance Evaluation in 45001: Monitoring, Measurement, and Assessment in Z10.0

This section of 45001 opens with a statement indicating that an organization shall determine what needs to be measured in relation to

- legal requirements;
- its health and safety provisions;

- progress made; and
- effectiveness of its controls.

It is also required that an organization determines the methods to be used in the required processes and when the data is to be analyzed.

In Z10.0, they say that measures of performance may include various indicators for system effectiveness. Examples of system elements may include the following:

- management leadership.
- worker participation.
- measurement of progress toward achieving objectives and targets.
- support provided.
- quality of hazard identification and control.
- quality and completeness of risk assessment and risk reduction activities.
- indicators from occupational health risk assessment and risk indicators.
- occupational injury and illness rates.
- workplace inspections and testing.

For both standards—using the data developed, the organization is to evaluate its health and safety performance, communicate on the results, and take the actions necessary for improvement in the processes. Proper documentation is to be retained.

In 45001, a subsection titled Evaluation of compliance follows the subsection on Monitoring, measurement, analysis, and performance evaluation. Its opening statement indicates that an organization shall have processes in place to assure that legal and other requirements are met.

Incident Investigation

Although incident investigation is the next subject addressed in Z10.0, comments on this subject are not made in 45001 until the requirements for Clause 10—Improvement—are defined.

Requirements in Z10.0 are that an organization establishes a process to report, investigate, analyze, document, and communicate on incidents to learn from the event, improve the management system, and address OHSMS nonconformances and other factors that may be causing or contributing to the occurrence of incidents. Investigations shall be performed by personnel demonstrating competence in investigation procedures, conducted in a timely manner and include worker participation.

Incident investigation processes should define what needs to be investigated and when, who should participate, and how recommendations to prevent recurrence should be generated and communicated. For incidents to be investigated, they must be reported. Organizations should have all barriers to incident reporting removed.

They say that findings from incident investigations should be used for developing and implementing corrective action plans. Lessons learned from these investigations can then be fed into the planning or corrective action processes.

It should be understood that incidents may be symptoms of deficiencies in the occupational health and safety management system. If the ideal could be reached, hazards and their underlying system deficiencies would be identified before any injury or illness occurs.

They say in Annex A for Z10.0, at E9.2, that "Many techniques are available for incident investigation. Only a few of the tools used in business management systems are mentioned here:

- The Five Why Problem Solving Technique.
- Fishbone Analysis.
- Critical Incident Technique or CIT.
- 8D Analysis.
- FMEA – Failure Mode and Effects Analysis.
- Bow-tie Analysis.
- CAST – Causal Analysis using Systems Theory."

Where appropriate, the investigation process used to identify causal factors for incidents should be the same as the organization uses for incidents that occur in other disciplines.

Understanding the gap between work as imagined and work-as-done helps in an incident investigation. A just culture is needed to insure that workers feel secure when they disclose how work is actually performed.

Incident investigation is the subject addressed in Chapter 19 in this book.

Audits in Z10.0: Internal Audits in 45001

A summation of what is said in both standards about audits is given here. An organization shall:

- define the criteria for audits.
- achieve an understanding that audits are management system audits rather than compliance audits.
- include management shortcomings with respect to relative legal requirements.
- have audits made periodically with respect to application of the provisions in the occupational health and safety management system.
- ensure that audits are made by competent persons not attached to the location being audited (but those persons can be in the same organization).
- have auditors communicate immediately on potentials for serious injuries, illnesses, or fatalities so that swift corrective action can be taken.
- document and communicate the results.
- have management action taken on deficiencies mentioned in audit reports.

9.0 EVALUATION AND CORRECTIVE ACTION IN Z10.0

Audits are to measure the organization's effectiveness in implementing the elements of the occupational health and safety management system. An expanded treatise on audits is presented in this book in Chapter 18, Audit Requirements.

Corrective Action

For Z10.0, this section defines requirements for processes to take corrective action when nonconformances, system deficiencies, hazards, and incidents that are not being controlled to an acceptable level of risk are found. Processes are to be in place that:

- identify and address new and residual risks associated with corrective actions that are not being controlled to an acceptable level of risk;
- expedite action on high-risk hazards that could result in fatality or serious injury/illness (FSII) that are not being controlled to an acceptable level of risk;
- account for worker participation; and
- review and ensure effectiveness of corrective actions taken.

In Note 1, organizations are informed that they are to address all identified system deficiencies and inadequately controlled hazards through the corrective action process, regardless of how those deficiencies and hazards were identified.

In Note 2, organizations are advised that:

- while risks cannot typically be eliminated entirely, they can be substantially reduced through application of the hierarchy of controls;
- they are to assure that the residual risks are acceptable, which are defined as the remaining risk after controls have been implemented.

Organizations should utilize common business processes to get corrective actions taken because that will be more effective. An example given is to use an existing maintenance work order system.

This standard, Z10.0, also says that involvement of other parts of the organization, such as engineering, procurement, maintenance, and R&D to correct occupational health and safety issues will foster relationships that will support the integration of OSH into the business.

It is also said that when corrective actions for an identified hazard will require a significant period of time to implement, immediate interim corrective action should be taken. They say that expedited actions should consider eliminating the exposure by removing the persons at risk, discontinuing the operation, or reducing the risk.

Management Review in 45001

In 45001, management review is a subset of Performance Evaluation. Reviews shall include consideration of:

- the status of open items from previous reviews.
- audit results.
- input from workers.
- risks and opportunities that should be discussed.
- changes in external and internal issues that are relevant to the occupational health and management system.
- extent to which the OH&S policy and objectives have been met.
- trending of performance.
- notable incidents that have occurred since the previous review.
- opportunities for continual improvement.

Outputs from management reviews shall include all the actions to be taken to reduce risk and for continual improvement.

10. MANAGEMENT REVIEW IN Z10.0: 10. IMPROVEMENT IN 45001

Just previously in this chapter, a summary was given of the requirements for Management Review as in 45001. Comments immediately following here pertain to the provisions for management reviews in Z10.0. They are similar to what was written for management reviews in 45001, and different.

Requirements for the Improvement category in 45001 have substantially different purposes, as will be seen later in this chapter.

Management Reviews as in Z10.0

They say in Z10.0 that management reviews should evaluate how well the occupational health and management system is fulfilling its purposes and how well it is integrated with other business systems. Reviews by top management are required because management has the authority to make the necessary decisions about actions to be taken and resources to be allocated.

To be effective, the review process should ensure that the necessary information is available for top management to evaluate the effectiveness and adequacy of the OHSMS. Management reviews should consider:

- the results of investigations of injuries and illnesses;
- performance monitoring and measurement; audit activities, and
- other relevant data such as the degree to which OHSMS is integrated into the business.

Results of management reviews should be summarized, specifying top management commitments and directives, as well as action items. Action items should designate the individuals responsible for activities and completion target dates.

As with action items from other parts of the business, the status of OHSMS management review action items should be periodically reported to top management until they are completed. Where applicable, using the same processes used in other parts of the business will support the integration of OH&S.

Improvement as in 45001

Immediately after a general statement about Improvement in 45001 indicating that an organization shall determine opportunities to improve its OH&S management system, there is a subsection titled **Incident, nonconformity, and corrective action**. Incidents are joined together with nonconformity.

When an incident or nonconformity occurs, the organization shall have processes in place to:

- react in a timely manner;
- take control and correcting action; and
- deal with the consequences.

With the participation of workers and other interested parties, the incident or nonconformity is to be investigated to determine its causes and whether similar incidents have occurred previously. As appropriate:

- existing assessments of risks are to be reviewed;
- additional action is to be taken if needed;
- risks that particularly relate to the incident are to be assessed;
- effectiveness of corrective actions is to be evaluated; and
- changes are to be made to the OH&S system.

Communication and documentation on the incident or nonconformity is to be appropriate in relation to the mature of the occurrence.

In 45001, there is another subsection to **Improvement** titled **Continual Improvement**. Requirements are that an organization is to continually improve its OH&S system by:

- enhancing performance;
- promoting a supporting culture;
- encouraging participation by workers;
- communicating on continual improvement; and
- doing the proper documentation.

This author finds "promoting a supporting culture" to be of particular interest. Chapter 2 in this book is titled **Organizational Culture, Management Leadership, and Worker Participation**.

ANNEXES, BIBLIOGRAPHIES, AND GUIDANCE MANUALS (NOT A PART OF THE STANDARDS)

Annex A in Z10.0 provides considerable explanatory and advisory text on the following subjects.

- Scope, Purpose, and Application.
- Context of the Organization: Strategic Considerations.
- Management Leadership and Worker Participation.
- Planning.
- Support.
- Implementation and Operation.
- Evaluation and Corrective Action.
- Management Review.

Annex B in Z10.0 is a Bibliography that consists of references pertaining to the subjects in the standard. It runs many pages.

Annex A in 45001 is titled "Guidance on the use of this document." It is the only annex. It gives guidance on all of the sections and subsections in the standard.

In the Bibliography for 45001, there is a list of 10 ISO standards, 2 ILO Guidelines, 2 publications pertaining to OHSAS 18001 and 18002, and IEC publication (International Electrotechnical Commission).

It is expected that the Z10.0 committee will have a Guidance Manual available consisting of chapters on each of the main sections of the standard.

For OHSAS 18001:2007, the British Standards Institute produced OHSAS 18002:2008, *Occupational health and safety management systems—Guidelines for the implementation of OHSAS18001: 2007*. Quite probably, a similar document will be produced by BSI for 45001.

OBSERVATIONS ON THE STANDARDS, INCLUDING SIGNIFICANT PROVISIONS THAT ARE MISSING

Mental Health

Mention was made earlier in this chapter that a requirement in 45001 indicates that organizations are responsible for a worker's mental health, a subject not defined. Placing a responsibility on management to protect the mental health of workers is unusual of itself. Perhaps the requirement is unachievable and quite probably unmeasurable by safety practitioners.

If management is to have that responsibility, the standard should have defined the term, provided guidance on how to achieve it and how to know that it is being achieved.

Safety practitioners—take care. How does one determine if an organization is protecting the mental health of workers, not knowing what is expected?

This requirement is also a problem for certifying auditors who would attest that a safety and health management system in place meets the requirements of 45001.

Design Review and Management of Change

These are the Design Review and Management of Change requirements in Z10.0 at Section 8.5.

The organization shall establish a process to identify and take appropriate steps to prevent or otherwise control hazards at the design and redesign stages, and for situations requiring Management of Change, to reduce potential risks to an acceptable level. The process for design and redesign and management of change shall include:

A. Identification of tasks and related health and safety hazards;
B. Recognition of hazards associated with human factors, including ergonomics, human errors induced by the design or design deficiencies and where design makes successful job completion unnecessarily difficult;
C. Review of applicable regulations, codes, standards, internal, and external recognized guidelines;
D. Application of control measures (hierarchy of controls—Section 8.4);
E. A determination of the appropriate scope and degree of the design review and management of change; and
F. Worker participation.

Design requirements are extensive in Z10.0. They recognize that the most effective and economical method to avoid, eliminate, or control hazards and their accompanying risks is to address them in the design and redesign processes.

In 45001, the comparable section is titled "Management of change" only. Although the word design appears in several places in 45001, the standard is considerably short on requirements for preventing or otherwise controlling hazards at the design and redesign stages.

Having said that, the reader is asked to hold judgment until the comments that follow under "Annex A(Informative) in 45001" are read.

Training

In Z10.0, under Education, Training and Competence, an organization is to provide for the competence of workers, supervisors, managers, engineers, contractors, etc. and ensure that they are competent through appropriate education, training, and other suitable means.

Training is to be ongoing and timely, and evaluations are to be made to assure that trainers are competent and that training is effective. Training requirements in Z10.0 are extensive.

In 45001, for Competence, an organization is required to assure that: workers are competent; that they are able to identify hazards through education, training or experience; and retain the necessary documentation on competence. There is no major section in 45001 that includes the word training. Although the word training appears in several sections of 45001, the standard is inadequate on training. This author, because of his experience, believes that training is a vital element within an operational risk management system.

For a reference on the proper place and guidance that should be given to training, see the *Recommended Practices for Safety and Health Programs* issued by OSHA in October 2016. Education and Training is a major section in the *Recommended Practices*.

Occupational Health

In the Background data for 45001, they say that "An organization is responsible for the occupational health and safety of workers and others who can be affected by its activities." (0.1). And the term OH&S is used throughout the standard. But there is no separate section for Occupational Health in 45001.

Occupational Health requirements in Z10.0 are broadly stated, fit an important need, and give good guidance. Writers of 45001 could say that their standard is all-inclusive and that recognizing the health requirements specifically is unnecessary.

Nevertheless, on this subject, 45001 is inadequate. It is easy to ignore the health exposures, which is too often the case.

Annex A (Informative) in 45001

There is one Annex in 45001. It briefly provides guidance on application of the standard's provisions. All major captions in the Annex are prefaced by the letter A.

With respect to the several excerpts from Annex A cited here, this question is asked:

> Could it be that leaders in the development of 45001 were aware that other elements should have been included in a current occupational risk management standard but could not get agreement on those additional provisions?

Provision A.6.1.2.1 is titled Hazard identification. Terminology following that caption is very close to the wording in the *Prevention through Design* standard, the designation for which is ANSI/ASSE Z590.3-2011 (R2016). This is an excerpt from A.6.1.2.1.

> The ongoing proactive identification of hazard begins at the conceptual design stage of any new workplace, facility, product or organization. It

should continue as the design is detailed and then comes into operation, as well as being ongoing during its full life cycle to reflect current, changing and future activities.

Supportive wording is included in A.6.1. Item (f) pertains to opportunities to improve OH&S performance. One of the examples given follows.

Integrating occupational health and safety requirements at the earliest stage in the life cycle of facilities, equipment or process planning for facilities relocation, process re-design or replacement of machinery and plant.

That is good guidance. Terminology in Annex A implies that those who adopt 45001 should apply prevention through design concepts. It would have been appropriate for such provisions to be included in the standard.

Had that been done and expanded, 45001 would not have been short on design, as was previously noted.

Subjects Pertaining to Both 45001 and Z10.0

A.8 in 45001 pertains to Operations. A General Statement on Operations is provided at A.8.1. It follows.

Operational planning and control of the processes need to be established and implemented as necessary to enhance occupational health and safety, by eliminating hazards or, if not practicable, by reducing the OH&S risks to levels as low as reasonably practicable for operational areas and activities.

ALARP is the well-known acronym for "as low as reasonably practicable." Important guidance would have been provided for users of 45001 and Z10.0 if the requirements implied by ALARP were included in the standards.

A.8.1 also provides examples of operational controls. Item (c) says "establishing preventive or predictive maintenance and inspection programmes." Having effective inspection programs provide several advantages.

There is no inspection provision in 45001 or in Z10.0. (For an example of an inspection provision, see the previously mentioned *Recommended Practices for Safety and Health Programs* issued by OSHA in October 2016.)

Preventive or predictive maintenance is vital for effective occupational risk management. Having effective maintenance programs are required to assure operational integrity. **There is no preventive maintenance provision in 45001 or Z10.0.**

It is of interest that OSHA's Voluntary Protection Program (VPP) includes requirements for periodic inspections and for maintenance activities.

CONCLUSION

This chapter presents an overview of 45001 and Z10.0 only. It is to advise briefly on the contents of the standards. Additional chapters follow on selected subjects for which only brief comments are made here.

REFERENCES

ANSI/ASSP Z10.0-2019. *Occupational Health and Safety Management Systems*. Park Ridge, IL: American Society of Safety Professionals, 2019.

ANSI/ASSP/ISO 45001-2018. *Occupational Health and Safety Management Systems—Requirements with Guidance for Use*. Park Ridge, IL: American Society of Safety Professionals, 2018.

ISO 14001:2015. *Environmental Management Systems—Requirements with Guidance for Use*. Geneva, Switzerland: International Organization for Standardization, 2015.

OHSAS 18001:2007. *Occupational Health and Safety Management Systems—Requirements*. London, UK: British Standards Institution (BSI), 2007.

OHSAS 18002:2008. *Occupational Health and Safety Management Systems—Guidelines for the Implementation of OSHSAS 18001*. London, UK: British Standards Institution (BSI), 2008.

OSHA. *Recommended Practices for Safety and Health Programs*. Washington, D.C.: OSHA, October 2016. Available at https://www.osha.gov/shpguidelines/. Accessed May 31, 2019.

FURTHER READING

ANSI/ISO/ASQ Q9001-2015. *American National Standard: Quality Management Systems—Requirements*. Milwaukee, WI: American Society for Quality, 2000.

ISO – Comments on 45001. Available at https://www.iso.org/iso-45001-occupational-health-and-safety.html. Accessed May 31, 2019. n.d.

OSHA's VPP Site-Based Participation Evaluation Report. Available at http://www.osha.gov/dcsp/vpp/vpp_report/site_based.html. Accessed May 31, 2019. n.d.

CHAPTER 2

ORGANIZATIONAL CULTURE, MANAGEMENT LEADERSHIP, AND WORKER PARTICIPATION

Safety is culture-driven. Everything that occurs or doesn't occur that relates to safety is a reflection of an organization's culture. Over time, every organization develops a culture, positive or negative, although it may not realize that it has one.

Culture is mentioned several times in 45001 and Z10.0. Slowly, recognition is emerging among safety practitioners that a positive organizational culture is vital for success. Section 10.3 in 45001 sets forth the requirements for continual improvement. This paraphrase of an element in that section that pertains to this chapter:

> The organization shall continually promote a culture that supports an occupational health and safety management system.

Continual improvement is also dominant in Z10.0. Three examples follow from the statements that make reference to continual improvement, all taken from Z10.0's Introduction:

- Sustainable growth encourages organizations to continually improve all facets of their business.
- The purpose of the standard is to provide organizations an effective tool for continual improvement of their occupational health and safety performance.
- The design of ANSI Z10.0 encourages integration with other management systems to facilitate organizational effectiveness using the elements of Plan-Do-Check-Act (PDCA) model as the basis for continual improvement (p. 7).

Advanced Safety Management: Focusing on Z10.0, 45001, and Serious Injury Prevention, Third Edition. Fred A. Manuele.
© 2020 John Wiley & Sons, Inc. Published 2020 by John Wiley & Sons, Inc.

Continual improvement initiatives cannot be successful unless an organization's culture demands continual improvement and that implies that sincere top management direction is to be applied.

Section 5 in Z10.0 is captioned "Management Leadership and Worker Participation." In 45001, the title of Section 5 is "Leadership and worker participation." These sections pertain to the most important element in a safety and health management system.

Safety professionals will surely agree that top management leadership is crucial and that success cannot be achieved without it. As management provides the leadership and makes decisions directing the organization, the outcomes of those decisions establish its safety culture.

Major improvements in safety will be achieved only if a culture change takes place—only if major changes occur in the reality of the *system of expected performance*.

Management owns the culture. An organization's culture is represented by the application of its goals, performance measures, and sense of responsibility to its employees, to its customers, and to its community—all of which are translated into a *system of expected performance*.

Over the long term, the injury and illness experience attained are a direct reflection of an organization's safety culture.

Discussions in this chapter relate to

- A System of Expected Performance.
- Roles of Safety Professionals with Respect to the Safety Culture.
- Absolutes for Management to Attain Superior Results.
- Solving an Immediate Hazard/Risk Problem May Be Only a Part of the Fix: A Culture Change May Also Be Needed.
- Positive Safety Culture Defined.
- Evidence of the Culture in Place.
- How an Organization's Positive Safety Culture Is Created.
- Characteristics of a Positive Safety Culture.
- Characteristics of a Negative Safety Culture.
- Several Safety Cultures May Exist at the Same Time.
- Safety Culture and Safety Climate.
- A Very Unusual Survey Instrument.
- Proposing an Internal Analysis of the Safety Culture.
- Management Leadership and Worker Participation as in Z10 and 45001.
- A Case Study.

A SYSTEM OF EXPECTED PERFORMANCE

Strong emphasis is given to the phrase *a system of expected performance* because it defines what the staff believes what management wants done. Assume that an organization issues desirable policies, manuals, and standard operating procedures. But the perception the staff has of what is really expected of them—*the system of expected performance*—may differ from what is written.

Colleagues remind this author of having written years ago that management is what management does, which may differ from what management says. What management does defines the actuality of an organization's safety culture and its commitment or noncommitment to safety. Having superior management leadership is an absolute requirement if the goal is to achieve stellar results.

Where there is a passion for superior results, management insists that its hazard and risk problems be identified and resolved. This insistence by management to be informed of the reality of management system deficiencies is vital to achieving stellar results.

ROLES OF SAFETY PROFESSIONALS WITH RESPECT TO THE SAFETY CULTURE

In an organization, where safety is a core value and management at all levels walks-the-talk and demonstrates by what it does that it expects the safety culture to be superior, the role of safety and health professionals is to give advice that supports and maintains the culture.

In a large majority of organizations, an advanced safety culture does not exist. Then, the role of safety and health professionals as culture-change agents has greater significance and requires more diligence as attempts are made to influence management to move toward achieving a superior culture.

Possibilities of being successful in such endeavors are enhanced if the safety professional attains the status of an integral member of the business team. That could result from the safety professional giving well-supported, substantial, and convincing risk reduction advice that serves the interests of the organization.

It must be recognized that convincing management that safety should be one of the organization's core values may not be easily achieved. Safety professionals should understand that as steps forward are taken by management to improve on management system deficiencies, the result in each instance is a culture change. And the requirements to achieve a permanent culture change should be intertwined into each proposal made to improve on a management deficiency. Chapter 3 in this book is titled "Safety Professionals as Culture Change Agents."

ABSOLUTES FOR MANAGEMENT TO ATTAIN SUPERIOR RESULTS

During a review of statements made in annual reports on safety, health, and environmental controls issued by five companies that consistently achieve outstanding results, a pattern became evident that defines the absolutes necessary to attain such results. Elements in that pattern are as follows:

- Safety considerations are incorporated within the company's culture, within its expressed vision and core values, and its *system of expected performance*.
- The governing board and senior management lead the safety initiative and make clear by what they do that safety is a fundamental within the organization's culture.
- There is a passion for and a sense of urgency for superior safety results.
- Safety considerations permeate all business decision-making, from the concept stage for the design of facilities and equipment, through to their disposal.
- An effective performance measurement system is in place.
- All levels of personnel are held accountable for results.

Whatever the size of an organization—10 employees or 100,000—the foregoing principles apply to achieve superior results. Safety is culture-driven, and the board of directors and senior management define the culture and the *system of expected performance*.

To achieve superior results, management must establish open communications so that knowledge about hazards and risks flows upward to decision-makers. Proof of management's wanting to know about hazards and risks is demonstrated by the actions they take to eliminate or control hazards and risks.

Safety professionals should be working toward influencing management that it is in their best interest to have processes in place to find, face, and act on hazards and the risks that derive from system shortcomings. Yet realism with respect to the management practices in some companies must be acknowledged. Unfortunately, what Whittingham (2004) wrote in *The Blame Machine: Why Human Error Causes Accidents* speaks of actuality, sometimes. He wrote

> Organizations, and sometimes whole industries, become unwilling to look closely at the system faults which caused the error. Instead the attention is focused on the individual who made the error and blame is brought into the equation. (Preface)

Readers will not find a statement in this book indicating that the role of the safety professional as a culture change agent is easily fulfilled. As many authors have written, achieving a culture change is difficult, takes time, and absolutely requires support and direction from senior management.

Yet the endeavor is worth undertaking and attaining positive results, perhaps in small steps, can be rewarding.

SOLVING AN IMMEDIATE HAZARD/RISK PROBLEM MAY BE ONLY A PART OF THE FIX: A CULTURE CHANGE MAY ALSO BE NEEDED

This subject is important. Assume that appropriate attention is given to the hazards/risk situation. Safety professionals must become aware that the fix is not complete unless the management deficiencies that allowed the hazard/risk problem to exist are also addressed. Consider the following example:

Safety professionals in an organization were elated about how they had brought prevention through design concepts to the shop floor. All categories of employment were involved. First-level workers became convinced that supervision would respond favorably to their comments and a large number of improvements were made that reduced the risks in that work setting. Almost all of the improvements made were related to workplace or work methods design.

But neither operations people nor safety practitioners had communicated information on the design changes made to the personnel who designed the workplace and the work methods. So designers would continue to design processes as they had in the past, not knowing that their designs included the hazardous situations that were corrected by operations personnel.

What was needed to complete the fix was a change in the *expected performance* at a senior level in the design process so that fewer hazards were included in their designs.

Experienced safety professionals could cite many similar situations in which fixing a hazards/risk situation, of itself, was not adequate unless the management deficiencies that related to the existence of the situation were also corrected.

This idea should be kept in mind as attempts are made to fix things because of OSHA citations. While it may be satisfactory for OSHA personnel if the subject of the citation is fixed, safety professionals would be aware that attention is also needed to fix the deficiencies in the management system that are relative to the subject of the citation.

POSITIVE SAFETY CULTURE DEFINED

Unfortunately, there are far too many descriptions of a safety culture, many of which indicate that personnel at all levels share values. Composite examples follow:

- Safety culture is defined as the shared values, beliefs, assumptions, and norms that prevail in an organization.
- Safety culture reflects attitudes, beliefs, perceptions, and values pertaining to safety that employees and management share.
- Safety culture is defined as the attitudes, values, norms, and beliefs which a particular group of people share.

As in the foregoing and in many other definitions of safety culture, terms such as the following appear shared beliefs, attitudes, values, and norms of behavior; shared assumptions and entrenched attitudes and opinions which a group of people share.

This author has used such terms in the past. But, on reflection, they may not relate to the reality of an existing safety culture. In some organizations, while management believes that operational risks are acceptable, employees may have a different opinion and conclude that risks are unacceptable. Thus, employees may not share the views and beliefs held by management with respect to safety and the operational risk levels. A safety culture can be positive or negative.

A safety culture is a subset of the overall organizational culture. To achieve stellar results in operational risk management, an organization must have a positive safety culture that meets the following composite definition:

> A positive safety culture exists when the decisions made by the governing body and senior management create a *system of expected performance* in which employees believe, and perform as though, safety is to be achieved in all operations.

Experience has shown that most often there is a difference between work as imagined—as described in standard operating procedures—and how the work is actually done. If supervisors and department heads approve of the way the work is actually done, that becomes the *system of expected performance* for which workers believe they will be held accountable.

Changes made in how the work is done may add to risk, be neutral, or be less risky. Safety professionals should be cognizant of the possible variations in outcomes of changes when risk assessments are made.

Entries are common in incident investigation reports indicating that the causal factor was that—the worker did not follow the standard operating procedure. Variations from written procedures are a norm and that is to be accepted.

In such situations, the culture should require inquiry to determine why the worker did not follow the written and the established procedure and whether supervision condoned the variation.

EVIDENCE OF THE CULTURE IN PLACE

With respect to safety, an indication of an organization's culture is demonstrated through the design decisions that determine what the facilities, hardware, equipment, tooling, materials, layout and configuration, the work environment, and the work methods, are to be.

Where the culture demands superior safety performance, the socio-technical aspects of operations are well balanced—design and engineering, management and operations, and the task performance requirements.

> For all facilities, the physical look and feel of the workplace send messages about what senior management really expects. You can see and feel the culture, positive or negative.

HOW AN ORGANIZATION'S POSITIVE SAFETY CULTURE IS CREATED

The first three items in *A Socio-Technical Model for an Operational Risk Management System* (Figure 6.1) are duplicated here. They represent the management decision-making that results in a positive safety culture. Those three elements are the following:

- The governing body and senior management establish a culture that requires having barriers and controls in place to achieve and maintain acceptable risk levels for people, property, and the environment.
- Management leadership, commitment, involvement, and the accountability system, establish that the performance level to be achieved is in accord with the culture established.
- To achieve acceptable risk levels, management establishes [appropriate] policies, standards, procedures, and processes.

What management does or does not do and the effectiveness of the accountability system determine what the culture is to be. Over the long term, the incident experience that the organization attains—injuries, illnesses, fatalities, and property and environmental damage—is a direct reflection of an organization's safety culture.

CHARACTERISTICS OF A POSITIVE SAFETY CULTURE

Those organizations in which actions taken by the governing bodies and senior management create a positive safety culture have effective communication systems—up and down—in which all levels of employment participate. A few of the characteristics of a positive safety culture are as follows:

- Senior management leadership, commitment, involvement, and the accountability system make clear that safety is a fundamental within the organization's vision, core values, and culture and demonstrate a passion and a sense of urgency for superior safety-related results.
- Management commitment is demonstrated continuously by what management does and by making decisions indicating that its *system of expected performance* is in accord with the positive culture established.
- Safety considerations permeate all business decision-making, from the concept stage for the design of facilities and equipment through to disposal.
- Risks are acceptable.
- Employees are encouraged to participate in safety considerations. Employee knowledge and experience are respected. Their viewpoints are seriously considered.
- Leaders often going to the work, intensely listening and interacting, asking for input and decisively acting on commitments made.

- Managers and supervisors understand that when employees raise a safety issue, it is another opportunity for improvement. This mindset allows them to respond positively to the employee that raised the issue and escalate the issue far enough to get a solution put in place in a timely and efficient manner.
- Training is extensive for all levels of employment.

CHARACTERISTICS OF A NEGATIVE SAFETY CULTURE

A depiction of the management shortcomings in a negative safety culture is provided by an internally produced report by BP Products, North America, pertaining to a fire and explosion that occurred on March 23, 2005, at an owned and operated refinery in Texas City, Texas, USA. As a result of that incident, 15 people were killed and over 170 were harmed.

It is important to note that the following excerpts from that report present a self-evaluation. They are taken from a paper which is on the Internet and titled Fatal Accident Investigation Report, Isomerization Unit Explosion Final Report, Texas City, Texas, USA. This report was approved for release by J. Mogford (2005), the Investigation team leader, and is known as the Mogford Report.

- Over the years, the working environment had eroded to one characterized by resistance to change, and lacking of trust, motivation, and a sense of purpose.
- Coupled with unclear expectations around supervisory and management behaviors this meant that rules were not consistently followed, rigor was lacking and individuals felt disempowered from suggesting or initiating improvements.
- Process safety, operations performance, and systematic risk reduction priorities had not been set and consistently reinforced by management.
- Many changes in a complex organization had led to the lack of clear accountabilities and poor communication, which together resulted in confusion in the workforce over roles and responsibilities.
- A poor level of hazard awareness and understanding of process safety on the site resulted in people accepting the levels of risk that are considerably higher than comparable installations.
- Given the poor vertical communication and performance management process, there was neither adequate early warning system of problems, nor any independent means of understanding the deteriorating standards in the plant.

This BP Report says, better than an outsider could say, how management decision-making can result in a culture that may lead to the occurrence of incidents having serious consequences.

Many organizations are members of associations that relate to their business and some of those associations collect OSHA type incident data which may be discussed at their meetings. That creates a sort of competition, and the organizations that continuously have incident experience falling in the top two quintiles may be boastful about their achievements and needle the representatives of organizations having experience in lower quintiles.

If an organization's incident statistics are always in the bottom quintiles and management condones those results, believing them to be satisfactory, the probability is that the safety culture in that organization is negative. In such a situation, an organization's safety culture may be more indicative of what management has not done—actions that have not been taken.

SEVERAL SAFETY CULTURES MAY EXIST AT THE SAME TIME

An organization may have several safety cultures—subcultures—and safety professionals should be aware of them and deal with opportunities as they may arise. Although the organization established the culture, its application is substantially influenced, positively or negatively, by the likes, dislikes, and personalities of department heads, supervisors, and line employees. It can be expected that each location or department has its own safety culture—a modification of the corporate safety culture.

Assume that a location manager is newly assigned, and she believes that safety should be a core value and important to her success. She notes that the risk levels considered acceptable at the location, reflecting on the safety culture in place are

- not in accord with corporate policy, and
- contrary to what she thinks is appropriate.

She finds a way to display that she has taken the leadership role to reduce injuries—such as by

- holding the people who report directly to her accountable,
- placing the review of incident reports at the top of the agenda for her management meetings,
- insisting on improved incident reports that include the reality of causal factors,
- requiring improved workplace order, and
- personally following some hazard/risk situations to a conclusion by going to the work.

She expects to encounter a time of doubt and the normal resistance to change. She is aware of the magnitude of her undertaking. Through what she does, personnel slowly come to believe that she is serious about reducing risks and the culture slowly shows improvement.

SAFETY CULTURE AND SAFETY CLIMATE

Often in safety-related literature, safety culture and safety climate are used as exchangeable terms. They may be used in the same sentence as though they had identical meanings. But the meanings of the two terms are separated here.

An organization's safety culture is an outcome of the decisions and actions taken by the governing board and senior management over a long term, and those decisions have an impact throughout all operations at all locations.

An organization's culture usually has a long life and is the dominant and governing factor with respect to the risk levels attained—acceptable or unacceptable.

Safety climate is specific to a given operation. A composite definition of safety climate is the beliefs and perceptions that employees have pertaining to safety in a given area of operations at a given time.

EVALUATING THE CULTURE OR THE CLIMATE IN PLACE

Culture and climate assessment have become a big business. Enter the Internet, and it will be apparent that there are a large number of suppliers who have developed culture/climate survey instruments, and they will be pleased to provide their services.

Most of what is being offered by independent culture and climate surveyors are really climate surveys. They can be valuable because climate surveys may contain indicators of culture shortcomings. And a safety climate can be measured with a reasonably good degree of accuracy.

Very importantly, if a climate survey is to be made, whether internally or externally, arrangements must be made for management to be supportive of the endeavor and be prepared to respond to the results of the survey. Since employees in different areas of an operation may have substantially different views, safety professionals should determine how broad the survey is to be, what they want to accomplish by having a survey made and how and to whom the outcome of the survey is to be communicated.

Perception surveys could be considered as being in the same class as climate surveys. They have also been used to provide measurements of the quality of safety management in place, using an interview method or the completion of written questionnaires.

Perception surveys are outcome surveys in which the perceptions people have of how safety is managed derive from their observations of what got done or did not get done with respect to the elements covered in the interview or questionnaire process.

Occasionally, situations arise in which it would advantageous to engage an independent provider to produce a climate or a perception survey and safety professionals should keep that in mind. An example follows.

In a former employment, a telephone call was received from a corporate safety director to arrange for an audit at one of his company's locations. And he recited all that he believed was to be found by the auditor. He was asked why he was arranging for an independent auditor to tell his company what he already knew. His response was that the report would be on an independent auditor's stationary and would get management attention. In his work environment, his reasoning was sound.

A VERY UNUSUAL SURVEY INSTRUMENT

All climate and perception survey instruments noted in the research done have the survey leader determine the opinions of employees on safety—except one.

This survey instrument was developed at Harvard's T. H. Chan School of Public Health, Center for Work, Health, and Well-being. It was funded by the National Institute for Occupational Safety and Health (NIOSH, n.d.). Its title is *Workplace Integrated Safety and Health (WISH) Assessment*.

What makes this survey tool unique is that it is meant to be completed by safety personnel. Completers of the survey form—safety professionals—would be recording their views of the safety management system in place at the locations chosen and then be in a position to establish culture-related priorities for further activities to be undertaken. Subject headings in the survey instrument are the following:

1. Leadership commitment
2. Participation
3. Policies, programs and practices focused on positive working conditions
4. Comprehensive and collaborative strategies
5. Adherence
6. Data-driven change

This survey instrument is recommended as a starting point from which safety professionals can and should develop their own culture survey instruments, with caution. Survey instruments should be structured particularly for the entity to be surveyed. It should also produce practicable and actionable outcomes. A caution is expressed about assuming that one survey instrument fits all locations. They do not.

Some safety professionals may say that the crafters of the Harvard survey instrument went beyond the areas of their influence on certain subjects. For example, safety professionals are to consider whether "The wages for the lowest paid employees in this organization seem to be enough to cover basic living expenses such as housing and food" (Item 5-e).

Nevertheless, this survey instrument provides a good base from which safety professionals can choose. Without substantial tweaking, the Harvard produced survey instrument could also be transformed into a self-evaluation tool to be used by senior management to evaluate its culture.

Safety professionals—think about that possibility. Getting a self-evaluation done by senior management, with the assistance of a safety professional, could result in major changes in what management does, thereby the culture would be improved.

PROPOSING AN INTERNAL ANALYSIS OF THE SAFETY CULTURE

Assume that management responded favorably to a suggestion made by a safety professional that an internally conducted survey of the organization's safety culture would be beneficial. Its purpose would be to gather the perceptions of all levels of employment on the quality of the safety management system in place.

Such a self-analysis provides data on the positive and negative effects of management leadership, the extent of employee participation, and whether an accumulation has developed of latent technical conditions and operating practices that could be the causal factors for low probability incidents that have severe consequences.

A survey mechanism is necessary for such a self-analysis. A basic survey guide follows. It is vigorously recommended that the outline not be used as presented.

For the survey mechanism to relate to the hazards and risks in a particular operation, it is necessary that management, assisted by a safety professional, add or delete items. Also, a scoring system for each item, compatible with practices applicable in the organization, should be included in the revision of the Guide so that a compilation of results can be more easily made. In many situations, a simple yes, no, and not applicable scoring system will suffice.

It must be emphasized—this guide is not offered as a one-size-fits-all instrument.

A SAFETY MANAGEMENT SYSTEM SURVEY GUIDE

1. Does management demonstrate by what it does that safety is a core value in our organization?
2. Is there a significant gap between what management says and what management does?
3. Has the staff reporting directly to the senior manager been held accountable, in reality, for a high level of safety decision-making?
4. Is this a safe place to work?
5. Are you asked to effectively participate in safety discussions and meetings?
6. Are you asked to provide input on safety matters that affect you directly?
7. Is your input on safety matters respected and considered valuable?
8. Do you believe that some of the equipment you operate or the work methods you are required to follow are hazardous and overly risky?

9. Do you believe you are free to report hazardous conditions and practices without reprimand?
10. Are you encouraged to report hazardous conditions and practices?
11. Does your supervisor effectively give safety a high priority?
12. Is accident investigation in sufficient depth to identify the reality of causal factors (organizational, cultural, design and engineering, technical, procedural)?
13. Is there a broadly held belief that unsafe acts of workers are the principle causes of accidents?
14. Is safety often relegated to a lower status and overlooked when there are production pressures?
15. Have you been given adequate training on hazards, risks, and safe operating procedures?
16. Does the organization's culture accept gradually escalating risk?
17. Does the organizational structure enhance or dissuade adequate safety decision making?
18. Are there organizational barriers that prevent effective communication on safety, up and down?
19. Have streamlining and downsizing conveyed a message that efficiency and being on schedule are paramount, and that safety considerations can be overlooked?
20. Is staffing adequate in your group adequate so that work can be done safely?
21. Are you discouraged to report injuries?
22. Have technical and operational safety standards been at a sufficiently high level?
23. Has it been the practice to accept safety performance at a lesser level than the prescribed standard operation procedures?
24. Have known safety problems, over time, been relegated to a "not of concern" status and, thereby, become "acceptable risks"?
25. Has safety-related hardware or software become obsolete?
26. Are certain operations continued with the knowledge they are unduly hazardous?
27. Have budget constraints had a negative effect on safety decision-making?
28. Has inadequate maintenance resulted in an accumulation of hazardous situations that have gone attended, e.g., is the detection equipment adequate, maintained, and operable; are basic safety-related repairs unduly postponed?
29. Has adequate attention been paid to "near miss" incidents that could, under other circumstances, result in a major accident?
30. Are safety personnel encouraged to be aggressive when expressing their views on hazards and risks, even though their views may differ from those held by others?

62 ORGANIZATIONAL CULTURE, MANAGEMENT LEADERSHIP, AND WORKER PARTICIPATION

31. Has there been an over-reliance on outside contractors (outsourcing) to do what they cannot do effectively with respect to safety?
32. Are purchasing and contracting procedures in place to limit bringing hazards into the workplace?

MANAGEMENT LEADERSHIP AND WORKER PARTICIPATION AS IN Z10.0 AND 45001

In Chapter 1, "An Overview of ANSI/ASSP Z10.0-2019 and ANSI/ASSP/ISO 45001-2018," comments were made on the provisions in Z10.0 and 45001 pertaining to management leadership and worker participation. In both standards, the provisions for these subjects are appropriately lengthy because of their importance.

Generally, but not absolutely, it can be said that the provisions in both standards for these subjects are similar. For simplicity and to avoid unnecessary duplication, a verbatim copy of the requirements in Z10.0 is duplicated here. They are extensive as are the provisions in 45001.

To repeat for emphasis: Management Leadership and Worker Participation as in Z10.0 and 45001 are the most important subjects in the standards.

Z10.0

5.0 Management Leadership and Worker Participation

This section defines the requirements for management leadership and worker participation. Top management leadership and effective worker participation are crucial for the success of an Occupational Health and Safety Management System (OHSMS).

5.1 Management Leadership

5.1.1 Occupational Health and Safety Management System Top management shall direct and support the organization to establish, implement, continually improve and maintain an OHSMS in conformance with the requirements of this standard that is appropriate to the nature and size of the organization and its OHS risks.

Note: Leadership begins with top management providing the directive for integrating OHS into the daily functions of the business and promoting a strong organizational culture that supports the OHSMS.

5.1.3 Responsibility and Authority Top management shall provide leadership and assume overall accountability for implementing, maintaining, and monitoring performance of the OHSMS; including the following:

A. Providing appropriate financial, human and, organizational resources to plan, implement, operate, check, correct, and review the OHSMS;
B. Defining roles, assigning responsibilities, establishing accountability, and delegating authority to implement an effective OHSMS for continual improvement. (See Note)
C. Integrating the OHSMS within the organization's other business activities systems and processes (strategic planning, operations, support functions) and assuring the organization's performance review, compensation, reward, and recognition systems are aligned with the OHS policy and the OHSMS objectives.
D. The significant OHSMS roles and responsibilities shall be documented and communicated.

Workers shall assume responsibility for aspects of OHS over which they have control, including adherence to the organization's OHS rules and requirements.

Note: Establishing a set of desired OHS leadership traits/actions for leaders at all levels may drive the organization toward the organization's desired OHSMS objectives (Section 6.3). Outcomes could include Leadership Accountability, Management System Process effectiveness, Risk Management, and Operational Excellence.

5.2 Worker Participation

The organization shall establish a process to ensure effective participation in the OHSMS by its workers at all applicable levels of the organization, including those working closest to the hazards by (see Notes 1, 2, 3, and 4):

A. Including worker input and involvement in determining acceptable level of risk and in other decision-making within the scope of their responsibility. (Note 1 and 4)
B. Understanding the reasons that work may be done differently than planned through periodic discussion. (See Note 4)
C. Providing worker and worker's representatives' with the ability to participate in, at a minimum, the processes of: (See Note 5 and 6)
   - Context of the Organization—Strategic Consideration (4.0);
   - Management Leadership and Worker Participation (5.0);
   - Planning (6.0);
   - Support (7.0);
   - Implementation and Operation (8.0);
   - Evaluation and Corrective Action (9.0);
   - Management Review (10.0)

Workers shall be provided with the time and resources necessary to participate in the OHSMS.

D. Providing workers, and worker's representatives, with timely access to information relevant to the OHSMS and access to the decision-making process; and
E. Identify and remove, or reduce obstacles or barriers to participation. (See Notes 7 and 8)

Note 1: Determining the acceptable level of risk is a collaborative process that accounts for the needs and expectations of all affected workers, the needs of the organization and includes the input and involvement of workers in decision-making.

Note 2: This provision is applicable to all workers, but gives special emphasis to participation by nonsupervisory workers because they are often those closest to the hazards, and often have the most intimate knowledge of workplace hazards and processes.

Note 3: One aspect of worker participation may include consultation where input and dialogue with workers occurs before decisions are made affecting health and safety. Another aspect may include full worker participation in the decision.

Note 4: Participation in OHS processes and in issues with potential OHS implications should be integrated into the business systems. This includes participation from "across" the organization's departments, in addition to "up and down" the various "levels" of an organization.

Note 5: The difference between how work is planned and how it is actually done is referred to as the gap between work as imagined and work as performed. The gap represents the adaptive actions of workers to adjust work to current circumstances to get the work done. (Guidance Manual Chapter 2)

Note 6: Some workers may be more involved in the decision-making than others depending on their work assignment and role.

Note 7: Examples of obstacles or barriers include lack of response to worker input or suggestions, reprisals (supervisory and/or peer), or any policy, practice or program that penalizes or discourages participation.

Note 8: If used, incentive programs, voluntary wellness programs, drug testing programs, and disciplinary mechanisms should be carefully designed and implemented to ensure workers are not discouraged from effective participation in the OHSMS.

5.3 OHS Management System

The organization shall establish, implement, maintain, and continually improve an OHS management system, including the processes needed and their interactions, in accordance with the requirements of this document.

ORGANIZATIONAL CULTURE AND WORKER PARTICIPATION

An organization's culture is represented by, among other things, actions taken that display management's sense of responsibility to its workers. A safety culture is unsound if workers do not have opportunities to participate in an organization's occupational health and safety management system and if they do not have the mechanisms, time, and the resources necessary to participate.

This author has expressed the view that if an employer does not take advantage of the knowledge, skill, and experience of the workers close to the work, and its hazards and risks, opportunities are missed to reduce injury and illness potential and sometimes to improve productivity and lower unit costs. Two examples of outstanding contributions to risk reduction made by hourly workers come to mind.

- In a maker of heavy machinery, the innovations of tool and die makers in redesigning work situations to reduce ergonomics risks were so creative that visitors to the plant were often shown their inventions as a matter of pride.
- In a space industry location, it became standard practice for design engineers to seek the opinions of hourly workers before proceeding to manufacture what was designed. They learned through experience that the suggestions made by hourly workers resulted in:
  - risk avoidance,
  - correction of human factors design errors, and
  - improved efficiency in the production process.

Effectiveness of a safety and health management system is improved if the culture makes it clear that the knowledge of workers is valued and respected and that they are to participate in the safety management system.

This author has also taken the position that while workers should be well trained and participate in the safety management system up to their capacities, they should not be expected to do what they are not capable of doing. This is a composite of what is said in 45001 (p. 10) and Z10 (p. 22) about worker responsibilities and authority.

Workers at each level of the *organization shall assume responsibility for those aspects of the OH&S management system over which they have control, including adherence to the organization's OHS rules and requirements.*

A CASE STUDY

A safety director in a very large organization with about 13,000 employees read an article written by this author in which emphasis was given to the necessity of having a positive safety culture to achieve superior performance levels. That organization's work is considered high hazard and fatalities and serious injuries occur.

The safety director had concluded that the senior executive in his organization, to whom he reported, was somewhat removed from the leadership necessary to reduce fatalities and serious injuries and that he did not hold the staff reporting to him accountable for their incident experience.

And, the safety director sought help. During our discussions, it was agreed that he would approach his boss to convince him that the organization would be well served if the opinions of the staff were solicited on the quality of the safety management system in place. He did so, and it worked.

A Safety Management System Survey Guide comparable to that just previously shown was sent to the safety director. He worked up a version of it that fit the high-hazard operation with which he is involved. The survey guide was sent to a statistically adequate sampling of the staff—at all employment levels. Over 70% of the receivers of the survey guide responded. They took the survey seriously.

When the safety director analyzed the results, he found that the same shortcomings in the safety management system were recorded, largely, by all levels of employment. And there were many shortcomings. But most important, some of the staff members reporting directly to the senior executive who authorized the survey said that, for the department as a whole, safety was not a high-level value.

During a meeting attended by this author with the safety director, his boss, and other interested persons, the senior executive was well prepared with questions about how superior safety results were achieved elsewhere. He was surprised by the results of the culture survey. He learned that every level of the organization said that they wanted safety to be given a higher status.

As the discussions proceeded, this author asked the senior executive to draw an organizational chart starting with his position and showing the positions of all who reported to him. He was quick and acknowledged that if he wanted risks of injury and fatalities to be reduced, he would have to provide the leadership and hold the staff reporting to him accountable for results.

Subsequently, the senior executive convened his staff and laid out what he expected of them. A bulletin was issued throughout the organization setting forth the safety policy and the procedures to implement it. Safety is now an agenda item for several levels of management meetings, division heads are holding the staffs reporting to them accountable for results, and management is encouraging input and involvement from all levels of employees.

Communication downward and upward has improved. Safety-related suggestions get quicker attention than formerly.

CONCLUSION

Safety is culture-driven, and leaders create the culture. It is the responsibility of leadership to change the safety culture when it is deficient. Top management must take responsibility for operational risk management by remaining alert to the effects their decisions have on the work system. Leaders are responsible for establishing the conditions and the atmosphere that lead to their subordinates' successes or failures.

As top management makes decisions directing the organization, its safety culture is established and that culture is translated into *a system of expected performance*. Where safety management systems are most effective, there is a commitment to

- finding the facts about hazards, risks, and deficiencies in management systems, regardless of any unpleasantness that may arise in the discovery process, and
- taking actions to achieve acceptable risk levels.

Nearly every author who writes on organizational change makes it clear that achieving permanent change is difficult. Safety professionals must be aware of the importance of giving attention to an organization's existing culture, how deeply certain practices are embedded within processes, who presumes to have ownership of them and the need to plan and communicate to achieve change.

For repetition and emphasis: Safety practitioners will not find a statement in this book indicating that a deficient safety culture can be easily changed. Yet, the endeavor is worth undertaking and attaining positive results, perhaps in small steps, can be rewarding.

Consider this premise: The entirety of purpose of those responsible for safety, regardless of their titles, is to manage their endeavors with respect to hazards so that the risks deriving from those hazards are at an acceptable level.

REFERENCES

ANSI/ASSP Z10.0-2019. *Occupational Health and Safety Management Systems*. Park Ridge, IL: American Society of Safety Professionals, 2019.

ANSI/ASSP/ISO 45001-2018. *Occupational Health and Safety Management Systems—Requirements with Guidance for Use*. Park Ridge, IL: American Society of Safety Professionals, 2018.

BP Mogford Report. Internet and titled Fatal Accident Investigation Report, Isomerization Unit Explosion Final Report, Texas City, Texas, USA. Available at https://people.uvawise.edu/pww8y/Supplement/-ConceptsSup/Work/WkAccidents/BPTxCityFinalReport.pdf. Accessed May 31, 2019. 2005.

NIOSH. Funded: "Workplace Integrated Safety and Health (WISH) Assessment". Harvard T.H. Chan School of Public Health, Center for Work, Health, and Well-being Assessment Introduction. Available at http://centerforworkhealth.sph.harvard.edu/resources/workplace-integrated-safety-and-health-wish-assessment. Accessed May 31, 2019. n.d.

Whittingham, R.B. *The Blame Machine. Why Human Error Causes Accidents*. Burlington, MA: Elsevier-Butterworth-Heinemann, 2004.

CHAPTER 3

SAFETY PROFESSIONALS AS CULTURE CHANGE AGENTS

When safety professionals give advice on improving operational risk management systems, their overarching role is that of a culture change agent. That premise applies universally. There are no exceptions. In their ordinary work, safety professionals are culture change agents, although they may not be aware of it.

When the second edition of this book was published in 2014, there were few promoters of the premise that safety professionals are culture change agents. That has changed. A journey into the Internet will reveal that several authors have written on the subject, there is a large amount of resource material, and that a goodly number of consultants are offering training courses on being effective change agents.

Also, presentations are being made at meetings of safety professionals on the function and the qualifications of a culture change agent. A few safety professionals say in their promotional material that they are skilled in culture change.

If safety professionals want to have additional recognition and visibility at a senior management level, to be viewed as a member of the management team, and to be considered as a contributor to achieving organizational goals, they could sharpen their culture change management skills and pursue the opportunities that are often presented or which they might create. This chapter

- Provides definitions of relative terms.
- Gives the characteristics of a change agent.
- Establishes that when solving a risk problem, a culture change may be needed to complete a fix.

Advanced Safety Management: Focusing on Z10.0, 45001, and Serious Injury Prevention, Third Edition. Fred A. Manuele.
© 2020 John Wiley & Sons, Inc. Published 2020 by John Wiley & Sons, Inc.

- Provides examples of safety professional involvement as culture change agents.
- Comments on opportunities and capabilities.
- Comments on the significance of management leadership.
- Explores the role of a safety professional with respect to the safety culture.
- Comments on drift.
- Discusses why change initiatives fail.
- Provides selected resources.
- Includes a basic guide.

WHAT DOES THE TERM "OVERARCHING" MEAN?

A composite of definitions for overarching, as found in dictionaries, is encompassing everything, embracing all else, including or influencing every part of something.

WHAT IS A CHANGE AGENT?

Definitions of a change agent are numerous. This definition is a composite that fits well with the safety professional's position.

> A change agent is a person who serves as a catalyst to bring about organizational change. Change agents assess the present, are controllably dissatisfied with it, contemplate a future that should be, and take action to achieve the culture changes necessary to achieve the desired future.

CHARACTERISTICS OF A CHANGE AGENT

Knowing that the following is close to the ideal, this list is offered for consideration by safety professionals who want to sharpen their skills. Successful change agents are

- Technically knowledgeable and capable.
- Good leaders and have notable management skills.
- Respectable as presenters.
- Forever inquisitive, seeking opportunities.
- Risk takers.
- Patient, yet persistent.
- Respectful of all levels of management, supervision, and the workforce.
- Thorough information gatherers and good at analysis.

- Knowledgeable about general business practices, financial terms, and language.
- Aware of the culture in place.
- Conscious of the need to acquire change management skills to be able to influence change.
- Capable when identifying what ought to be, deciding how to get there and influencing decision-makers to adopt their ideas.
- Able to tactfully inform management when they believe decisions made and actions taken may have negative results.
- Successful, most of the time, in their attempts to influence decision-makers.

Notice that some of these skills and attributes are business-related. Safety professionals, to be successful as agents of change, must operate within the business framework in the entities to which they give counsel. They could benefit from "A Short Course on Financial Management" as in Chapter 26 in the fourth edition of *On the Practice of Safety* (Manuele, 2013).

A reader could say that the qualifications previously listed are the same for the success of any safety professional. But those skills have to be more sharply honed for the role of a change agent.

SOLVING A HAZARD/RISK PROBLEM MAY BE ONLY A PART OF THE FIX: A CULTURE CHANGE MAY ALSO BE NEEDED

Good references for the premise of this section are ANSI/ASSP Z10-2019, the American National Standard titled *Occupational Health and Safety Management Systems* and the MORT User's Manual issued by the U.S. Department of Energy in 1992. MORT is an acronym for Management Oversight and Risk Tree.

In the definitions for Z10.0, they say that OHSMS Issues are: Hazards, risks, management system deficiencies, legal requirements, and opportunities for improvement (p. 3). There is a Note following the definition that reads: The term "system deficiency" includes both nonconformances and weaknesses.

MORT is unusual in that its premise is that where there are deficiencies in operation controls that contribute to an incident there will be related deficiencies in management systems (p. 11). That statement implies that insufficient barriers and controls that are causal factors for incidents may derive from deficiencies in management systems. As the practice of safety evolves, that premise has become more significant.

Now, comments follow on the real-world instance that relates to "deficiencies in management systems; and to achieving a culture change."

Safety professionals in an organization were elated about how they had brought prevention through design concepts to the shop floor. All categories of employment were involved. First-level workers became convinced that supervision would respond favorably to their comments and a large number of improvements were made that reduced the risks in that work setting. Almost all of the improvements made were related to workplace or work methods design and were in the ergonomics field. Obviously, there were deficiencies in the design management system.

But neither operations people nor safety professionals had communicated information on the design changes made to the personnel who designed the workplace and the work methods. So designers would continue to design processes as they had in the past, not knowing that their designs included the hazardous situations that were corrected by operations personnel.

What was needed to complete the fix was a change in the design standards, *the expected performance level* in the design process, so that fewer hazards were included in their designs. Doing so would result in a culture change—how ergonomics hazards would be addressed.

Are safety professionals capable of acting on a situation of that sort? If safety professionals have acquired a good base of technical and managerial skills, they are well equipped to identify—the relative management system deficiencies and propose the changes necessary for a complete fix—and achieve a culture change.

Safety professionals should understand that solving a hazard/risk problem may be only a part of the fix and that deficiencies in operation systems must also be addressed. And that requires a culture change—a change in how things are done. Their overarching role is that of a culture change agent.

To make this chapter closer to complete and have it be a stand-alone essay, comments are repeated that appeared earlier in this book. They pertain to the Significance of Management Leadership: Positive Culture Defined, the Role of Safety Professionals with Respect to a Positive Safety Culture, the Role of Safety Professionals with Respect to a Negative Safety Culture, and the Concept of Drift.

SIGNIFICANCE OF MANAGEMENT LEADERSHIP: POSITIVE CULTURE DEFINED

From whence does the culture derive which safety professionals are to influence as they promote continual improvement? It was said in Chapter 1 in this book that as management provides the leadership, makes decisions, and takes actions directing the organization, the outcome of those decisions establishes the reality of its safety culture which is a subset of its overall culture. Safety is culture-driven and management establishes and owns the culture.

To achieve stellar results in operational risk management (Figure 3.1), an organization must have a positive safety culture that is in accord with the following definition:

> A positive safety culture exists when the decisions made by the governing body and senior management create a *system of expected performance* in which employees believe, and perform as though, safety is to be achieved in all operations.

SIGNIFICANCE OF MANAGEMENT LEADERSHIP: POSITIVE CULTURE DEFINED

The governing board and senior management establish a culture that requires having barriers and controls in place to achieve and maintain acceptable risk levels for people, property, and the environment.

Management leadership, commitment, involvement, and the accountability system, establish that the performance level to be achieved is in accordwith the culture established.

To achieve acceptable risk levels, management establishes policies, standards, procedures, and processes with respect to:

- Risk assessment, prioritization, and management
    - Applying a hierarchy of controls
- Prevention through design
    - Inherently safer design
- Providing adequate resources
- Competency and adequacy of staff
    - Capability—skill levels
    - Sufficiency in numbers
- Training and motivation
- Maintenance for system integrity
- Barrier management
- Management of change/pre-job planning
- Procurement—hazard and risk specifications
- Emergency planning and management
- Risk-related processes
    - Organization of work
    - Employee participation
    - Information—communication
    - Permits
    - Inspections
    - Incident investigation and analysis
    - Providing personal protective equipment
- Third party services
    - Relationships with suppliers
    - Contractors—on premises
- Conformance/compliance assurance reviews

Periodic performance reports are prepared pertaining to the adequacy of systems and controls in place to maintain acceptable risk levels.

FIGURE 3.1 A socio-technical model for an operational risk management system.

THE ROLE OF SAFETY PROFESSIONALS WITH RESPECT TO A POSITIVE SAFETY CULTURE

Assume that there is a positive safety culture that safety is a core value in an organization and the governing board and senior management are determined to achieve and maintain acceptable risk levels in all operations. Usually, the environmental, health, and safety professionals in those organizations are well qualified, they have stature, and the advice they give is well received and seriously considered.

But, even in the organizations where safety is a core value and have has superior safety management systems, change—favorable or unfavorable—is a constant. Information will be developed through the established communication systems indicating that improvements can be made in certain safety-related processes.

Then, acting from a sound professional and technical base, the advice given by safety professionals in their role as culture change agent is welcomed, mostly.

THE ROLE OF SAFETY PROFESSIONALS WITH RESPECT TO A NEGATIVE SAFETY CULTURE

But all organizations do not have superior safety management systems. In such entities, when the safety culture is negative, the role of the safety professional as a culture change agent can be more difficult, particularly if senior management takes the position that all is well and believes there is little need to make changes.

Skill levels required to be a successful culture change agent in such situations can be exceptionally demanding. Patience is required. Satisfaction may derive principally from small steps forward. But the goal remains the same—to try to positively influence the safety culture as continual improvement is proposed.

THE CONCEPT OF DRIFT

Assume that the organization's culture has been either positive or negative and is "drifting" into a more hazardous state for a variety of reasons. Because of this author's observations, the concept of "drift" needs discussion in this chapter. Jens Rasmussen (1997) writes in "Risk Management in a Dynamic Society: A Modelling Problem" that

> The scale of industrial installations is steadily increasing with a corresponding potential for large-scale accidents. Companies today live in a very aggressive and competitive environment which will focus the incentives of decision makers on short term financial and survival criteria rather than long term criteria concerning welfare, safety and environmental impact. (p. 186)

The word "drift" has been attached to Rasmussen's premise. For example, a book by Sidney Dekker is titled *Drift into Failure*. He references Rasmussen's work. Dekker (2011) writes:

Drift occurs in small steps. This can be seen as decrementalism, where continuous adaptation around goal conflicts and uncertainty produces small, step-wise normalizations of what was previously judged as deviant or seen as violating some safety constraint. (p. 15)

When a safety professional senses "drift" resulting from the decisions made that result in "violating some safety constraint" or more likely, several safety constraints, the challenges faced as a culture change agent require an approach that attempts to deter or slow down the pace of "drift." With diligent data gathering and analysis to identify process shortcomings and risk prioritization counsel can be given to management on the facts of and the pace of the "drift to danger."

Their purpose should be to have management become aware that the organization is adding to the potential for a "large scale accident" to occur. Prioritizing risks and emphasizing the growing potential for the occurrence of incidents that result in severe consequences achieves major importance.

Safety personnel at locations where the safety culture has drifted into a negative state or the culture has always been negative, most likely have to deal with situations in which resources are greatly limited. That has to be emphasized. Drift occurs often when money is short.

This author does not say that safety professionals will always achieve success in such situations. But the probability of being successful as a culture change agent is enhanced if a safety professional attains the status of an integral member of the business team. That will result from giving well-supported, substantiated, and convincing technical and managerial risk management advice that is perceived as serving the business interests.

EXAMPLES OF SAFETY PROFESSIONALS BEING INVOLVED AS CULTURE CHANGE AGENTS

Three examples of real-world situations in which safety professionals were involved as culture change agents concerning minor to major culture changes:

- Assume that a safety professional is trying to help resolve what is obviously a misuse of a ladder. And the outcome should be an understanding of the risks taken and a change in the views the supervisor and the worker have of how the work is to be done. If the change is successful and lasting (repeat for emphasis—lasting), a modest culture change has been achieved.
- Now assume safety professionals are brought into a discussion of a serious injury that resulted from the misuse of a ladder and senior management expresses dissatisfaction and wants to know why and what is to be done to prevent future accidents. This situation is seen by the safety professionals as an opportunity to propose major, major changes as respects the existence and use of ladders and is willing to take the risk in proposing those changes.

Studies made by safety professionals indicated that the first step in ladder safety is to eliminate them by designing processes so that they can be operated and maintained at floor level. Through colleagues, they obtained data on a manufacturing location in which the design engineers did that—there are no ladders in the plant.

Safety professionals developed a presentation on their idea, the target being the head of design and her staff. Its gist was that a study be made of all ladder usage to determine where it was practicable to redesign so that the work could be done at floor level and ladders would not be needed.

After a presentation to department heads and supervisors, in which concurrence was obtained, a meeting was arranged with the head of design and her staff who design new facilities and alterations in existing facilities. A presentation was made, and there were many questions, some of which were answered by department heads who spoke of economies to be attained in operations and maintenance.

Conclusions of the meeting were that for new facilities, designers would try to eliminate the use of ladders, and for all existing locations, studies would be made to determine where it was practicable to redesign so that ladders would not be necessary.

Through the work of safety professionals, major risk reductions had been achieved and important changes had been made in how things would be done. That represents a major culture change. Word got around. Senior management was pleased.

- An organization's president attended a trade association meeting at which there was, among many other things, a review of OSHA type statistics. Her organization's record had been historically in the bottom quintile and her co-executives needled her about it, as they had done in the past.

She decided that she had enough of the needling and that major changes would have to be made. She arranged a meeting with the person to whom the safety director reported and the safety director to tell them what had happened and that she wanted her OSHA type statistics improved and moved up two quintiles—in two years, and she assigned that responsibility to the safety director.

Fortunately, the safety director was experienced and knowledgeable. Tactfully, his response was that to accomplish her goal she would have to take the leadership and convince all personnel who reported to her that major changes were necessary in the level of risks that were acceptable, that their level of accountability for accidents and illnesses was now at a higher level, and that she expected to see results in a reasonable term.

He added that his role was that of an advisor. He then informed the president of the elements in their operations in which there were notable shortcomings. If she wanted, he said, he would prepare an action plan for her, with priorities.

As the meeting closed, he suggested that she arrange to have the corporate policy on safety updated and widely distributed so that all employees would be aware that significant changes were being made.

He also said that if she commenced reviewing incident investigation reports with her department heads during regularly scheduled weekly meetings, she would become aware, quickly, of what was being condoned.

This safety director rose to the occasion and placed himself in a position to have a major impact on the organization's culture, on how things are to be done.

OPPORTUNITIES AND CAPABILITIES

Readers may say that opportunities of the sort described in the last example given arise infrequently. True, but consider this: with respect to comparative OSHA type statistics, there are always 60% of organizations in the bottom three quintiles in relation to similar organizations. That spells opportunities for safety professionals who can get excited about creating opportunities to influence decision-makers.

In most enterprises, change is a constant, and staying relevant is a necessity. If safety professionals have acquired a good base of technical and managerial skills, they are well equipped to become culture change agents and continue their professional development.

SELECTED RESOURCES ON SAFETY PROFESSIONALS AS CULTURE CHANGE AGENTS

Books, articles, and Internet entries on culture change management are now much abundant. Those listed here interested the author.

- Chapter 4 in the book *Safety Through Design* (Christensen and Manuele, 1999) is titled "Achieving the Necessary Culture Change." Steven I. Simon was the author. Its opening paragraph reads as follows: "A full explanation of what culture change is and is not, who is involved, why it is necessary and can achieve world class safety through design, and how to make it happen is provided in this chapter." Although the focus of the chapter is on safety through design, it is largely generic with respect to attaining a culture change. Simon is a good writer.
- In Volume 41, Issue 1, February 2003 of the magazine *Safety Science*, there is an article titled "The safety adviser/manager as agent of organisational change: a new challenge to expert training." Paul Swuste and Frank Arnoldy (2003) are the authors. They are attached to the Safety Science Group, Delft University of Technology. This is a paraphrasing of the Abstract for their article.

Swuste and Arnoldy say there is a great need for health and safety advisers to act as agents of change with respect to the organization of a company's safety and health management system. They say that they have developed a training program to meet the growing needs in a postgraduate master's degree course.

This course, they say, is for safety advisors who are effective in their attempts to influence and stimulate others. Their paper describes the development of the training course on change management and gives indications of its content.

- On the Internet, Pat Naughtin has a lengthy downloadable book review on *Leading Change*, a book written by John P. Kotter who is a Harvard-Professor Emeritus. He says that Kotter has been observing the process of change for 30 years. Maughtin notes that Kotter believes that there are critical differences between change efforts that have been successful, and change efforts that have failed. What interests Kotter is why some people are able to get their organizations to change dramatically—while most do not.

  Note the phrase "most do not" and be aware that similar statements are made by several authors. Getting a significant culture change done successfully requires a great deal of planning, communication, and management. Naughtin's book review is downloadable without charge.

- This entry is also about John P. Kotter's (1996) book *Leading Change*. His book has been well received, positively critiqued, and available for purchase. Many authors and consultants refer to Kotter's work. It should be considered foundational. I was much impressed with the thought pattern in Leading Change, a good part of which pertains to what was learned by Kotter from practical applications.

- This next reference is not an article or a book. It is a slide presentation pertaining to patient safety developed for the Department of Defense. Its title is "Change Management: How to Achieve a Culture of Safety." (n.d.) It is solid, downloadable, recommended, and free. I suggest that safety professionals not be turned off because this slide presentation pertains to patient safety.

- Under the banner of the Center for Chemical Process Safety, there is an article on the Internet titled "Safety Culture: 'What Is at Stake.'" A sub caption is "Building Process Safety Culture Tool Kit." Early in this paper, the following sentences appear: "Every organization has a safety culture, operating at one level or another; The challenges to the leadership of an organization are to: 1) determine the level at which the safety culture currently functions; 2) decide where they wish to take the culture; and 3) chart and navigate a path from here to there." This article is good read.

- Jim Spigener is an executive at Dekra. On the Internet, he has an article titled "Staying Relevant: The Emerging Role of the Safety Personnel—Part 2—Becoming a Change Agent." This is an important paper. It is obvious—the role of the safety professional is changing and professionals should be aware of the changes taking place and stay relevant and adopt to them. Spigener gives this definition of a change agent:

- A change agent is in the business of advancing performance by identifying how to get there and enticing others in that endeavor. Change agents in safety do not leave their technical expertise behind; they simply leverage it to develop strategies for sustainability.
- Shawn M. Galloway is the author of a paper on the Internet titled "Characteristics of Effective Safety Culture Change Agents." As a beginning, he says:

  > Who are the change agents in your organization? Who influences others to make improvements in performance without force? Cultures do not naturally resist all change. People change all the time, and so do the cultures they are a part of. It is not the change that is resisted; it is being changed by force.

- "Understanding safety culture" is an eight-page paper issued by Workplace Health and Safety in Queensland, Australia. It is a good read. These are its opening sentences.

  > Understanding what influences the culture of your organization can make a significant contribution to changing employee attitudes and behaviors in relation to workplace health and safety.
  >
  > For a safety culture to be successful it needs to be led from the top—that is, safety culture needs to be embraced and practiced by the CEO and senior managers.

- Because Jim Spigener has also been a promoter of the need for safety professionals to become change agent, reference is made here to another of his paper which is on the Internet.

  Its title is "Changing the game in safety performance: the leader's role." This is an eight-page treatise in which, after an Introduction, the next caption is "Changing the Game Means Changing the Culture."

  Spigener writes about several barriers that can interfere with attempts to change a culture and then discusses the leadership competencies "that will change the game."

WHY CULTURE CHANGE INITIATIVES FAIL

As culture change agents, safety professionals should be well informed on how change initiatives succeed and fail and how successes and failures are measured. There are many articles and offers of consulting servicers pertaining to failures of change initiatives on the Internet. Examples follow. All were accessed on June 20, 2019

- At http://EzineArticles.com/26088, you will find "Five reasons why leaders fail to create successful change."
- At https://www.inc.com/lee-colan/10-reasons-change-efforts-fail.html, you will find "Ten Reasons Why Change Efforts Fail."

SOLVING A HAZARD/RISK PROBLEM MAY BE ONLY A PART OF THE FIX

- At https://seapointcenter.com/why-most-change-efforts-fail/ there is "Why Most Change Efforts Fail and 7 Guidelines to Ensure Your Team Succeeds."
- At https://daniellock.com/kotters-8-step-change-model/ John Kotter has "Kotter's 8 Step Change Model: 8 Reasons for Change Failure."

Reflecting on this author's experience, a few of the reasons why change initiatives fail are recorded here, and the first is the most important:

1. People leading the change largely ignored the culture in place.
2. The leadership and commitment necessary at sufficiently high levels to achieve the change may not in reality exist because the change agent has not invested the time necessary to achieve the commitment.
3. Decision-makers are not seriously enthused about the change proposed because the supporting data is shallow and not convincing.
4. The importance of becoming aware of the power structure and determining how to work within it has not been sufficiently recognized.
5. Team building—vital to success—has been inadequate.
6. Preparing for the typical resistance to change at all levels comes up short.
7. Communication to all personnel levels that would be affected by the change is not as thorough as needed.
8. Management personnel who are assigned responsibility for the change may not be held accountable for progress by the people to whom they report, and in time, the urgency and importance for the change that may have been established diminishes.
9. Assumptions are made that a change in a process or a system has occurred without determining that it has. Others refer to this as declaring victory too soon. Some authors say that a change in a system or process should not be considered a success as a culture modification until the change has been in place for at least a year. It occurs too often that operators revert to previous methods if the supervision allows such a reversion.
10. Change agents are not sufficiently aware that achieving a culture change may take several years.

A BASIC GUIDE

This second edition of *Environmental Management Systems: An Implementation Guide for Small and Medium-Sized Organizations* (n.d.) was prepared by NSF International with funding through a cooperative agreement with the U.S. Environmental Protection Agency. It is on the Internet and downloadable. Although the text is devoted to environmental management, many of its parts are generic.

This author thinks the contents of the Guide are basic to almost all change initiatives. This publication can be a good addition to a section of a library titled Safety Professionals as Culture Change Agents. A few excerpts follow.

A BASIC GUIDE FOR ACHIEVING A CULTURE CHANGE

Objectives and Targets

Developing objectives and targets help an organization translate purpose into action. These environmental goals should be factored into your strategic plans. This can facilitate the integration of environmental management with your organization's other management processes.

You determine what objectives and targets are appropriate for your organization. These goals can be applied organization-wide or to individual units, departments or functions—depending on where the implementing actions will be needed.

In setting objectives, keep in mind your significant environmental aspects, applicable legal and other requirements, the views of interested parties, your technological options, and financial, operational, and other organizational considerations.

There are no "standard" environmental objectives that make sense for all organizations. Your objectives and targets should reflect what your organization does, how well it is performing and what it wants to achieve.

Setting objectives and targets should involve people in the relevant functional area(s). These people should be well positioned to establish, plan for, and achieve these goals. Involving people at all levels helps to build commitment.

- Get top management buy-in for your objectives. This should help to ensure that adequate resources are applied and that the objectives are integrated with other organizational goals.
- In communicating objectives to employees, try to link the objectives to the actual environmental improvements being sought. This should give people something tangible to work toward.
- Measurable objectives should be consistent with your overall mission and plan and the key commitments established in your policy (pollution prevention, continual improvement, and compliance). Targets should be sufficiently clear to answer the question: "Did we achieve our objectives?"
- Be flexible in your objectives. Define a desired result, then let the people responsible determine how to achieve the result.
- Objectives can be established to maintain current levels of performance as well as to improve performance. For some environmental aspects, you might have both maintenance and improvement objectives.
- Communicate your progress in achieving objectives and targets across the organization. Consider a regular report on this progress at staff meetings.
- To obtain the views of interested parties, consider holding an open house or establishing a focus group with people in the community. These activities can have other payoffs as well.
- How many objectives and targets should an organization have? Various EMS implementation projects for small- and medium-sized organizations indicate that it is best to start with a limited number of objectives (say, three to five)

and then expand the list over time. Keep your objectives simple initially, gain some early successes, and then build on them.
- Make sure your objectives and targets are realistic. Determine how you will measure progress toward achieving them.

In every situation, the planning, communication, and consideration of people relationships that are a part of a good culture change process must be at a high level.

Repeating for emphasis: Many culture change initiatives fail because people leading the change effort largely ignore the culture in place.

CONCLUSION

A case can be soundly made in support of the idea that the overarching role of a safety professional is that of a culture change agent. In my role as the managing director for a safety and fire protection consultancy, it became evident early on that the only product we had to sell was advice. And that our success would be determined by whether clients believed that our advice provided adequate value. Think about it. Are most safety professionals not primarily providers of advice to achieve change?

Safety professionals are most often in staff positions and their role is to provide advice to decision-makers on hazards, risks, legal requirements, and deficiencies in management systems so that modifying actions can be taken to achieve acceptable risk levels. Most often, actions taken to attain acceptable risk levels require culture changes. If safety professionals recognize the reality of their role, they should become more competent in achieving changes in systems and processes.

Nearly every author who writes on organizational change makes it clear that achieving permanent change is difficult. Safety professionals must become aware of the importance of giving attention to an organization's existing culture, how deeply certain practices are embedded within processes, who presumes to have ownership of them, and the need to plan and communicate to achieve change.

Safety professionals must be perceived as members of the management team. Change methods they adopt should, if practicable, be in concert with the procedures with which the management team is comfortable so that major conflict is avoided.

Good places to start in applying culture change agent skills are the subjects of the following chapters in this book:

- Management of Change, Chapter 14; and
- Incident Investigation, Chapter 19.

REFERENCES

ANSI/ASSP Z10.0-2019. *Occupational Health and Safety Management Systems*. Park Ridge, IL: American Society of Safety Professionals, 2019.

"Change Management: How to Achieve a Culture of Safety". U.S. Department of Defense. Available at https://www.ahrq.gov/sites/default/files/wysiwyg/professionals/education/curriculum-tools/teamstepps/longtermcare/module8/igltcchangemgmt.pdf. Accessed May 31, 2019. n.d.

Christensen, W. and F. Manuele, Co-editors. *Safety Through Design*. Itasca, IL: National Safety Council, 1999.

Dekker, S. *Drift into Failure*. Burlington, VT: Ashgate Publishing Company, 2011.

Environmental Management Systems: An Implementation Guide for Small and Medium-Sized Organizations, Second Edition. Available at https://www.epa.gov/sites/production/files/2015-07/documents/ems_an-implementation-guide-for-small-and-medium-sized-organizations_2nded.pdf. Accessed May 31, 2019. n.d.

Kotter, J.P. *Leading Change*. Boston, MA: Harvard Business Review Press, 1996. A book review. Available at http://www.metricationmatters.com/docs/LeadingChangeKotter.pdf. Accessed May 31, 2019.

Manuele, F.A. *On the Practice of Safety*, Fourth Edition. Hoboken, NJ: John Wiley & Sons, 2013.

MORT User's Manual. DOE, 1992. On the Internet. Downloadable free. Available at https://www.osti.gov/servlets/purl/5254810. Accessed May 31, 2019.

Rasmussen, J. "Risk management in a dynamic society: a modelling problem." *Safety Science*, Vol. 27, No. 2/3, pp. 183–213, 1997. *Elsevier Science Ltd*. Printed in Great Britain.

Swuste, P. and F. Arnoldy. "The safety adviser/manager as agent of organizational change: a new challenge to expert training." *Safety Science*, Vol. 41, No. 1, pp. 15–27, February 2003. Rotterdam, The Netherlands: Elsevier. An Abstract for their article is available at https://www.academia.edu/4979755/The_safety_adviser_manager_as_agent_of_organisational_change_a_new_challenge_to_expert_training. Accessed May 31, 2019.

FURTHER READING

ANSI/ASSP/ISO 45001-2018. *Occupational Health and Safety Management Systems—Requirements with Guidance for Use*. Park Ridge, IL: American Society of Safety Professionals, 2018.

Common Topic 4: Safety Culture. Health and Safety Executive, UK. Available at http://www.hse.gov.uk/humanfactors/topics/common4.pdf. Accessed May 31, 2019.

Promoting a Positive Culture: A Guide to Health and Safety Culture. 12 pages. https://www.highspeedtraining.co.uk/hub/health-and-safety-culture/. Accessed May 31, 2019.

Vecchio-Sadus, A.M. Enhancing Safety Culture Through Effective Communication, 9 pages. Available at https://www.digicast.com.au/hs-fs/hub/59176/file-15741271-pdf/docs/enhancing_safety_culture_through_effective_communication.pdf. Accessed May 31, 2019.

Schein, E.H. *Organization Culture and Leadership*, Third Edition. San Francisco: Josey-Bass, 2004—437 pages.

CHAPTER 4

ESSENTIALS FOR THE PRACTICE OF SAFETY

Creation in 2018 of a Council of Academic Affairs and Research at the American Society of Safety Professionals gives encouragement with respect to the possible development of an agreement on the essentials for the practice of safety. To be recognized as a profession, the practice of safety must have a sound theoretical and practical base which, if applied, will be *effective* in hazard avoidance, elimination, or control, in achieving acceptable risk levels, and in reducing the occurrence of events that result in harm or damage to people, property, or the environment.

Safety professionals take a variety of approaches in the advice they give to decision-makers and the advice is based on substantively different premises. They can't all be right or equally effective. On two previous occasions, this author wrote what he thought were the theoretical and practical bases for the practice of safety. In this third attempt, a markedly different approach is taken because of what has been learned.

This chapter presents general principles, statements, and definitions that are believed to be at the core of what safety professionals should be taught and be at the base of what they do. Surely, there can be debate on what is presented here and additions or deletions. This presentation is not complete.

But it is a certainty—the practice of safety will not be recognized as a profession until the people involved agree on a generic base for what they do and on a core curriculum for safety degree programs.

Advanced Safety Management: Focusing on Z10.0, 45001, and Serious Injury Prevention, Third Edition. Fred A. Manuele.
© 2020 John Wiley & Sons, Inc. Published 2020 by John Wiley & Sons, Inc.

ON ORGANIZATIONAL CULTURE

Safety is culture-driven. Everything that occurs or does not occur that relates to safety is a reflection of an organization's culture. Every organization develops a culture, positive or negative, although it may not realize that it has one.

1. An overarching role for a safety professional is that of a change agent who influences the cultures in the organizations to which advice is given so that those entities can achieve and maintain safety in their operations.
2. Safety is defined as a state that exists when the risks are acceptable. [A composite definition of state is characteristics of something at a specific time.] Or safety is defined as freedom from unacceptable risks.
3. Fixing an immediate hazards/risk problem may be required but safety professionals must be aware that the fix is not complete unless the management deficiencies that allowed the hazard/risk problem to exist are also fixed.
4. Management creates the culture for operational risk management—positive or negative.
5. A positive safety culture is achieved as a result of the outcomes of decisions made by the governing body and senior management which create a *system of expected performance* in which employees at all levels believe and perform as though safety is to be achieved in all operations.
6. A negative culture is developed when the decisions and actions taken by senior management create a *system of expected performance* in which a large percentage of the employees believe that management establishes and condones risk levels that are unacceptable—thus, the climate at the working level is doubtful about the reality of management's intent.
7. Major improvements in safety will be achieved only if a culture change takes place—only if major changes occur in the reality of the *system of expected performance*.
8. An organization's culture determines management's commitment or lack of commitment to safety and the level of safety to be attained. Commitment is doubtful when
   a. The accountability system does not include safety performance measures that impact on the well-being of those responsible for results.
   b. Adequate resources to maintain acceptable risk levels throughout the organization are not provided.
9. Principle evidence of an organization's culture with respect to operational risk management is demonstrated through the design decisions for the facilities, hardware, equipment tooling, materials, processes, configuration and layout, work environment and work methods. [You can see and feel evidence of the culture.]
10. Where the culture demands superior safety performance, the socio-technical aspects of operations are well balanced—design and engineering, management and operations, and the task performance aspects.

ON HAZARDS, RISKS, AND DEFICIENCIES IN MANAGEMENT SYSTEMS—THE BASICS

A citation taken from the American National Standard for *Occupational Health and Safety Management Systems*, ANSI/AIHA/ASSE—Z10.0—2019, appears here because it contains a particularly basic and sound premise pertaining to safety-related issues. In the standard's Definitions, OHSMS Issues are defined as "hazards, risks, management system deficiencies, legal requirements, and opportunities for improvement" (p. 3). **That is a seminal definition. It is basic in the practice of safety.** All issues with which safety practitioners are involved relate to risks that derive from hazards that may exist because of management system deficiencies.

On Hazards

1. A hazard is defined as the potential source of harm to people, property or the environment.
   Hazards include the characteristics of things (e.g., equipment, technology, processes, dusts, fibers, gases, materials, and chemicals) and the actions or inactions of persons.
2. Hazards are the generic base of, the justification for the existence of, the entirety of the practice of safety. If there are no hazards, safety professionals need not exist.
3. If a hazard is not avoided, eliminated, or controlled, its potential may be realized and a hazards-related incident or exposure may occur that will likely result in harm or damage, depending on exposures.
4. By definition, all risk controversies concern the risks associated with some hazard ... the term "hazard" is used to describe any activity or technology that produces risk (Fischhoff, 1989, p. 217).
5. Two considerations are necessary in determining whether a hazard exists.
   a. Do the characteristics of the things or the actions or inactions of people present the potential for harm or damage?
   b. And can people, property, or the environment be harmed or damaged if the potential is realized?
6. Every activity undertaken as a part of an operational risk management system should serve to avoid, eliminate, or control hazards so that the risks deriving from those hazards are acceptable.
7. Hazard analysis is the most important safety process in that, if that fails, all other processes are likely to be ineffective (Johnson, 1980, p. 245).
8. In the hazard analysis process, the exposure must be assessed. An exposure assessment would determine the number of people that could be affected and how often, and the extent of the probable property damage and environment damage.
9. If hazard identification and analysis do not relate to actual causal factors, corrective actions will be misdirected and ineffective.

10. Hazards and risks are most effectively and economically avoided, eliminated, or controlled in the design and redesign processes.
11. Hazards-related incidents or exposures, even the ordinary and frequent, may have multiple and interacting causal factors.

Defining Risk and Acceptable Risk

1. Risk is defined as an estimate of the probability of a hazard-related incident or exposure occurring and the severity of harm or damage that could result.
2. Probability is defined as an estimate of the likelihood of an incident or exposure occurring that could result in harm or damage for the selected unit of time, events, population, items, or activity being considered.
3. Severity is defined as an estimate of the magnitude of harm or damage that could reasonably result from a hazard-related incident or exposure. (Severity considerations include injury and illness to people, damage to property and the environment, business down time, loss of business, etc.)
4. Acceptable risk is that risk for which the probability of an incident or exposure occurring and the severity of harm or damage that may result are as low as reasonably practicable (ALARP) in the setting being considered.
5. ALARP—as low as reasonably practicable—is that level of risk that can be further lowered only by an expenditure that is disproportionate in relation to the resulting decrease in risk.
6. All risks to which the practice of safety applies derive from hazards: there are no exceptions.
7. It is impossible to attain a risk-free environment. Even in the most desirable situations, there will still be residual risk after application of the best, practical prevention methods.
8. Residual risk is the risk remaining after risk reduction measures have been taken. Residual risk must be acceptable.
9. Setting a goal to achieve a zero-risk environment may seem laudable, but doing so requires chasing a myth.
10. In the professional practice of safety, consideration is required of the two distinct aspects of risk:
    a. Avoiding, eliminating, or reducing the probability of a hazard-related incident or exposure occurring.
    b. Reducing the severity of harm or damage if an incident or exposure occurs.
11. For an operation to proceed, its risks must be judged to be acceptable.

Deficiencies in Management Systems

1. Awareness has emerged that if a hazard exists its source most often is a deficiency in some management system.

2. That applies whether the hazard derives from the features of things (e.g., equipment, technology, processes, dusts, fibers, gases, materials, and chemicals) or the actions or inactions of persons.
3. Thus, when seeking causal factors, it is necessary to probe into the related management systems for deficiencies, which if not corrected will lie dormant and possibly be causal factors for additional incidents.
4. Only one source is cited here in support of the significance of deficiencies in management systems (there are many) because it relates to outcomes of decisions that impact on the physical environment and human errors. Excerpts follow from *Guidelines for Preventing Human Error in Process Safety* (1994). Although written for the process industry, the following excerpts are generic.
    - It is readily acknowledged that human errors at the operational level are a primary contributor to the failure of systems. It is often not recognized, however, that these errors frequently arise from failures at the management, design, or technical expert levels of the company (p. xiii).
    - Almost all major accident investigations in recent years have shown that human error was a significant causal factor at the level of design, operations, maintenance, or the management process (p. 5).
    - One central principle presented in this book is the need to consider the organizational factors that create the preconditions for errors, as well as the immediate causes (p. 5).
5. Failures—deficiencies in management systems—were at the management, design, operations, and maintenance levels, or the management processes. Those failures affect the design of the workplace and the work methods—the operating system.

RISK ASSESSMENT

1. As the practice of safety has evolved, it has become apparent that safety professionals will be expected to have the capability to do risk assessments.
2. Many standards now require that risk assessments be made. For example, both Z10 and 45001 require that risk assessment processes be established.
3. Risk assessment is a process that commences with hazard identification and analysis that produces an estimate of the severity of harm or damage that may result if an incident or exposure occurs, followed by an estimate of the probability of an incident or exposure occurring and concluding with a risk category (e.g., Low, Moderate, Serious, or High).
4. Risk assessment is the cornerstone of initiatives to prevent harm or damage to people, property, and the environment.
5. In August of 2008, the European Union gave importance to risk assessment through a bulletin that is now available in "Prevention and Control Strategies" issued by OSHWiki, the Health & Safety Laboratory, UK. This is what the European Union (n.d.) said:

> Risk assessment is the cornerstone of the European approach to prevent occupational accidents and ill health. It is the start of the health and safety management approach. If it is not done well or not at all the appropriate preventative measures are unlikely to be identified or put in place (pages not numbered).

6. For emphasis: If risk assessments—the first step in operational risk management—are not done well or not done at all, the appropriate preventive measures are unlikely to be identified and put in place.
7. In producing the measure that becomes a statement of risk, it is necessary that determinations be made for the
   a. Existence of a hazard or hazards.
   b. Exposure to the hazard.
   c. Frequency of endangerment of that which is exposed to the hazard.
   d. Severity of the consequences should the hazard's potential be realized (the extent of harm or damage to people, property, or the environment).
   e. Probability of the hazard's potential being realized.
8. A successful communication with management on risks is not possible until an understanding has been reached on the meaning of the terms used in the communication.
9. Risk assessments can be considered as the first step in problem-solving in that they can be used to identify the potential precursors of incidents that could result in serious injuries and fatalities—before such incidents occur.
10. It will not be unusual when risk decisions are made that the decision process is influenced by elements of fear and dread, and the perceived risks of employees, the immediate community, a larger public, and management personnel.
11. Risk assessments would be made for all aspects of operational risk management: occupational safety, occupational health, damage to property or the environment, product safety, all aspects of transportation safety, safety of the public, health physics, system safety, fire protection engineering, business interruption avoidance, etc.

DEFINING THE PRACTICE OF SAFETY

For the practice of safety to be recognized as a profession, it must:

- Serve a declared and understood societal purpose and
- Clearly establish what the outcome of applying the practice should be.

The Practice of Safety

1. Serves the societal need to prevent or mitigate harm or damage to people, property, and the environment that may derive from hazards.
2. Is based on knowledge and skill for each of the following:
    a. Applied engineering
    b. Applied sciences
    c. Human factors
    d. Management principles
    e. Information and communication
    f. Legal and regulatory affairs
3. Is accomplished through
    a. Anticipating, identifying, and evaluating hazards and assessing the risks that derive from them; and
    b. Taking actions—having adequate barriers and controls—to avoid, eliminate, or control those hazards
4. Has as its ultimate purpose attaining a state for which the risks are judged to be acceptable.
5. While safety professionals may undertake many tasks in their work, the underlying purpose of each task is to have the attendant risks be acceptable.
6. There are four major stages in operational risk management to which this definition of the practice of safety applies.
    a. *Preoperational stage*: in the initial planning, design, specification, prototyping, and construction processes, where the opportunities are greatest and the costs are lowest for hazard and risk avoidance, elimination, reduction, or control.
    b. *Operational stage*: where hazards and risks are identified and evaluated and mitigation actions are taken through redesign initiatives or changes in work methods before incidents or exposures occur.
    c. *Postincident stage*: where investigations are made of incidents and exposures to determine the causal factors which will lead to appropriate interventions and acceptable risk levels.
    d. *Postoperational stage*: when demolition, decommissioning, or reusing/rebuilding operations are undertaken.

HIERARCHY OF CONTROLS

1. A hierarchy of controls provides a systematic way of thinking, considering steps in a ranked and sequential order, to choose the most effective means of avoiding, eliminating, or reducing hazards and their associated risks.

2. Acknowledging that premise—that risk reduction measures should be considered and taken in a prescribed order—represents an important step in the evolution of the practice of safety.
3. In all of the four stages in operational risk management, the following hierarchy of controls is to be applied to achieve acceptable risk levels.
4. With respect to the seven levels of control shown in the following hierarchy, it should be understood that the first through the fourth steps are most effective because they:
   a. Are preventive actions that eliminate or reduce risk by design, elimination, substitution, and engineering measures.
   b. Rely the least on human behavior—the performance of personnel.
   c. Are less likely to be defeated by managers, supervisors, or workers.
5. Actions in the fifth, sixth, and seventh levels are contingent actions that rely greatly on the performance of personnel for their effectiveness—thereby, they are less reliable.
6. It must be emphasized that lower levels of controls may not reduce the risks sufficiently.
7. This hierarchy of controls differs from others. It reflects this author's emphasis on designing out or controlling hazards in the design process and, thus, achieving acceptable risk levels. Its difference is in the first item—Risk Avoidance: Prevent entry of hazards into a workplace by selecting and incorporating appropriate technology and work methods criteria during the design processes (Figure 4.1).

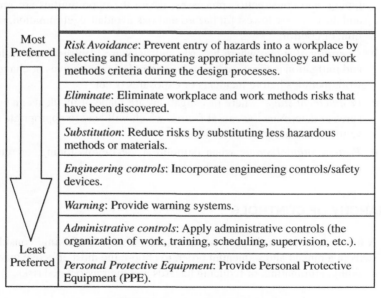

FIGURE 4.1 Hierarchy of controls.

CONCERNING LEADERSHIP AND TRAINING

1. Effective leadership, training, communication, persuasion, and discipline are vital aspects of safety management, without which superior results cannot be achieved.
2. But, training, persuasion, etc., are often erroneously applied as solutions to problems, with unrealistic expectations. Such personnel actions have limited effectiveness when causal factors derive from work place and work methods design decisions. (It is recognized that, in certain situations, personnel actions are the only preventive actions that can be taken.)
3. Heath and Ferry (1990) wrote this in *Training in the Workplace: Strategies for Improved Safety and Performance*:

   > Employers should not look to training as the primary method for preventing workplace incidents that result in death, injury, illness, property damage, or other down-grading incidents. They should see if engineering revisions can eliminate the physical safety and health hazards entirely. (p. 6)

4. As an idea, the substance of, but not the precise numbers of, Deming's 85-15 Rule applies to all aspects of the practice of safety. This is from *The Deming Management Method* by Mary Walton (1986):

   > Deming's 85-15 Rule holds that 85 percent of the problems in any operation are within the system and are the responsibility of management, while only 15 percent lie with the worker. (p. 242)

5. In *Out of the Crisis* by W. Edwards Deming, this is how the subject just previously mentioned is treated:

   > I should estimate in my experience most troubles and most possibilities for improvement add up to proportions something like this: 94% belong to the system (responsibility of management); 6% special (p. 315).

6. Deming's premise is valid—a large majority of the problems in any operation are systemic, deriving from the workplace and the work methods created by management, and can be resolved only by management. Responsibility for only the relatively small remainder lies with the worker.
7. Problems that are in the system can only be corrected by a redesign of the system or the work methods and management is responsible for the system and the work methods.
8. If system design and work methods design are the problems, the capability of employees to help is principally that of problem identification.

9. This is from *Out of the Crisis* in which W. Edwards Deming (1986), referencing Juran, speaks of workers being "handicapped by the system":

> The supposition is prevalent the world over that there would be no problems in production or in service if only our production workers would do their jobs in the way that they were taught. Pleasant dreams. The workers are handicapped by the system, and the system belongs to management. It was Dr. Joseph M. Juran who pointed out long ago that most of the possibilities for improvement lie in action on the system and that contributions of production workers are severely limited. (Deming, p. 134)

10. While employees should be trained and empowered up to their capabilities and encouraged to make contributions to safety, they should not be expected to do what they cannot do.
11. While safety is a line responsibility, it should be understood that achievements by management at an operating level are limited by previously made regulations and workplace and work methods design decisions.
12. If the design of the system presents excessive operational risks for which the cost of retrofitting is prohibitive—administrative controls, which perhaps may be the only actions that can be taken, will achieve less than superior results.

HUMAN ERRORS—UNSAFE ACTS: REVISED VIEWS

1. Views of many safety professionals concerning human errors—the so-called "unsafe acts"—have changed remarkably in recent years. Those changes pertain to how human errors occur, human error avoidance, and the corrective actions to be taken for human errors. Those changes indicate that a transition is on progress which gives greater emphasis to prevention through design, hazards analyses, risk assessment, and risk-based decision-making.
2. Speakers on human error prevention more often comment on human errors that result from workplace and work methods design errors and the best solution in such cases is to redesign the workplace or the work methods.
3. Reference is now being made more often to applying the top four elements of a hierarchy of controls system as the preferred method of limiting human errors. There is less reference of "getting into the heads" of employees for prevention purposes.
4. Bolder speakers say, in summation, it is management's responsibility to anticipate errors and to have systems and work methods designed so as to reduce error potential and to minimize severity of injury potential when errors occur.
5. Although training is often the default corrective action recommended when incident investigation reports cite human error/unsafe acts as the causal factor,

recognition has developed that training is not effective if error potential is designed into the work.

6. Sidney Dekker's (2006) comments on human error, as found in *The Field Guide to Understanding Human Error*, are pertinent to this subject. A few excerpts follow:
   - Human error is not a cause of failure. Human error is the effect, or symptom, of deeper trouble. Human error is systematically connected to features of people's tools, tasks, and operating systems. Human error is not the conclusion of an investigation. It is the starting point (p. 15).
   - Sources of error are structural, not personal. If you want to understand human error, you have to dig into the system in which people work (p. 17).
   - Error has its roots in the system surrounding it; connecting systematically to mechanical, programmed, paper-based, procedural, organizational, and other aspects to such an extent that the contributions from system and human error begin to blur (p. 74).

7. Particular attention is given here to the *Guidelines for Preventing Human Error in Process Safety*, as was done previously. Although "process safety" appears in the book's title, the first two chapters provide an easily read primer on human error reduction. Safety practitioners should view the following highlights as generic and broadly applicable. They advise on where human errors occur, who commits them and at what level, the effect of organizational culture and where attention is needed to reduce the occurrence of human errors. These highlights apply to organizations of all types and sizes.
   - It is readily acknowledged that human errors at the operational level are a primary contributor to the failure of systems. It is often not recognized, however, that these errors frequently arise from failures at the management, design or technical expert levels of the company (p. xiii).
   - Almost all major accident investigations in recent years have shown that human error was a significant causal factor at the level of design, operations, maintenance, or the management process (p. 5).
   - One central principle presented in this book is the need to consider the organizational factors that create the preconditions for errors, as well as the immediate causes (p. 5).

8. Since "failures at the management, design or technical expert levels of the company" affect the design of the workplace and work methods—that is, the operating system—it is logical to suggest that safety practitioners focus on system improvement to avoid human error and to attain acceptable risk levels rather than principally on affecting worker behavior.

PREVENTION THROUGH DESIGN

1. It is often said that hazards are most effectively and economically anticipated, avoided, or controlled in the initial design process.

2. For the practice of safety, the terms "design and redesign processes" apply to
   a. Facilities, hardware, equipment, tooling, selection of materials, operations layout, and configuration.
   b. Work methods and procedures, personnel selection standards, training content, management of change procedures, maintenance requirements, and personal protective equipment needs, etc.
3. The goal to be achieved in the design and redesign processes is acceptable risk levels.
4. Design and engineering applications that determine the workplace and work methods are the preferred measures of prevention since they are more effective in avoiding, eliminating, and controlling risks.
5. Over time, the level of safety achieved will relate directly to the caliber of the initial design of the workplace and work methods, and their subsequent redesign in a continuous improvement endeavor.
6. Requirements to achieve an acceptable risk level in the design and redesign processes can usually be met without great cost if the decision-making takes place early enough upstream.
7. When that does not occur, and retrofitting to eliminate or control hazards is proposed, the cost may be so great as to be prohibitive.
8. Those engaged in the professional practice of safety should be aware of the provisions in ANSI/ASSE Z590.3-2011 (R2016), the title of which is *Prevention through Design: Guidelines for Addressing Occupational Hazards and Risks in Design and Redesign Processes*.

SETTING PRIORITIES AND UTILIZING RESOURCES EFFECTIVELY

1. These principles are postulated:
   a. All hazards do not present equal potential for harm or damage.
   b. All incidents that may result in injury, illness, or damage do not have equal probability of occurrence, nor will their adverse outcomes be equal.
   c. Some risks are more significant than others.
   d. Resources are always limited. Staffing and money are never adequate to attend to all risks.
   e. The greatest good to employees, to employers, and to society is attained if available resources are *effectively and economically* applied to avoid, eliminate, or control hazards and the risks that derive from them.
2. Since resources are always limited, and since some risks are more significant than others, safety professionals must be capable of distinguishing the more significant from the lesser significant.
3. The professional practice of safety requires that the potentials for the greatest harm or damage be identified for the decision-makers and that a ranking system be applied to proposals made to avoid, eliminate, or control hazards.

4. Safety professionals must, therefore, be capable of using hazard analysis and risk assessment methods and of rating risks.
5. Causal factors for incidents resulting in severe harm or damage may be different from the causal factors for incidents that occur more frequently. Such low-probability incidents often involve unusual or nonroutine work, nonproduction activities, sources of high energy, and certain construction situations.
6. Thus, safety professionals must undertake a distinct activity to seek those hazards that present the most severe injury or damage potential so that they can be given priority consideration.

ON INCIDENT CAUSATION

1. For most all hazards-related incidents, even those that seem to present the least complexity, there may be multiple causal factors that derive from *less than adequate* workplace and work methods design, management and operations, and personnel task performance.
2. In MORT *Safety Assurance Systems*, Johnson wrote succinctly about the multifactorial aspect of incident causation, as in the following:

   > Accidents are usually multifactorial and develop through relatively lengthy sequences of changes and errors. Even in a relatively well-controlled work environment, the most serious events involve numerous errors and change sequences, in series and parallel (p. 74).

3. In the hazards-related incident process, deriving from those multiple causal factors:
   a. There are unwanted energy flows or exposures to harmful environments.
   b. A person or thing in the system or both are stressed beyond the limits of tolerance or recoverability.
   c. The incident process begins with an initiating event in a series of events; and
   d. Multiple interacting events may occur, sequentially or in parallel, over time and influencing each other, to a conclusion that is likely to result in injury or damage.
4. *Severity potential* should determine whether hazards-related incidents are considered significant, even though serious harm or damage did not occur.
5. Heinrich's 88-10-2 ratios indicate that among the direct and proximate accident causes, 88% are unsafe acts, 10% are unsafe mechanical or physical conditions, and 2% are unpreventable (p. 174).
   a. The methodology used in arriving at those ratios cannot be supported.
   b. Current causation knowledge indicates the premise to be invalid.

c. Among all the Heinrichean premises, application of the 88-10-2 ratios has had the greatest impact on the practice of safety, and has also done the most harm—since they promote preventive efforts being focused on the worker rather on improving the operating system.

d. Those who continue to promote the idea that 88% of all industrial accidents are caused primarily by the unsafe acts of persons do the world a disservice.

6. The Foundation of a Major Injury, the 300-29-1 ratios (Heinrich's triangle) is the least tenable of his premises (p. 27).
   a. It is impossible to conceive of incident data being gathered through the usual reporting methods in 1926 (when his postulation was made) in which 10 out of 11 reports would pertain to accidents that resulted in no injury.
   b. Conclusions pertaining to the 300-29-1 ratios were revised from edition to edition, without explanation, thus presenting questions about which version is valid.
   c. Heinrich's often-stated belief that the predominant causes of no-injury accidents are identical with the predominant causes of accidents resulting in major injuries is not supported by convincing statistical evidence, and is questioned by several authors.
   d. Application of the premise results in misdirection since those who apply it may presume, inappropriately, that if they concentrate on reducing the types of accidents that occur frequently, the potential for severe injury will also be addressed (p. 33).

7. Investigation of numerous accidents resulting in fatality or serious injury by modern-day safety professionals leads to the conclusion that their causal factors are different and that they may not be linked to the causal factors for accidents that occur frequently and result in minor injury.

8. No documentation exists to support Heinrich's 4-1 ratio of indirect injury costs to direct costs. Further, arriving at a ratio that is applicable universally is implausible.

9. In Heinrich's (1959) Principles of Accident Prevention, an inordinate emphasis is placed on the unsafe acts of individuals as causal factors, and insignificant attention is given to systemic causal factors. It is this author's belief that many safety practitioners would not agree with Heinrich's premise that "man failure is the heart of the problem and the methods of control must be directed toward man failure" (p. 4).

10. Incident investigation, initially, should address the work system, applying a concept that
    a. Commences with inquiries to determine whether causal factors derive from workplace and work methods design decisions.
    b. Promotes ascertaining whether the design of the work methods was overly stressful or error-provocative, or whether the immediate work situation encouraged riskier actions than the prescribed work methods.

PERFORMANCE MEASURES

1. If the practice of safety professionals is based on sound science, engineering, human factors engineering, and management principles, it follows that safety professionals should be able to provide measures of performance that reflect the outcomes of the risk management initiatives they propose with some degree of accuracy. That has been proven difficult to do.
2. Understanding the validity and shortcomings of the performance measures used is an indication of the maturity of the practice of safety as a profession.
3. Safety practitioners must understand that the quality of the management decisions made to avoid, eliminate, or control hazards and the risks that derive from them are impacted directly by the validity of the information they provide. Their ability to provide accurate information to be used in the decision-making is a measure of their effectiveness.
4. Since safety achievements in an organization are a direct reflection of its culture, and since it takes a long time to change a culture, short-term performance measures should be examined cautiously as to validity.
5. Except for low probability/high consequence incidents, as the exposure base represented by the number of hours worked increases, the historical incident record has an increasing degree of confidence as follows:
   a. A measure of the quality of safety in place; and
   b. A general, but not hazard specific, predictor of the experience that will develop in the future.
6. But no statistical, historical performance measurement system can assess the quality of safety in place that encompasses low probability/high consequence incidents since such events seldom appear in the statistical history. (Example a risk assessment concludes that a defined catastrophic event, one that has not happened and is not represented in the statistical base, has an occurrence probability of once in 200 plant operating years.)
7. Even for the large organization with significant annual hours worked, in addition to historical data, hazard-specific and qualitative performance measures (risk assessments, safety audits, perception surveys are also necessary, particularly to identify low probability/severe consequence risks).
8. Statistical process controls (Cause-and-Effect Diagrams, Control Charts, etc.), as applied in quality management, can serve as performance measures for safety, if the database is large enough and they are used prudently and with caution.
9. Incidents resulting in severe injury or damage seldom occur and would rarely be included in the plottings on a statistical process control chart. Although such a chart may indicate that a system is in control, it could be deluding if it was presumed that incidents that occur infrequently which could result in severe harm or damage were encompassed in the plottings.
10. Having achieved stellar OSHA type incident statistics does not assure that barriers and controls are adequate with respect to the potential for having serious injuries or fatalities.

11. Since the language of management is finance, safety practitioners must be able to communicate incident experience in financial terms.
12. Much interest has developed in what are being called leading and lagging indicators for the measurement of safety performance. Statistics traditionally gathered are considered lagging indicators. Leading indicators would be such as the number of training programs conducted, inspections made, safety discussions held, etc. In the long term, management will still want to know whether application of the leading indicators has been successful, and the success will be measured largely by the trending in the trailing indicators—the incident experience and costs.

ON SAFETY AUDITS

1. Safety audits must meet this definition to be effective: A safety audit is a structured approach to provide a detailed evaluation of safety effectiveness, a diagnosis of safety problems, a description of where and when to expect problems, and guidelines concerning what should be done about the problems.
2. The paramount goal of a safety audit is to influence favorably the organization's culture. Kase and Wiese (1990) concluded properly in *Safety Auditing: A Management Tool* that:

> Success of a safety auditing program can only be measured in the terms of the change it effects on the overall culture of the operation, and enterprise that it audits (p. 36).

3. Since evidence of an organization's culture and its management's commitment to safety is first demonstrated through its upstream design and engineering decisions, safety audits that do not evaluate the design processes are incomplete and fall short of the definition of an audit.
4. Safety audits must also properly measure management commitment, primary evidence of which is a results-oriented accountability system. If such an accountability system does not exist, management commitment is questionable.
5. Safety audits must also determine whether adequate resources are provided to achieve and maintain acceptable risk levels.

CONCLUSION

This chapter does not contain a complete listing of the principles for the practice of safety. Others may add to it or subtract from it. But the intent here is to produce a document that would encourage dialogue. To some extent, that has occurred and continues.

REFERENCES

ANSI/ASSP/ISO 45001-2018. *Occupational Health and Safety Management Systems—Requirements with Guidance for Use.* Park Ridge, IL: American Society of Safety Professionals, 2018.

ANSI/ASSP Z10.0-2019. *Occupational Health and Safety Management Systems.* Park Ridge, IL: American Society of Safety Professionals, 2019.

ANSI/ASSE Z590.3-2011 (R2016). *Prevention through Design: Guidelines for Addressing Occupational Hazards and Risks in Design and Redesign Processes.* Park Ridge, IL: American Society of Safety Engineers, 2011 (R2016).

Dekker, S. *The Field Guide to Understanding Human Error.* Burlington, VT: Ashgate Publishing Company, 2006.

Deming, W.E. *Out of the Crisis.* Cambridge, MA: Center for Advanced Engineering Study, Massachusetts Institute of Technology, 1986.

European Union statement on risk assessment. "Prevention and Control Strategies," UK: OSHWiki, The Health & Safety Laboratory. Available at https://oshwiki.eu/wiki/Prevention_and_control_strategies. Accessed May 31, 2019. n.d.

Fischhoff, B. *Risk: A Guide to Controversy.* Appendix C, Improving Risk Communication. Washington, DC: National Academy Press, 1989.

Guidelines for Preventing Human Error in Process Safety. New York: Center for Chemical Process Safety of the American Institute of Chemical Engineers, 1994.

Heath, E.D. and T. Ferry. *Training in the Work Place: Strategies for Improved Safety and Performance.* Goshen, NY: Aloray, 1990.

Heinrich, H.W. *Industrial Accident Prevention*, Fourth Edition. New York: McGraw Hill, 1959.

Johnson, W.G. *MORT Safety Assurance Systems.* New York: Marcel Dekker, 1980.

Kase, D.W. and K.J. Wiese. *Safety Auditing: A Management Tool.* New York: John Wiley & Sons, 1990.

Walton, M. *The Deming Management Method.* New York: Putnam Publishing Company, 1986.

FURTHER READING

European Agency for Safety and Health at work. Available at https://oiraproject.eu/en/what-risk-assessment. Accessed May 31, 2019. n.d.

Manuele, F.A. *On the Practice of Safety*, Fourth Edition. Hoboken, NJ: John Wiley & Sons, 2013.

CHAPTER 5

A PRIMER ON SYSTEMS/MACRO THINKING

"Systems thinking" is a term now used more often by some safety professionals. Occasionally, the term is found in safety-related literature and promoting systems/macro thinking is progressive and commendable and should be encouraged.

Why? Many safety professionals would increase their effectiveness if they adopted the premises on which systems thinking is based in a form that is practicable and effective in the organizations to which they give advice. To adopt systems/macro thinking, safety professionals must have an understanding of its bases. This chapter

- Comments on the various definitions of systems thinking.
- Makes clear that agreement has not been reached on a definition of systems thinking.
- Explores complexity, to which some writers on system thinking refer.
- Describes the status quo in the practice of safety.
- Encourages recognition of the enormity of the culture change necessary in some organizations to adopt systems/macro thinking concepts.
- Connects systems/macro thinking to having a socio-technical balance in operations.
- Encourages the use of the Five Why Problem-Solving Technique in the early stages of applying systems/macro concepts.

Advanced Safety Management: Focusing on Z10.0, 45001, and Serious Injury Prevention, Third Edition. Fred A. Manuele.
© 2020 John Wiley & Sons, Inc. Published 2020 by John Wiley & Sons, Inc.

WHAT IS SYSTEMS THINKING?

Practitioners in systems thinking use a variety of definitions for what they do, although they have a similarity—and that is for problem-solving, thinking macro and considering the entirety of the system being discussed and the interrelationship of the components. Some definitions follow:
Systems thinking was posted on the Internet by Margaret Rouse (n.d.).

Systems thinking is a holistic approach to analysis that focuses on the way that a system's constituent parts interrelate and how systems work over time and within the context of larger systems.

Peter M. Senge (n.d.). What Is the Fundamental Rationale of Systems Thinking?

[The fundamental rationale of systems thinking] is to understand how it is that the problems that we all deal with, which are the most vexing, difficult, and intransigent, come about, and to give us some perspective on those problems [in order to] give us some leverage and insight as to what we might do differently.

Ciaran John (n.d.). Thinking in the Workplace. Updated September 26, 2017.

Traditionally, managers and business owners have attempted to resolve issues or improve efficiency by breaking down the production process into sections and addressing problems in each segment of the business. Some academics argue that you should apply systems thinking to your business when you are confronted with issues. Systems thinking works on the premise that you must understand how the different components of your business interact and that you can only solve problems by studying the relationship between these components.

Several books relating to systems thinking have been published. One name that appears often in the literature is Peter M. Senge (1990), a professor at MIT. His book *The Fifth Discipline: The Art and Science of the Learning Organization* was first printed in 1990 and revised in 2006.

Donella H. Meadows (2008) wrote *Thinking in Systems: A Primer*. It is popular, in paperback, and not costly. Several quotations from her book follow.

Systems analysts use overarching concepts and have many fractious schools of systems thought. (Under is a note from the author, page unnumbered.)

As our world continues to change rapidly, and become more complex, systems thinking will help us manage, adapt, and see the wide range of choices we have before us. It is a way of thinking that gives us the freedom to identify root causes of problems and see new opportunities (p. 1).

The most marvelous characteristic of some complex systems is their ability to learn, diversify, complexify, evolve. The capacity of a system to make its own structure more complex is called self-organization (p. 79).

As Donella Meadows wrote, "there are many fractious schools of systems thought" among those who offer themselves as skilled in the field. If safety professionals are to adopt systems thinking, a definition suitable to the practice of safety must be written.

COMPLEX SYSTEMS AND SELF-ORGANIZATION

Meadows also speaks of complex systems and self-organization as do other authors. Both subjects require comment and very cautious consideration. Is the mention of those subjects predictive of the knowledge that safety professionals are expected to have?

Definitions of complexity, as on the Internet and in some books, are thought-provoking and challenging. They may apply to the work of a very small percentage of safety practitioners. But this author has great difficulty relating them to organizations that make a product or provide a service.

This definition appears on the Internet and is provided by "Volume 3—Complexity Theory, TU Delft."

> A complex system is defined as one in which many independent agents interact with each other in multiple (sometimes infinite) ways. This variety of actors also allows for the "spontaneous self-organization" that sometimes takes place in a system. This self-organization occurs without anyone being in charge or planning the organization. Rather, it is more a result of organisms/agents constantly adapting to each other. The complex systems are also adaptive (i.e., they always adapt in a way that benefits them).

Nancy Leveson, in *Engineering a Safer World: Systems Thinking Applied to Safety*, speaks of complexity and says that we build "systems that are beyond our ability to intellectually manage." She says:

> Complexity comes in many forms, most of which are increasing in the systems we are building. The operation of some systems is so complex that it defies the understanding of all but a few experts, and sometimes even they have incomplete information about the system's potential behavior. The problem is that we are attempting to build systems that are beyond our ability to intellectually manage; increased complexity of all types makes it difficult for the designers to consider all the potential system states or for operators to handle all normal and abnormal situations and disturbances safely and effectively. In fact, complexity can be defined as intellectual unmanageability. (p. 4)

Sidney Dekker implies that we should understand complexity theory and systems thinking to better understand how major incidents occur. In the Preface to his book *Drift into Failure*. He says:

> In this book, I explore complexity theory and systems thinking to better understand how complex systems drift into failure. (p. xiii)

Dekker distinguishes between failures that occur linearly in predictable ways and complex failures in which the sequence of events is unpredictable. He wrote this:

> Linear interactions among component are those (that occur) in expected and familiar production or maintenance sequences, and those that are visible and understandable even if they were unplanned.
>
> But complex interactions produce unfamiliar sequences, or unplanned and unexpected sequences, that are either not visible or not immediately comprehensible. An electric power grid is an example of an interactively complex system. (p. 128)

Dekker's example of an electric power grid being an interactively complex system is easily understood. When major power failures have occurred, there have been unforeseen and varied consequences. However, not many organizations have similar exposures.

Dekker (2006) distinguishes between complex systems and complicated systems in the following excerpt from *Drift into Failure*.

> Complex is not the same as complicated. A complicated system can have a huge number of parts and interactions between parts, but it is, in principle, exhaustively describable. We can, again in principle, develop all the mathematics to capture all the possible states of the system.
>
> Complicated systems often (if not always) do rely on an external designer or group or company of designers. The designers may not beforehand know how all their parts are going to work together (this is why there are lengthy processes of flight testing and certification), but in due time, with ample resources, in the limit, it is possible to draw up all the equations for how the entire system works, always.
>
> Reductionism, then, is a useful strategy to understand at least large parts of complicated systems. We can break them down and see how parts function or malfunction and in turn contribute to the functioning or malfunctioning of super-ordinate parts or systems. (p. 149)

A composite of definitions of reductionism is—a theory that all complex systems can be completely understood in terms of their components; the analysis of complex things into simpler constituents.

Terminology in the literature, such as the following, is difficult to comprehend and is somewhat scary. It may apply in a few organizations, but a rare few:

- The most marvelous characteristic of some complex systems is their ability to learn, diversify, complexify, and evolve;
- Self-organization occurs without anyone being in charge or planning the organization; and
- Intellectual unmanageability.

For a hugely huge share of the risk situations with which safety professionals are involved, the use of reduction concepts is sufficient in applying systems/macro thinking. They are rarely involved in operations that are self-organizing and unmanageable. For emphasis, it is said again that it is difficult to envision the characteristics of complexity existing in an organization that makes a product or provides a service.

PROMOTING SYSTEMS/MACRO THINKING: REQUIREMENTS FOR

Even though, as Donella Meadows wrote, there are "many fractious schools of systems thought" among those who offer themselves as skilled in the field, this author promotes adoption of the premises on which systems/macro thinking is based. However, the macro thinking model adopted must be practicably applicable in the organizations in which safety practitioners give advice. In its application, systems/macro thinking applied to the general practice of safety requires:

- Taking a macro view of the situation being considered and promoting collaborative discussion.
- Looking at the whole of the interrelationships, interdependencies, and interconnectedness of units within the processes and the human interactions within those systems.
- Looking at the forest and the trees at the same time.
- Being truly diagnostic.
- Recognizing that there may be several causal factors for a given situation.
- Determining how data can be obtained and used to evaluate the processes and the human interactions within those processes.
- Recognizing the need for communication feedback loops.
- Being willing to champion interventions that may not be popular.

COMMENTS ON THE STATUS QUO

By implication, the promoters of systems thinking—macro thinking—would have a large majority of safety professionals convert the fundamentals of their practice from being person centered (in which unsafe acts of workers are dominant as causal factors) to being systems centered (in which it is recognized that causal factors may be multiple and systemic and largely derive from the workplace and the work methods created by management).

That worker unsafe acts are the principal causes of occupational injuries and illnesses is deeply embedded in many organizations. And focusing on the unsafe acts of workers as causal factors is condoned by management and is a part of the culture in those organizations. Safety professionals may be among the condoning group.

That is a hindrance to applying macro thinking. Unfortunately, it defines the status quo (see Manuele (2014)). Reviews made by this author of over 1,950 incident investigation reports indicate that a large proportion of safety professionals are willing participants as the personnel involved take a narrow view—a micro view—in determining the causal factors for incidents. Examples follow:

1. This author was a speaker at a session arranged by ORCHSE, a consulting organization whose members represent Fortune 500 companies. When the 85+ attendees were asked for a show of hands, indicating whether identifying worker unsafe acts dominated the incident investigation systems in their organizations, over 60% responded.
2. At a meeting of about 42 safety practitioners who had gathered for technical discussions on a standard, about 50% raised their hands when asked whether unsafe acts of workers were the focus of incident investigation reports.
3. At a meeting of 121 safety personnel employed by a very large manufacturing company with many locations, the participants were asked this question: About what percentage of the incident reports at your location identify unsafe acts as the primary causal factor? Attendees could select from the percentages shown in Table 5.1.

TABLE 5.1   Percentages–Unsafe Acts

Percentage of Incidents	Percentage Indicating Unsafe Acts
100	3
75	33
50	37
25	12
<25	15

Recordings by participants indicate that for 50% or more of incidents, identification of worker unsafe acts as the primary accident cause total 73%. As the colleague who conducted this survey said, "We've got work to do."

THE FOCUS MUST CHANGE

To achieve a broad adoption of systems/macro thinking in operational risk management, the focus must change from unsafe acts of workers being the principal causal factors for incidents and illnesses to a focus principally on the work systems and work methods as the principal sources for causal factors.

What occurs now for incident investigation in many organizations is micro thinking. If that was replaced by macro thinking, a major and beneficial step forward will have been taken. That will require a major culture change in a huge percentage of organizations. Figure 5.1 depicts micro and macro thinking.

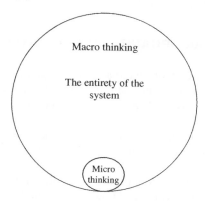

FIGURE 5.1 Macro/micro thinking.

While providing encouragement, the promoters of systems/macro thinking should be aware of the enormity of the task they undertake. Figure 5.1 depicts the enormity of the difference between macro thinking and micro thinking.

SYSTEMS OR MACRO THINKING

Erik Hollnagel (2004) implies, in his book *Barriers and Accident Prevention*, that the term "system" is overly used and may not be sufficiently descriptive to convey the thoughts and purposes intended. Hollnagel wrote this:

> The term [system] is ubiquitous in technical (and popular) writing today and is generally used on the assumption that it is so well understood by everyone that there is no need to define its meaning. (p. 6)

While supporting the premises on which systems thinking is based, this author now believes that the term "systems" may not communicate the breadth of what is intended. If the goal is to have safety professionals think broadly about hazards and risks, use of other terms would be advantageous—such as macro thinking, micro thinking, and collaborative thinking. Composite definitions of those terms follow:

Macro thinking: very large in scale, scope, or capability, taking a broad and holistic approach to the interdependent and integrated relationships between all aspects of processes and humans.

Collaborative thinking: in real time, interfacing, and discussion with colleagues who may have substantially differing views to achieve plausible and actionable conclusions.

Micro thinking: small and narrow in scope, centering on the unsafe acts of employees and immediately apparent physical conditions as causal factors.

SIGNIFICANCE OF AN ORGANIZATION'S CULTURE

Safety professionals must understand that how an organization encourages or does not encourage avoiding, eliminating, and controlling hazards and risks is established within its culture. They must also be aware that the culture is created and controlled by management.

In far too many organizations, taking a narrow view—a micro view—of causal factors is the acceptable practice. That identifies an element of the culture that has been created.

As culture change agents, safety professionals should be well informed on how change initiatives succeed and fail and how success and failure are measured.

So for management to adopt the concept of macro thinking, a major change may be required that results in broadening the view it has taken of the impact its decisions and actions have with respect to hazards, risks, legal requirements, and possible deficiencies in management. Doing so will often require a major culture change.

It cannot be overemphasized that an organization's culture will be the major determinant if a safety professional tries to have an entity adopt macro thinking concepts. There is a relative and all-too-truthful paragraph that supports that premise in *Guidelines for Preventing Human Error in Process Safety* (1994). It follows.

Cultural Aspects of Data Collection System Design

A company's culture can make or break even a well-designed data collection system. Essential requirements are minimal use of blame, freedom from fear of reprisals, and feedback which indicates that the information being generated is being used to make changes that will be beneficial to everybody. All three factors are vital for the success of a data collection system and are all, to a certain extent, under the control of management.

Authors of the book go on to say that such guarantees may not be obtained in organizations that maintain a traditional view of accident causation (p. 259).

RELATING MACRO THINKING TO A MODEL FOR A BALANCED SOCIO-TECHNICAL OPERATION

Promoting macro thinking should also promote the benefits of having a balanced socio-technical operation. That should be a basic for applied macro thinking. In this author's article titled "Preventing Serious Injuries & Fatalities"—the subtitle is

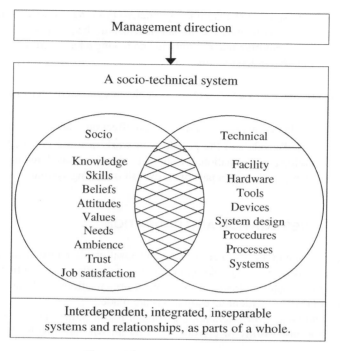

Figure 5.2 A primer on systems.

"Time for a Socio-technical Model for an Operational Risk Management System" (Manuele, 2013).

Definitions of a socio-technical system vary. This author's depiction of a socio-technical system is shown in Figure 5.2. A composite definition follows. Note the similarity with the definitions previously given of systems thinking.

> A socio-technical system stresses the holistic, interdependent, integrated and inseparable interrelationship between humans and machines. It fosters the shaping of both the technical and the social aspects of work in such a way that both the output goal of the system and the needs of workers are accommodated.

When safety professionals offer a recommendation to improve a facet of an operational risk management system, it is suggested that they apply macro thinking to determine how application of the recommendation may also affect other operational aspects.

Applied macro thinking takes a holistic approach to analysis that focuses on the whole of a system and its parts at the same time and the way a system's parts interrelate. Macro thinking contrasts with an analytical process that addresses a technical or social aspect of a system separately—micro thinking—without considering the relationship of that aspect to the system as a whole.

112 A PRIMER ON SYSTEMS/MACRO THINKING

To further understand systems thinking—macro thinking—safety professionals need to be aware of and include the human interface in the process for which they give advice When promoting macro thinking and having a balanced socio-technical operation, it would be understood that the

- Technical and social systems are inseparable, integral, and interrelated parts of a whole.
- Changes made in one process may have an effect on others.
- Organizational needs and the people who work in it are not well served if, when resolving a risk situation, the subject at hand is considered narrowly and in isolation rather than as a part of an overall operating system.

INCIDENT INVESTIGATION AND CAUSATION

Why have a section in this chapter on incident causation? Because, if safety professionals start applying macro thinking, incident investigation is a function in which they can be immediately influential. Also, most organizations badly need to apply macro thinking when incident investigations are made.

In *The Field Guide to Understanding Human Error,* Sidney Dekker makes this astute observation—worthy of consideration by all who are involved in incident investigations.

> Where you look for causes depends on how you believe accidents happen. Whether you know it or not, you apply an *accident model* to your analysis and understanding of failure. An accident model is a mutually agreed, and often unspoken, understanding of how accidents occur. (p. 81)

Safety practitioners must recognize as fact that they apply an accident model as they participate in and give advice for incident investigations. They are also obligated to have their advice be based on a sound thought process that takes into consideration the reality of the causal factors.

Unfortunately, there is an overabundance of incident causation models. A sound causation model for hazards-related incidents must take into consideration the entirety of the socio-technical system—applying a holistic approach to both the technical aspects and the social aspects of operations. It must be understood that those aspects are interdependent and mutually inclusive.

As safety practitioners become change agents and promote macro thinking, they should be certain that the causation model they endorse appropriately encompasses the premises that many incidents have multiple causal factors and that a majority of the causal factors are systemic. For emphasis, it is again said that causal factors may derive from decisions made at the upper management level.

This author purposely avoids proposing a complicated causal factor identification system. Such a system would not likely be accepted by management It is strongly recommended that safety practitioners become familiar with The Five Why System (n.d.) for Problem-Solving and use that system, **initially**, as they apply macro thinking in their work, perhaps commencing with incident investigations. (See Chapter 21—The Five Why Problem-Solving Technique).

CONCLUSION

The author promotes adoption of the premises on which systems/macro thinking is based for the practice of safety. Doing so would add immensely to the content and effectiveness of what safety professionals do. Now that more entities are initiating activities to further improve on the prevention of serious injuries and fatalities, macro thinking about processes and the interrelations between the elements within a process would additionally identify the reality of causal factors.

For all incidents, preferably, but particularly for incidents that result in serious injuries, illnesses, or fatalities, systems/macro thinking should be applied concerning the interaction and interplay of elements within a process before adverse events occur.

Emphasis is given throughout this chapter to the importance of considering the entirety of the processes and the probable human interfaces within those processes and their interdependence when risk-related problems are being discussed and resolved.

Safety practitioners must understand that if they promote adoption of systems/macro thinking concepts, they should be prepared to participate in a major culture change. Also, they should be aware that they have a self-imposed responsibility to bring to the attention of executives the deficiencies in management systems that may be uncovered, some of which could relate to decisions made at upper levels.

REFERENCES

Dekker, S. *The Field Guide to Understanding Human Error*. Burlington, VT: Ashgate Publishing Company, 2006.

Guidelines for Preventing Human Error in Process Safety. New York: Center for Chemical Process Safety of the American Institute of Chemical Engineers, 1994.

Hollnagel, E. *Barriers and Accident Prevention*. Barrington, VT: Ashgate Publishing Limited, 2004.

John, C. Thinking in the Workplace. Updated September 26, 2017. Available at https://bizfluent.com/info-12124336-advantages-disadvantages-systems-thinking-workplace.html. Accessed May 31, 2019. n.d.

Manuele, F.A. "Incident investigation: our methods are flawed." *Professional Safety*, October 2014.

Manuele, F.A. "Preventing serious injuries & fatalities: time for a socio-technical model for an operational risk management." *Professional Safety*, May 2013.

Meadows, D.H. *Thinking in Systems: S Primer*. White River Junction, VT: Chelsea Green Publishing, 2008.

Senge, P.K. *The Fifth Discipline: The Art and Science of the Learning Organization*. New York: Doubleday, 1990 and revised in 2006.

Senge, P. What Is The Fundamental Rationale Of Systems Thinking? Available at http://www.mutualresponsibility.org/science/what-is-systems-thinking-peter-senge-explains-systems-thinking-approach-and-principles. Accessed May 31, 2019. n.d.

Systems thinking, posted by Margaret Rouse. Available at https://searchcio.techtarget.com/definition/systems-thinking. Accessed May 31, 2019. n.d.

The Five Why System. Available at https://www.mindtools.com/pages/article/newTMC_5W.htm; http://www.moresteam.com/toolbox/5-why-analysis.cfm. Accessed May 31, 2019. n.d.

FURTHER READING

ANSI/ASSP Z10.0-2019. *Occupational Health and Safety Management Systems*. Park Ridge, IL: American Society of Safety Professionals, 2019.

ANSI/ASSP/ISO 45001-2018. Occupational Health and Safety Management Systems—Requirements with Guidance for Use. Park Ridge, IL: American Society of Safety Professionals, 2018.

Manuele, F.A. "Culture change agent: the overarching role of OSH Professionals.". *Professional Safety*, December 2015.

Manuele, F.A. "Root-causal factors: uncovering the hows & whys.". *Professional Safety*, May 2016.

CHAPTER 6

A SOCIO-TECHNICAL MODEL FOR AN OPERATIONAL RISK MANAGEMENT SYSTEM

Several writers on socio-technical systems say that the term "socio-technical" was coined in the 1960s by Eric Trist and Fred Emery. They were working as consultants at the Tavistock Institute in London. Personnel at Tavistock said, based on their research and from their experience, that a good fit between the technical subsystems and the social subsystems in operations was needed to achieve superior results.

Trist and Emery, the writers say, postulated that stellar performance can be obtained only if the interdependency of the technical and social subsystems is unequivocally recognized. Thus, in the design process, the decision-makers must recognize the impact each subsystem has on the other and design accordingly to assure that the subsystems are working in harmony.

Even though the term—"socio-technical systems"—is not prominent in the current literature about the organization of work, the idea it conveys is dominant in conventional thinking about the interrelationship between the technical and social aspects of operations.

Sidney Dekker (2006), in *The Field Guide to Understanding Human Error*, spoke of "the socio-technical system" as follows:

> The Systemic Accident Model focuses on the whole [system], not [just] the parts. It does not help you much to just focus on human errors, for example, or an equipment failure, without taking into account the socio-technical system

---

Advanced Safety Management: Focusing on Z10.0, 45001, and Serious Injury Prevention, Third Edition. Fred A. Manuele.
© 2020 John Wiley & Sons, Inc. Published 2020 by John Wiley & Sons, Inc.

that helped shape the conditions for people's performance and the design, testing and fielding of that equipment. (p. 90)

Dekker makes it plain that operating systems should be considered and examined as a whole to properly determine solutions to problems.

DEFINING A SOCIO-TECHNICAL SYSTEM AND MACRO THINKING

Definitions of a socio-technical system vary in detail, although they maintain the substance of what the originators of the term intended. This author's current definition is

> A socio-technical system stresses the holistic, interdependent, integrated and inseparable interrelationship between humans and machines. It fosters the shaping of both the technical and the social conditions of work in such a way that both the output goal of the system and the needs of workers are accommodated.

This author believes that, when safety professionals offer a recommendation to improve an aspect of an operational risk management system, that macro thinking—an extension of systems thinking—be applied and that they encourage decision-makers to determine how the improvement may affect other operational aspects. What is macro thinking? This author's definition of applied macro thinking follows.

> Applied macro thinking takes a holistic approach to analysis that focuses on the whole and its parts at the same time and the way a system's parts interrelate.

Macro thinking contrasts with an analytical process that addresses a technical or social aspect of a system separately—micro thinking—without considering the relationship of that aspect to the system as a whole.

In taking a socio-technical systems approach, a macro thinking approach, it would be understood that the:

- Technical and social systems in an organization are inseparable parts of a whole.
- Parts are interrelated and integrated.
- Changes made in one system may have an effect on others.
- Organizations and the people who work in them are not well served if, when resolving a risk situation, the subject at hand is considered in isolation rather than as a part of an overall system.

EMPHASIZING THE WHOLE AND INTERDEPENDENCE

It is the intent here to encourage the decision-makers to integrate the technical aspects with the social aspects of business practice as an interdependent and integrated whole not only for good operating results but also for the benefit of all.

The technical aspects of a system include the facility, hardware, tools, devices, system design, physical surroundings, and the prescribed procedures.

The social system consists of the knowledge and skills of employees from the most senior management level down to the most newly hired, the attitudes that derive from their beliefs, their values, their needs, job satisfaction, respect, trust, relations with each other, the spirit and ambience of the work place, authority structures, the reward system, and whether there is an open communications system through which the views of all can be heard.

In an effective socio-technical system, management recognizes the interdependence of the technical and social aspects of operations and integrates them. And a feedback process would be created to monitor alignment.

SIGNIFICANCE OF AN ORGANIZATION'S CULTURE

It all starts with the culture created by the governing body and senior management. In the ideal situation, decision-makers recognize the interrelationship of the social and technical aspects of operations.

Safety is culture driven. Everything that occurs or doesn't occur that relates to safety is a reflection of an organization's culture. Over time, every organization develops a culture, positive or negative, although it may not realize that it has one.

Only the first three statements in *A Socio-Technical Model for an Operational Risk Management System*—Figure 6.1—are duplicated here. They represent the decision-making that results in a positive safety culture. They are

- The governing body and senior management establish a culture that requires having barriers and controls in place to achieve and maintain acceptable risk levels for people, property, and the environment.
- Management leadership, commitment, involvement, and the accountability system, establish that the performance level to be achieved is in accord with the culture established.
- To achieve acceptable risk levels, management establishes [appropriate] policies, standards, procedures, and processes.

An organization's governing body and senior management establish and own the culture. What management does or does not do and the effectiveness of the accountability system determine what the culture is to be.

Over the long term, the incident experience attained—injuries, illnesses, fatalities and property, and environmental damage—are a direct reflection of an organization's safety culture.

118 A SOCIO-TECHNICAL MODEL FOR AN OPERATIONAL RISK MANAGEMENT SYSTEM

The governing board and senior management establish a culture that requires having barriers and controls_in place to achieve and maintain acceptable risk levels for people, property, and the environment.

Management leadership, commitment, involvement, and the accountability system, establish that the performance level to be achieved is in accord with the culture established.

To achieve acceptable risk levels, management establishes policies, standards, procedures, and processes with respect to:

- Risk assessment, prioritization, and management
    - Applying a hierarchy of controls
- Prevention through design
    - Inherently safer design
- Providing adequate resources
- Competency and adequacy of staff
    - Capability—skill levels
    - Sufficiency in numbers
- Training and motivation
- Maintenance for system integrity
- Barrier management
- Management of change/pre-job planning
- Procurement—hazard and risk specifications
- Emergency planning and management
- Risk-related processes
    - Organization of work
    - Employee participation
    - Information—communication
    - Permits
    - Inspections
    - Incident investigation and analysis
    - Providing personal protective equipment
- Third party services
    - Relationships with suppliers
    - Contractors—on premises
- Conformance/compliance assurance reviews

Periodic performance reports are prepared pertaining to the adequacy of systems and controls in place to maintain acceptable risk levels.

FIGURE 6.1   A socio-technical model for an operational risk management system.

CHARACTERISTICS OF A POSITIVE SAFETY CULTURE

Those organizations in which actions taken by the governing bodies and senior management create a positive safety culture have effective communication systems—up and down—in which all levels of employment participate. A few of the characteristics of a positive safety culture follow. There are others.

- Senior management leadership, commitment, involvement, and the accountability system make clear that safety is a fundamental within the organization's vision, core values and culture and demonstrate a passion and a sense of urgency for superior safety-related results.
- Management commitment is demonstrated continuously by what management does and by making decisions indicating that its *system of expected performance* is in accord with the positive culture established.
- Safety considerations permeate all business decision-making, from the concept stage for the design of facilities and equipment through to disposal.
- Risks are acceptable.
- Employees are encouraged to participate in safety. Employee knowledge and experience are respected. Their viewpoints are seriously considered.
- Leaders often going to the work, intensely listening and interacting, asking for input and decisively acting on commitments made.
- Managers and supervisors understand that when employees raise a safety issue, it's another opportunity for improvement. This mindset allows them to respond positively to the employee that raised the issue and escalate the issue far enough to get a solution put in place in a timely and efficient manner.
- Training is extensive for all levels of employment.

Whatever the size of an organization—10 employees or 100,000—the foregoing principles apply. Where there is a passion for superior results, management insists that it be informed about significant hazards/risk problems and that they be resolved.

EVIDENCE OF THE CULTURE IN PLACE

With respect to safety, an indication of an organization's culture is demonstrated through the design decisions that determine what the facilities, hardware, equipment, tooling, materials, layout and configuration, the work environment (the climate), and the work methods are to be.

Where the culture demands superior safety performance, the socio-technical aspects of operations are well balanced—design and engineering, management and operations, and the task performance requirements.

For all facilities, the physical look and feel of the workplace sends messages and creates expectations about what senior management really expects. You can see and feel the culture and the climate created, positive or negative.

An operating system—the system of expected performance—whatever the product made or the service provided, is a reflection of an organization's culture, its values and its sense of responsibility to its employees and to its community, all of which determine its design decisions for hardware, facilities, and the work environment; the work methods; and the management aspects of operations. Where hazards are given the required consideration in the design and engineering processes, a foundation is established that gives good probability to avoiding hazards-related incidents.

When what is acceptable to the governing body and to senior management concerning safety is less-than-adequate with respect to attaining and maintaining acceptable risk levels, causal factors for incidents may derive from the organization's culture impacting on both the social and technical aspects of operations.

MANAGEMENT COMMITMENT OR NONCOMMITMENT TO SAFETY

An organization's culture is translated into a system of expected performance. Management commitment or noncommitment to safety is an expression of the culture. All aspects of safety, favorable or unfavorable, derive from that commitment or noncommitment. When an organization's culture is less-than-adequate, it commonly occurs that the reality of the system of expected performance deviates from good operational procedures—deviates from an entity's own internally established standard operating procedures.

Where management commitment to and leadership for safety are solid, management achieves an understanding—because of what it does—that all in the organization are to manage their endeavors with respect to hazards so that the risks deriving from those hazards are acceptable.

SAFETY POLICIES, STANDARDS, PROCEDURES, AND THE ACCOUNTABILITY SYSTEM

Management commitment or noncommitment to safety determines the appropriateness or nonappropriateness of safety policies, standards, procedures, and the accountability system, and their implementation with respect to acceptable risk levels. When such instructions and procedures are *Less Than Adequate*, causal factors may originate from them in any of the following operational categories.

ON RISK ASSESSMENTS

Reviews of several texts on problem-solving techniques indicate that their authors agree on at least one vitally important premise—that the first step in problem-solving is to define the problem. To prevent accidents, management must identify the potentials of hazards and their related risks. To know of those potentials, hazards identification and analyses and risk assessments must be made. *Thus, making risk assessments should be considered an identification activity for potential problems.*

Risk assessments made in the design process provide opportunities to avoid bringing hazards into the workplace. Risk assessments made in existing facilities can identify the potentials of discovered hazards and their accompanying risks so that preventive action can be taken in the form of effective barriers and controls.

It is to the credit of those who wrote ANSI/AIHA/ASSE Z10.0-2019 and ISO 45001 that provisions are included in the standards requiring that risk assessments be made. For Z10.0, the requirement appears at Section 8.3 (p. 18). For 45001, the requirement is at Section 6.1.2.2 (p. 13).

This author believes that risk assessments should be the core of an Operational Risk Management System. If employees at all levels had more knowledge and awareness of hazards and risks, there would be fewer serious injuries, illnesses, and fatalities.

In August 2008, the European Union gave importance to risk assessment in an undated bulletin that is no longer available on the Internet. But it is available in "Prevention and Control Strategies" issued by OSHWiki, the Health & Safety Laboratory, UK (2017). This is what the European Union said:

> Risk assessment is the cornerstone of the European approach to prevent occupational accidents and ill health. It is the start of the health and safety management approach. If it is not done well or not at all the appropriate preventative measures are unlikely to be identified or put in place (pages not numbered).

This statement issued by the European Union is foundational. Huge benefits can derive if an organization makes risk assessments a cornerstone of its operational risk management system.

When the outcome of a risk assessment indicates that risks are not at an acceptable level, reductions would be achieved through the application of a hierarchy of controls which is discussed in Chapter 11.

PREVENTION THROUGH DESIGN

This is the definition of prevention through design as given in ANSI/ASSE Z590.3-2011 (R2016), the American National Standard for *Prevention through Design: Guidelines for Addressing Occupational Hazards and Risks in Design and Redesign Processes.*

> The integration of hazard analysis and risk assessment methods early in the design and redesign stages and taking the necessary actions so that the risks of injury or damage are at an acceptable level. This definition applies to all functions that are hazard based: occupational safety, environmental safety, product safety, fire protection, security, public safety and quality.

This author has written that the greatest strides forward with respect to safety, health, and the environment are being made through the design and redesign processes. What is proposed for prevention through design is an agreed-upon and well-understood concept, a way of thinking that effectively addresses hazards and risks as far "upstream" as possible in the design and redesign processes. The goals of applying prevention through design concepts are to

- Achieve acceptable risk levels.
- Prevent or reduce injuries, illnesses, or damage to people, property or the environment.
- Reduce the cost of retrofitting that becomes necessary to mitigate hazards and risks that were not sufficiently addressed in the design or redesign processes.

Reviews of incident investigation reports indicate that often hazards and risks were not properly addressed in the design and redesign processes. And design decisions that result in less-than-adequate outcomes are often the source of incident causal factors.

A commonly faced problem is that there is a major difference between what the designer perceives and how the work is done. In a few organizations, this problem is largely resolved by having skilled operations personnel take part in the design development process. It is understood that doing so is not always easy.

Since it is well established that the most effective and economical method of achieving and maintaining acceptable risk levels it to address hazards and risks in the design and redesign processes, prevention through design must be given prominence in a socio-technical model.

Because from 40 to 65% of occupational injuries are musculoskeletal, particular mention is being made here of the importance of encompassing ergonomics design guidelines within the design processes.

Adopting and extending from the work of Alphonse Chapanis (1980), causal factors will derive directly from design decisions or operations practices if the design of the workplace or the work methods is error-provocative. A workplace or the work methods are error-provocative if it

- Is overly stressful.
- Induces fatigue.
- Promotes riskier behavior than the prescribed work methods.
- Is unnecessarily difficult or unpleasant.
- Is unnecessarily dangerous.
- Requires "jerry rigging" for job accomplishment.
- Does not allow easy access for the work to be done. (p. 99)

To be most effective, intervention to achieve acceptable risk levels has to begin with the upstream decision-making out of which a socio-technical system can be created. An appropriate goal is to have designers sign off on finished plans, indicating that appropriate consideration has been given to avoiding serious injuries, illnesses, fatalities, property damage, and environmental damage.

PROVIDING ADEQUATE RESOURCES: COMPETENCY AND ADEQUACY OF STAFF

Resources to be provided include adequate staffing in numbers and in skill levels, and for facilities, equipment, tooling, hardware, processes, layout, work stations and the environment, and, very importantly, the design of the operating methods—the procedures for getting the work done.

Resources, then, include all aspects that impact on the socio-technical processes in which operations take place. Over time, if decisions are made that result in not having adequate sources and which have a negative impact on safety—on acceptable risk levels—the shortcomings accumulate and become the causal factors of many incidents that result in serious injury or fatality.

OPERATING PROCEDURES: ORGANIZATION OF WORK

This subject is important as an attempt is made to achieve a good socio-technical balance. It takes a reader back to a statement made at the beginning of this chapter indicating that a good fit between the technical subsystems and the social subsystems in operations was needed to achieve superior results.

It also requires that the system as a whole be considered to determine what risks may exist in the technical and social aspects of the operation as assignments are distributed among departments and the work tasks are assigned among individuals. That requires coordination to effectively and economically make a product or provide a service.

Experience requires the caution that the people who make the relative decisions often get it wrong, both as to hazards and risks and efficiency. It is common for the way work is actually done and the sequence of it differs from the prescribed methods.

That could be advantageous or disadvantageous, depending on whether risks are increased, neutral, or decreased. Getting it right requires an open communication system through which line workers and all levels of supervision can comment on their observations and make suggestions.

TRAINING AND MOTIVATION

Too often, phrases such as the following appear in incident investigation reports as causal factors:

- did not know the proper procedure.
- was not aware of the requirements of the permit system.

- was put on this job without going through the training process.
- we no longer have the time to do the training that our standard operating procedure says we should do.
- obviously, the training given was inadequate.

Unfortunately, training is much talked and written about as being vital to avoid incidents but the delivery may be poorly done or not done. Adequate or inadequate training, in a sense, derives from decisions made at the senior management level. Employees at all levels cannot be expected to follow safe work practices if they have not been instructed in the proper procedures and be required to follow them.

Too much emphasis cannot be given to the priority that must be given to training at all operational levels.

Much has been said in this book about an organization's safety culture. For the workers to be motivated, there must be a positive culture.

MAINTENANCE FOR SYSTEM INTEGRITY

Incident investigation reports reviewed indicate that, because of expense reductions, the maintenance staff had been adversely affected and maintenance deteriorated—the result being serious injuries and illnesses.

In *Managing the Risks of Organizational Accidents*, James Reason (1997) wrote about how the effects of decisions accumulate over time and become the causal factors for incidents resulting in serious injuries or major damage when all the circumstances necessary for the occurrence of a major event fit together. Maintenance failures were included among the latent conditions to which he referred, and rightly so (p. 10).

Failure to maintain facilities, equipment, and processes is one of the immediate and outward indications of an organization's safety culture being inadequate or deteriorating. Also, because of the severity of incidents that occur, particular consideration is necessary to avoiding, eliminating, or controlling hazards and risks for those who do maintenance work. This need was addressed in *ANSI/ASSE Z590.3—Prevention Through Design: Guidelines for Addressing Occupational Hazards and Risk in the Design and Redesign Processes* as follows:

> Consideration shall be given to accessibility for maintenance personnel (clear, unobstructed pathways and adequate clearance) and having systems that require minimum maintenance, to reduce exposures for maintenance personnel. (p. 47)

This section was included in the standard because it often occurs that the needs of maintenance personnel are not adequately addressed in the design processes. As a result, job titles—such as maintenance mechanic or maintenance supervisor—appear often in investigation reports on serious injuries.

MANAGEMENT OF CHANGE/PREJOB PLANNING: PRESTARTUP REVIEW

Because of the significance of this subject, a separate chapter is devoted to it in this book. Why so? Because many incidents that have severe outcomes occur when changes are made.

Management of change (MOC) is a process to be applied before modifications are made and continuously throughout the modification activity to assure that

- hazards are identified and analyzed and risks are assessed.
- appropriate avoidance, elimination, or control decisions are made so that acceptable risk levels are achieved and maintained during the change process.
- new hazards are not knowingly created by the change.
- the change does not impact negatively on previously resolved hazards.
- the change does not make the potential for harm of an existing hazard more severe.

Consideration would also be given to the safety of

- employees making the changes.
- employees in adjacent areas that may be affected.
- employees who will be engaged in operations after changes are made.
- the environment and the public.
- avoiding property damage and business interruption.

Reviews made by the author of incident investigation reports support the need for and the benefit of having MOC systems. They showed that a significantly large share of incidents resulting in serious injury and fatalities occur

- when unusual and nonroutine work is being performed
- in nonproduction activities
- in at-plant modification or construction operations (replacing a motor weighing 800 pounds to be installed on a platform 15 feet above the floor)
- during shutdowns for repair and maintenance, and startups
- where sources of high energy are present (electrical, steam, pneumatic, and chemical)
- where upsets occur: situations going from normal to abnormal

Additional and substantial support for having a Management of Change/Pre-Job Planning system in place comes from a study led by Dr. Thomas Krause, who was then Chairman of the Board at BST. (This data was provided in personal communication with Dr. Krause.) Seven companies participated in a study made in 2011 during which incidents that had serious injury or fatality potential were separated from the remainder of the reports collected.

Pre-Job Planning shortcomings were noted in 29% of the incidents that had serious injury or fatality potential. For the nonserious injury potential group, Pre-Job Planning inadequacies were identified in 17%. Pre-Job Planning is another name for Management of Change.

Data clearly establishes that the potential for serious injuries and fatalities occurring can be diminished by having a MOC/Pre-Job Planning System in place. But the data also shows that many organizations do not have effective MOC systems in place.

It was previously said in this chapter that serious injuries and fatalities often occur when unusual and nonroutine work is being performed. Startups are in that category. OSHA's process safety standard requires a startup review. So should every organization.

PROCUREMENT: RELATIONSHIPS WITH SUPPLIERS

To reduce the probability of bringing hazards and their accompanying risks into a workplace, safety specifications should be included in procurement papers such as purchase orders and contracts. Although the requirements for procurement in 45001 and Z10.0 are plainly stated and easily understood, they are brief in relation to the enormity of what will be required to implement them.

An Addendum C in *ANSI/ASSE Z590.3-2011 (R 2016)* provides a good guide on this subject for prevention through design.

Almost all of the requirements that companies have published for suppliers pertain to such as their qualifications, ability to deliver goods and services, compliance with laws and standards, and their financial procedures for payment. There are other subjects that may be covered.

A few companies have included detailed safety requirements in their agreements with suppliers. They may be titled such as Environmental, Health, and Safety Requirements for All Contracts, or Design Specifications to Be Met by Suppliers.

There is no one answer to what should be required of suppliers. Safety professionals should make a needs and opportunities assessment, the goal of which is to have requirements established to avoid purchasing equipment or services that result in hazards and their accompanying risks being brought into the workplace.

The premise is simply stated: If management systems do not serve to avoid bringing hazards into the workplace, employees are exposed to those hazards and retrofitting activity to achieve acceptable risk levels can be difficult to apply, and costly.

EMERGENCY PLANNING AND MANAGEMENT

Emergency planning and preparedness relates to the potential for harm or damage to employees, members of the public, property not under the control of an organization, product safety and quality, the environment, and the organizations assets and continuity of business.

Most importantly, an organization should have the ability to detect the beginning of an incident process and to marshal the trained personnel and the equipment necessary to restrict its impact.

Because of the broad range of harm or damage that can result from an incident, good community relations are necessary with organizations such as fire and police departments, governmental environmental control organizations, ambulance services, hospitals, etc. Having good emergency planning can be the difference between minor personal injury and damage and a catastrophe.

RISK-RELATED PROCESSES

Employee Participation

An organization that does not encourage employee participation and involvement and does not seek the advice of employees who have experience with the work being done do not benefit from the contributions they can make with respect to hazard identification and risk reduction, and improved efficiency.

Failure to involve employees sometimes results in known hazardous situations to remain in existence over time and later become incident causal factors.

Employees should be trained and empowered up to their capabilities, and procedures should be established for employees to make contributions to safety. But employees should not be expected to do what they cannot do. Nor, should the focus be on improving employee behavior when the causal factors for hazards-related incidents derive principally from less-than-adequate technical or social aspects of operations.

While the focus here is on the operations system created by management and which can be changed only by management, it is not intended that the significance of inappropriate employee performance be diminished.

Information—Communication

This is a broad ranged subject which, too often, is not well managed. Organizations that are the best in their class usually have good communication systems—upward and downward. Workers are encouraged to be communicators about their work. Shortcomings that have been noted and to be avoided are

- inadequate documentation and recordkeeping.
- incomplete communication to employees on work to be done.
- no communication on changes that have been or about to be made.
- inaccurate information being given.
- not communicating at all on situations that impact on maintaining acceptable risk levels.

Permits

Special mention is given to permit systems in this model because of the author's experience. As an example, far too many fires that occur as an outcome of welding, cutting, or burning have resulted in major property damage, personal injury, environmental damage, and business interruption. Either the permit system was inadequate or it was not properly managed. Similar comments apply to other permit systems—such as confined space entry, for example.

It is easier to write a good permit system than it is to have it applied. Effective permit systems are vital to safe operations. Whether a permit system is well managed or not is a culture matter.

Inspections

It was said in Chapter 1 that there was no better way to establish the sincerity of management concerning safety than for leaders going to the work, intensely listening and interacting, asking for input and decisively acting on commitments made. That includes leaders participating in inspections.

The quality of inspections made, as reflected by their thoroughness and the actions taken with respect to findings, is another immediate and outward evidence of an organization's safety culture.

Hazards and risks may be overlooked when inspections are made because that is what is expected. Or actions may not be taken on hazards identified in the inspection process. If that occurs, employees determine precisely that, no matter what management says or writes, management does not really intend to have a safe operation.

Those unidentified or unattended hazards appear regularly as important elements in incident investigation reports.

Incident Investigation and Analysis

As is also said elsewhere in this book, incident investigation and analysis to determine the causal factors for incidents are a vital element in an operational risk management system. High scores on incident investigation are not often achieved. Chapter 19 is devoted to this subject.

Failure to detect causal factors and their elimination or control results in their lying dormant and becoming the causal factors for other incidents. Very often, the deterrents to good incident investigation are cultural and management based, meaning that what is expected and condoned is less than adequate.

Providing Personal Protective Equipment

In hierarchies of control, personal protective equipment is usually the last element because the use of such equipment is the least effective method of control for hazards and risks. But some operations require the use of personal protective equipment. And that presents a management problem. Injuries and illnesses occur because

- the wrong type of equipment is provided.
- fitting to individual needs may be difficult.
- the work environment (heat, cold, humidity) make it difficult to wear the equipment.
- maintenance is inadequate.
- a deficient culture may condone nonuse of personal protective equipment.

One of the goals in the design process should be to avoid the use of such equipment in so far as is practicable. In most situations, it should be recognized that getting the use of personal protective equipment right requires considerable management time and effort.

CONTRACTORS—ON PREMISES

Safety professionals have struggled with the risks presented by having contractors doing work on their employer's premises for well over 60 years. Rarely will safety professionals say that their companies have the subject under control exceptionally well.

Work that third parties do on an organization's premises may be hazardous to the employees of those contractors and may be hazardous to an organization's employees while the work is being done. It may also introduce hazards and risks to an operation that must be dealt with after the work is completed. It is easy to suggest that

- diligent selection criteria be applied.
- the organization's insurance requirements be met.
- an assessment be made of a contractor's capability before engagement.
- agreement be reached with the contractor on safety requirements.

But, usually, this activity is not done exceptionally well. Representatives of several large employers agreed in discussion that even though they had wrestled with the subject, about 50% of the fatalities that occurred on their organization's premises over a ten-year period occurred to third-party contractor employees.

An organization should assess its situation management and decide whether it is satisfied with the status quo or it is to be changed. If improvements are to be made, the bulleted items above should be considered.

CONFORMANCE/COMPLIANCE ASSURANCE

While next to last in this chapter, compliance with government regulations is usually, but not always, treated as important at a corporate level and in most situations the needed attention to regulations percolates down through the organization.

Compliance programs do not determine operating standards. It is common in the best situations for government regulations to be considered basic standards to be considered, with actual design and operating requirements exceeding them.

PERFORMANCE MEASUREMENT

Scheduled audits are performed, internally or externally, to provide indications that expectations with respect to the subjects mentioned in this chapter are or are not being met. Also, data produced by incident recording and analysis systems is a principal aspect of performance measurement. That could include OSHA statistics and workers compensation costs. Some organizations also provide performance measures for leading and lagging indicators.

Communication, up and down, about situations encountered and resolved or not resolved is vital to achieving stellar results.

Many companies benchmark with others in similar businesses, formally and informally. They may exchange statistics on incident experience through their trade associations and, and that may produce competition.

CONCLUSION

This author believes that for the practice of safety to become recognized as a profession, one of the requirements is to reach a general agreement on a soundly based model for a socio-technical operational risk management system. It is also proposed, as is presented here, that the model must relate to the following truisms—which are repeated in this conclusion for emphasis.

- An organization's culture is the primary determiner with respect to the avoidance, elimination or control of hazards and whether acceptable risk levels are achieved.
- Management commitment or noncommitment to operations risk management is an extension of the organization's culture.
- Causal factors may derive from decisions made at the management level with respect to adequate or less than adequate policies, standards, procedures, provision of adequate resources, the accountability system, or their implementation.
- A large majority of the problems in any operation are systemic. They derive from the decisions made by management that establish the socio-technical system—the workplace, the work methods, and the governing social atmosphere-environment.
- The focus in operational risk management should be moved away from improving the behavior of workers to encompass the impact made by management decisions that create and control the socio-technical operations system.

- A sound operational risk management model must take into consideration the entirety of the socio-technical system—applying a holistic approach to both the technical aspects and the social aspects of operations. It must be understood that those aspects are interdependent and mutually inclusive.

REFERENCES

ANSI/ASSE Z590.3-2011 (R 2016). American National Standard. *Prevention through Design: Guidelines for Addressing Occupational Hazards and Risks in Design and Redesign Processes*. Park Ridge, IL: American Society of Safety Professionals, 2011 (R2016).

ANSI/ASSP/ISO 45001-2018. American National and ISO Standard. *Occupational Health and Safety Management Systems—Requirements with Guidance for Use*. Park Ridge, IL: American Society of Safety Professionals, 2018.

ANSI/ASSP Z10.0-2019. American National Standard—*Occupational Health and Safety Management Systems*. Park Ridge, IL: American Society of Safety Professionals, 2019.

Chapanis, A. "The error-provocative situation." *The Measurement of Safety Performance*. William E. Tarrants, ed. New York: Garland Publishing, 1980.

Dekker, S. *The Field Guide to Understanding Human Error*. Burlington, VT: Ashgate Publishing Company, 2006.

European Union statement on risk assessment. "Prevention and Control Strategies," UK: OSHWiki, The Health & Safety Laboratory, 2017. Available at https://oiraproject.eu/en/what-risk-assessment. Accessed June 1, 2019.

Reason, J. *Managing the Risks of Organizational Accidents*. Burlington, VT: Ashgate Publishing Company, 1997.

FURTHER READING

Frey, W. "EPUB is an electronic book format that can be read on a variety of mobile devices." Socio-Technical Systems in Professional Decision Making, n.d. Available at http://cnx.org/content/m14025/latest/?collection=col10396/latest. Accessed June 1, 2019.

Heinrich, H.W. *Industrial Accident Prevention*. New York: McGraw-Hill, 1950.

Johnson, W.G. *MORT Safety Assurance Systems*. New York: Marcel Dekker, 1980.

Manuele, F.A. *Reviewing Heinrich: Dislodging Two Myths from the Practice of Safety*. Park Ridge, IL: Professional Safety, October 2011.

Manuele, F.A. *On the Practice of Safety,* Fourth Edition. Hoboken, NJ: John Wiley & Sons, 2013.

OSHA's Rule for Process Safety Management of Highly Hazardous Chemicals. 29 CFR 1910.119. Washington, DC: U.S. Department of Labor, 1992.

Reason, J. *Human Error*. New York, NY: Cambridge University Press, 1990.

Trist, E.L. *The Evolution of Socio-technical Systems: A Conceptual Framework and an Action Research Program*. Toronto, Canada: Ontario Quality of Working Life Center, Occasional Paper no. 2, 1981.

CHAPTER 7

INNOVATIONS IN SERIOUS INJURY, ILLNESS, AND FATALITY PREVENTION

Preventing occupational incidents and illnesses that have serious results has now become a subject that speakers address in conferences, authors write about and for which knowledgeable safety professionals seek information.

While incident frequency has been substantially reduced, the greater part of the reduction has been for incidents resulting in lesser severity. Also, statistics indicate that the incident rate for fatalities has plateaued.

This chapter sets forth the actions that can be taken to achieve additional reductions in severity. It is understood that other safety professionals may propose other actions because of their experience. First, though, illustrative statistical and supportive data is presented.

EMPLOYEE INJURIES HAVE DECREASED

Reductions achieved in employee injuries and illnesses have been stellar. In the National Council on Compensation Insurance (NCCI, 2011) publication "2015 State of the Line," the Chief Actuary, Kathy Antonello (2015), said that "Workers compensation claim frequencies have dropped more than 50% over the last 20 Years" (p. 1).

NCCI has issued several bulletins indicating that while claims frequencies have been reduced, the reductions were more prominent for injuries resulting in lesser severity. That trend is displayed Table 7.1.

Advanced Safety Management: Focusing on Z10.0, 45001, and Serious Injury Prevention,
Third Edition. Fred A. Manuele.
© 2020 John Wiley & Sons, Inc. Published 2020 by John Wiley & Sons, Inc.

TABLE 7.1   Private Industry: Trending—Percent of Days-Away-from-Work Cases

	Number of Days Lost						
	1	2	3–5	6–10	11–20	21–30	31 or more
1995	16.9	13.4	20.9	13.4	11.3	6.2	17.9
2014	13.9	10.7	17.1	11.8	11.3	6.3	29.0
% Change from 1995	−17.8	−20.1	−18.2	−1.2	0.0	+0.2	+62.0

Obviously, there was a significant reduction in the percent of lost work day cases for incidents resulting in injuries with lesser severity—those having from 1 day through 6 to 10 days lost.

For incidents resulting in 11 to 20 and 21 to 30 lost work days, not much change occurred. However, incidents that were more serious and resulted in a loss of 31 or more days are a much larger share of the total.

There is a signal in the 62% increase in the cases resulting in 31 or more days of lost time for which safety professionals should be attentive. There is opportunity here.

Sources for Table 7.1. BLS—Lost-Worktime Injuries or Illnesses: Characteristics and Resulting Time Away from Work, 1995. Nonfatal Occupational Injuries and Illnesses Requiring Days Away from Work, 2014. Table 9.

FATALITY HISTORY

Each year the Bureau of Labor Statistics (BLS) produces reliable statistics on occupational fatalities. Data in Table 7.2 is taken from their reports.

In these BLS reports, fatality rates are computed for the number of fatalities per 100,000 equivalent full-time employees. In the years from 2011 through 2017, it is plain to see that

- While the numbers of fatalities have varied from 4,693 to 5,147, and increased 9.7%;
- Fatality rates have plateaued and remained in a very narrow range—from 3.3 to 3.6, with a numerical average of 3.44.

TABLE 7.2   U.S. Number of Fatalities and Fatality Rates

Year	Number of Fatalities	Fatality Rates	Workers in Millions
2011	4,693	3.5	134
2012	4,628	3.4	136
2013	4,585	3.3	139
2014	4,821	3.4	142
2015	4,836	3.4	142
2016	5,190	3.6	144
2017	5,147	3.5	147

Sources for Table 7.2 are the "Census of Fatal Occupational Injuries," 2011–2017, issued by the U.S. Department of Labor, Bureau of Labor Statistics.

IMPLICATIONS

For management to get off the plateau and achieve reductions in incidents that result in more serious injuries, illnesses, and fatalities, changes in the focus of their safety management systems are necessary.

And how can safety professionals help the organizations, to which they give advice, determine what changes in operational risk management systems should be undertaken? A list of such suggested actions follows. Safety professionals should choose the subjects that they believe to be pertinent and achievable while keeping the number of subjects selected manageable.

ACTIONS TO BE CONSIDERED BY SAFETY PROFESSIONALS

1. Study the principles on which their practice is based to determine what is sound and not sound. Ask—is it time to recognize evolving principles and practices?
2. Develop meaningful and convincing data on the activities in which serious injuries, illnesses, and fatalities have occurred in your operations and for the industry of which it is a part.
3. Evaluate the culture in place.
4. Recognize that for most of the revisions to be made that a culture change will be necessary.
5. Convince management that having good OSHA type incidence rates may not provide assurance that barriers and controls are adequate to prevent serious injuries and fatalities.
6. Educate management on the inappropriateness of Heinrich's principle indicating that 88% of occupational accidents are caused by employee unsafe acts.
7. Persuade decision-makers that focusing on reducing incident frequency may not result in an equivalent reduction in serious injuries.
8. Be aware of the changes in approach that have taken place in some circles with respect to addressing human errors/unsafe acts and determine how they might affect what you do.
9. Appreciate the importance of barriers in safety management.
10. Encourage all engineers to recognize that they are also safety engineers and that they should be adept in making risk assessments both in original designs and in alterations.
11. Promote defining potential problems through making risk assessments.
12. Recognize the beneficial impact of applying prevention through design principles.

13. Encourage the effective application of a Management of Change (MOC) system.
14. Analyze the incident investigation system in place and propose improvements as necessary.
15. Understand that *tweaking existing systems will not achieve the substantial improvements desired*.

To repeat, these subjects pertain to this author's experience. They result particularly from his having reviewed over 1,950 incident investigation reports. This list is not all inclusive. Other safety professionals may cite other measures to be taken.

SAFETY PROFESSIONALS REVIEWING THE PRINCIPLES ON WHICH THEIR PRACTICE IS BASED

Why would an author suggest that professionals in his field study the principles on which their practice is based to determine what is sound and not sound? As the practice of safety evolved over the past 60 years, some safety practitioners who sought beneficial effects from what they proposed, revised their views as analyses were made of what they did. Some safety professionals are still adhering to outdated concepts.

If safety professionals undertake a review of their basics and beliefs, they could consider Chapter 4 in this book—Essentials for the Practice of Safety—a significant reference.

Into the 1960s, this professional was a devoted advocate of Heinrich's principles as stated in *Industrial Accident Prevention: A Scientific Approach* (1959). Heinrich wrote that 88% of occupational accidents were caused by employee unsafe acts and that if you focus on reducing incident frequency, there would be an equivalent reduction in serious injuries (pp. 21 and 33).

Studies commenced in the late 1960s show that neither premise could be upheld. (See Manuele (2011), "Reviewing Heinrich: Dislodging Two Myths from the Practice of Safety.")

As a case in point, it was a common practice when trying to assist management to prevent back injuries for safety professionals to recommend training programs for workers on how to lift safety. But analyses show that such training did not measurably reduce back injuries.

Having visited several locations which had reported back injuries, it became obvious that the design of the workplace and the work methods were the problem. Often, a large percentage of the workers were required to handle excessive workloads as they were doing what they were expected to do.

Studies of the causes of back injuries led, progressively, into ergonomics, risk assessments, prevention through design, serious injury and fatality prevention, inadequacies in incident investigation, and the benefit of having a socio-technical workplace in which the technical aspects of the work and the social aspects are well balanced.

Over time, several writers influenced this author's evolving thought process. Alphonse Chapanis, who was prominent in ergonomics, coined the phrase error-provocative. His position was that if the workplace is designed to be error-provocative, it is nearly certain that errors will occur (Chapanis, 1980, p. 111). His work is still cited, particularly by researchers and writers on human error prevention.

Ergonomics is design based, as is all of safety. Having become involved in ergonomics, it was easy to be a promoter of safety professionals becoming participants in the design of the workplace and work methods.

W. Edwards Deming (1982), who was world renown in quality management, proposed quality achievement principles that also applied to safety, such as—if you want to achieve superior (safety) quality, you must design a system in which talented people could achieve superior (safety) quality (*Out of the Crisis*—p. 49). It became obvious that the overall principles to be applied to achieve superior quality were the same as needed to achieve superior safety.

Other authors whose work was influential were Willie Hammer (1972) who *wrote Handbook of System and Product Safety*; Trevor Kletz (1991) who was the author of several notable texts, one of which was *An Engineer's View of Human Error*; a Center for Chemical Process Safety publication—*Guidelines for Preventing Human Error in Process Safety* (1994); and William Johnson whose 1980 book *MORT Safety Assurance Systems* is still referenced. So is the 1992 version of the "MORT User's Manual: For Use with the Management Oversight and Risk Tree" issued by the U.S. Department of Energy.

As will be evident in the remainder of this chapter, the writings of numerous other and more recent authors have also been influential as this author's concepts of operational risk management evolved.

ACTIVITIES IN WHICH SERIOUS INJURIES AND FATALITIES OCCUR

A statistical history supports proposing that safety professionals pay particular attention to the characteristics of incidents resulting in serious injuries, particularly concerning the nature of the work being done and the job titles of injured personnel.

Reviews made by this author of incident investigation reports, mostly for serious injuries and fatalities, showed that a significantly large share of incidents resulting in serious injuries, illnesses, and fatalities occurs:

- When unusual and nonroutine work is being performed.
- In non-production activities.
- In at-plant modification or construction operations (replacing a motor weighing 800 pounds to be installed on a platform 15 feet above the floor).
- During shutdowns for repair and maintenance, and startups.
- Where sources of high energy are present (electrical, steam, pneumatic, chemical).

- Where upsets occur: situations going from normal to abnormal.

These observations were also made by this author as the research progressed:

- Causal factors for low probability/serious consequence events are seldom represented in the analytical data on accidents that occur frequently. (Some ergonomics-related incidents are the exception.)
- Many incidents resulting in serious injury are unique and singular events, having multiple and complex causal factors that may have organizational, technical, operational systems, or cultural origins.

Dan Petersen was an early promoter for giving particular attention to serious injury prevention. In the second edition of *Safety Management*, Petersen (1998) wrote:

If we study any mass data, we can readily see that the types of accidents that result in temporary total disabilities are different from the types of accidents resulting in permanent partial disabilities or in permanent total disabilities or fatalities. The causes are different. There are different sets of circumstances surrounding severity. Thus, if we want to control serious injuries, we should try to predict where they will happen. Today, we can often do just that. (p. 12).

The key to Petersen's message is prediction. He implies that studies should be made of serious injuries within an operation to anticipate and "predict where they will happen."

While this author listed the types of activities that research found to be prominent in serious injury and fatality reports, safety professionals should do their own analyses and develop their own lists relating to an entity's history and inherent risks.

OBTAINING INTEREST IN FATALITY PREVENTION

Fatality data for the manufacturing industry was selected for this discussion to illustrate that the statistical probability of an organization having a fatality is low and that specially crafted methods to achieve interest in fatality potential may be necessary.

Consider the data in Table 7.3 for which the sources are reports issued by the Bureau of Labor Statistics. (Select any industry and comparable data can be produced.)

Fatality rates, obviously, fall within a very narrow range. Data contained in a once-in-five-years report issued by the U.S. Census Bureau indicates that there were 292,825 manufacturing establishments in the United States in 2015.

If it is assumed that no location reported more than one fatality in 2015 (which is unlikely), the ratio is about 12 fatalities per 10,000 establishments. Some manufacturing locations have not ever experienced a fatality.

TABLE 7.3 Manufacturing Year. Number of Fatalities, and Rates

Year	Number of Fatalities	Fatality Rates
2011	322	2.2
2012	314	2.1
2013	304	2.0
2014	341	2.2
2015	353	2.3
2016	318	2.0
2017	303	1.9
	Average rate over 7 years—	2.17

As was said previously, the causal factors for serious injuries and illnesses are the same as for fatalities. Thus, if persuasive data is produced on serious injuries and management devotes additional attention to their prevention, reducing the causal factors for fatalities will also be encompassed. For the development of such data, three possible courses of action are presented here. Safety professionals may conclude that other possibilities should be considered.

1. All incident investigation reports for a three-year period should be collected for a sorting operation. The purpose is to select out those reports describing situations for which, under slightly different circumstances, the results could have been or were serious injuries or fatalities. That could lead to additional analyses of operations in which the incidents occurred and furthering the idea that serious potential needs special consideration.
2. Have a computer run made of all workers compensation claims valued at US$25,000 or more for three years. Why selects this level? This author's experience has been that using a US$25,000 cutoff level results in print-outs of 6 to 8% of the total number of claims and 60 to 80% of total claims values. The number of cases in the computer outputs, even for very large companies, has been manageable and they provided good bases for analysis.

   There have been outliers. For a manufacturing company that also had a mining operation, 25% of cases valued at US$25,000 or more represented 75% of total claims value. In another organization, 90% of the total claims value came from 5% of the claims valued at US$25,000 or more. In each case, valuable data was produced.
3. Engage employees in an information gathering system that continuously reports on hazardous situations that have serious injury potential. That system should include near misses that could have resulted in serious consequences under slightly different circumstances. For such a process to succeed, it must be understood that employees who are encouraged to provide input are recognized as valuable resources because of their extensive knowledge about how the work gets done. Also, they must be respected for the knowledge and skills that they have. Feedback on employee input is a must.

The data collected should be mined for information that will support a proposal that additional attention be given to serious injury and fatality potential. Specifically, reviews should be made of job titles, units, or departments that are prominent in the data, and the types of operations in which the injuries occurred.

For example, it was found in several companies that 60 to 80% of injuries valued at US$25,000 or more occurred to employees that were not making products.

Other methods can be used to produce meaningful and convincing data relating to the inherent risks in an organization. This subject presents opportunity for safety professionals to be creative.

AN EVALUATION OF THE CULTURE IN PLACE

Safety professionals should consider making a culture assessment, from the top down, to determine how much creative destruction and reconstruction are needed to achieve a mindset that gives a proper place to reducing the potential for serious injury (see Chapter 2).

Assessing the culture in a place provides information from which goals and priorities can be set. They will probably find that an organization's safety management systems concentrate largely on the personal aspects of safety and do not include activities to anticipate and identify the causal factors for low-probability/severe-consequence accidents.

James Reason (1997) made similar comments in *Managing the Risks of Organizational Accidents* that are pertinent here. He observed that occupational safety approaches directed largely on the unsafe acts of persons have limited value with respect to the prevention of accidents having serious consequences (p. 239). Reason (1990) also says that:

> Workplaces and organizations are easier to manage than the minds of individual workers. You cannot change the human condition, but you can change the conditions under which people work. (p. 223)

CHANGES IN AN ORGANIZATION'S CULTURE WILL BE NECESSARY

Reference is made several times in this book to an organization's culture and how it impacts, favorably or unfavorably, on the injury experience attained. Causal factors for incidents resulting in serious injury are largely systemic and their presence is a reflection of the organization's culture.

Since such causal factors are not properly addressed in many organizations, safety professionals should anticipate that culture changes may be necessary to have additional attention given to the prevention of incidents resulting in severity.

They should expect that considerable time will be spent as they assist management in achieving those culture changes.

All who undertake an inquiry to determine what additional actions may be taken to reduce serious injury potential will learn from R.B. Whittingham's (2004) *The Blame Machine: Why Human Error Causes Accidents.*

In his Introduction, Whittingham makes a significant observation that derives from his investigative experience—one that safety professionals should think about as they attempt to give counsel to their clients on improving serious injury prevention. He states that in many organizations, and sometimes whole industries, there is an unwillingness to look closely into error-provocative system faults (p. xii).

In organizations where there is a reluctance to explore systemic causal factors, the incident investigation stops after addressing the individual human error—the so-called "unsafe act." Thus, a more thorough investigation that looks into the reality of the systemic causal factors is avoided.

That occurs. In some organizations technical, organizational, management systems, and cultural causal factors for incidents are glossed over when incident investigations are made. Where avoidance of the reality of incident causal factors is deeply embedded into the system of expected performance, opposition to culture changes should be expected.

CONVINCING MANAGEMENT THAT HAVING GOOD OSHA STATISTICS MAY BE DECEIVING

A major educational undertaking will be necessary to convince management, and subsequently all personnel, that achieving low OSHA incident rates does not indicate that controls are adequate with respect to serious injury and fatality potentials. For over 40 years, emphasis has been placed on achieving such rates—resulting in competition within companies and among companies within an industry group.

Many companies that have achieved stellar OSHA type statistics, nevertheless, have occasional serious injuries and fatalities. A relative case is mentioned in the final report published by the U.S. Chemical Safety and Hazards Investigation Board (CSB, n.d.) in April 12, 2016 on an explosion and fire that occurred at the Macondo Deepwater Horizon rig in the Gulf of Mexico on April 20, 2010. That incident resulted in 11 fatalities, 17 injuries, and extensive environmental damage.

BP and Transocean were the two principle operators. Both companies celebrated their good OSHA type incident rates while decisions made in operations resulted in an accumulation of hazardous conditions and practices—and the event occurred.

Management should be made aware that having good incident rates may be deceptive with respect to the adequacy of controls and barriers to avoid the occurrence of serious injuries and fatalities.

ON HEINRICH'S PRINCIPLES: FOCUSING ON UNSAFE ACTS AND INCIDENT FREQUENCY

Heinrich's 88-10-2 Ratios

H.W. Heinrich has had more influence on the practice of safety than any other author. His premises have been adopted by many safety practitioners as certainty. They permeate safety literature. Four editions of his book *Industrial Accident Prevention* were printed, the last being in 1959. Some of Heinrich's premises are questionable.

Heinrich's 88-10-2 ratios indicate that among the direct and proximate occupational accident causes, 88% are unsafe acts, 10% are unsafe mechanical or physical conditions, and 2% are unpreventable (p. 174).

Current causation knowledge indicates the premise to be invalid. Heinrich's 88-10-2 premise conflicts with the work of others, such as W. Edwards Deming, whose research finds that root causes derive from shortcomings in management systems. Among all the Heinrichean premises, application of the 88-10-2 ratios have had the greatest impact on the practice of safety, and has also done the most harm—since the ratios promote preventive initiatives being focused on worker performance rather on improving the operating system.

Those who continue to promote the idea that 88% of all occupational accidents are caused primarily by the unsafe acts of persons do the world a disservice. (Comments are made later in this chapter on "Consideration of Unsafe acts—Human Errors: A Different Approach.").

Safety practitioners must convince management of the inappropriateness of Heinrich's principle indicating that 88% of occupational accidents are caused by employee unsafe acts.

Heinrich's Emphasis on Frequency

Heinrich's often-stated belief that the predominant causes of no-injury accidents are identical with the predominant causes of accidents resulting in major injuries is not supported by convincing statistical evidence. Application of the premise results in misdirection since those who apply it may presume, inappropriately, that if they concentrate on reducing the types of accidents that occur frequently, the potential for severe injury will also be addressed (p. 33).

Heinrich's premise derives from his "Foundation of a Major Injury," the 300-29-1 ratios (Heinrich's triangle). His premise follows:

> Analysis proves that, in the average case, for every mishap resulting in an injury there are many other accidents that cause no injuries whatever. From review of data now available concerning the frequency of potential-injury accidents, it is estimated that in a unit group of 330 accidents of the same kind and involving the same person, 300 result in no injuries, 29 in minor injuries, and 1 in a major or lost-time case. (p. 26)

Conclusions pertaining to the 300-29-1 ratios were revised from edition to edition, without explanation. Thus, questions arose about which version is valid or whether any version is valid. It is impossible to conceive of data being gathered through the usual reporting methods on 330 incidents which were of the same type and involving the same person. And of those 330 incidents, 10 out of 11 resulted in no injury.

Investigation of numerous accidents resulting in fatality or serious injury by modern-day safety professionals leads to the conclusion that their causal factors are different and that they may not be linked to the causal factors for accidents that occur frequently and result in minor injury.

Safety practitioners must persuade decision-makers that, as the statistics previously given indicate, focusing on reducing incident frequency may not result in an equivalent reduction in serious injuries.

CONSIDERATION OF UNSAFE ACTS—HUMAN ERRORS: A DIFFERENT APPROACH

In *Human Error*, James Reason wrote that latent failures may not be immediately apparent but can serve both to promote unsafe acts and to weaken defense [barrier] systems." (p. 197).

Reason was, sort of, sympathetic for the person who allegedly committed an unsafe act which was considered the causal factor for an incident. He wrote, importantly, that

> Preconditions or psychological precursors are *latent* states (emphasis added). They create the potential for a wide variety of unsafe acts. (p. 205)

Sidney Dekker's (2006) book *The Field Guide to Understanding Human Error* supports the view taken by Reason about employee unsafe acts being influenced by the latent conditions and practices that have developed in an organization over time. (p. 17)

Typically, the actions taken when the causal factor for an accident is identified as an employee unsafe act would be such as—retraining, reposting the written Standard Operating Procedure, or having a group meeting during which the unsafe act was reviewed and the correct way to do the job was discussed.

Dekker, and others, soundly propose that a different approach be taken when errors—unsafe acts—occur. A few excerpts from Dekker follow:

- If you want to understand human error, you have to assume that people were doing reasonable things given the complexities, dilemmas, trade-offs, and uncertainty that surrounds them. Just finding and highlighting mistakes people make explains nothing. Saying what people did or not do, or what they should have done, does not explain why they did what they did (p. 13).

- Human error is not a cause of failure. Human error is the effect, or symptom, of deeper trouble. Human error is not random. It is systematically connected to features of people's tools, tasks, and operating systems. *Human error is not the conclusion of an investigation. It is the starting point* (emphasis added, p. 15).
- Sources of error are structural, not personal. If you want to understand human error, you have to dig into the system in which people work. You have to stop looking for people's shortcomings (p. 17).

A comment made by Dekker in the previous citation is important in the professional practice of safety and is repeated: "If you want to understand human error, you have to dig into the system in which people work" (p. 17). If you dig into the system in which people work, it is logical as the digging takes place to explore the decision-making that resulted in barriers being nonexistent, or inadequate, or disengaged.

In *Human Error*, Reason implied similarly, that employees are exposed to that which management creates through its decision-making and, thus, improving the work system to have sufficient and effective barriers should be the focus in the practice of safety.

Rather than being the main instigators of an accident, operators tend to be the inheritors of system defects created by poor design, incorrect installation, faulty maintenance and bad management decisions. Their part is usually that of adding the final garnish to a lethal brew whose ingredients have already been long in the cooking (p. 173).

Reason may have exaggerated slightly. But what does all this mean? In the practice of safety, the emphasis is slowly moving:

- From trying to change the behavior of workers which, for quite some time, was the emphasis of many safety practitioners;
- To reducing operational risks through digging "into the system in which people work" and promoting prevention in the design processes and the existence of adequate physical and administrative barriers.

One of the best references on human error prevention is the book *Guidelines for Preventing Human Error in Process Safety*. Although written for the process industry, the following excerpts are generic. They advise on where human errors occur, who commits them and at what levels, the influence of organizational culture, and where attention is needed to reduce the occurrence of human errors. The author suggests that safety practitioners seriously consider the following statements and relate them to the facilities in which they work.

- It is readily acknowledged that human errors at the operational level are a primary contributor to the failure of systems. It is often not recognized, however, that these errors frequently arise from failures at the management, design, or technical expert levels of the company (p. xiii).

- Almost all major accident investigations in recent years have shown that human error was a significant causal factor at the level of design, operations, maintenance, or the management process (p. 5).
- One central principle presented in this book is the need to consider the organizational factors that create the preconditions for errors, as well as the immediate causes (p. 5).

Note particularly that "failures [were] at the management, design, or technical expert levels of the company" and that "human error was a significant causal factor at the level of design, operations, maintenance, or the management process." Those failures and human errors affect the design of the workplace and the work methods—the operating system.

Thus, it is reasonable to suggest that the focus in the practice of safety should be on improving the operating system by stressing prevention through design and the need for having adequate and effective barriers so that acceptable risk levels can be achieved and maintained.

Also, with respect to human error—the unsafe act—attention must be given specifically to "the organizational factors that create the preconditions for errors." Then, in the design process, attempts would be made to anticipate and avoid "preconditions for error." That calls for including the necessary barriers.

BARRIERS DEFINED

What is a barrier? In Erik Hollnagel's (2004) book *Barriers and Accident Prevention*, Chapter 3 is titled "Barrier Functions and Barrier Systems" and Chapter 4 discusses "Understanding the Role of Barriers in Accidents." Hollnagel's coverage of barriers is extensive and informative. His writings are good references on barriers and are recommended. Hollnagel acknowledges that

> Despite the importance of barriers in accident analysis and prevention, there are surprisingly few systematic studies in the available literature. (p. 75)

Hollnagel then refers to two particular resources: William Haddon's unwanted energy release theory, which some 50 plus years ago promoted the use of barriers to protect people and property (p. 79); and the Management Oversight and Risk Tree (MORT) (pp. 79, 80, 83, 86–89). His references prompted this author to refresh his knowledge and history with respect to barriers. Hollnagel says:

> An accident can be described as one or more barriers that have failed, even though the failure of a barrier only rarely is a cause in itself. A barrier is, generally speaking, an obstacle, an obstruction, or a hindrance that may either
>
> 1. Prevent an event from taking place, or
> 2. Thwart or lessen the impact of the consequences if it happens nonetheless.

In the former case, the purpose of the barrier is to make it impossible for a specific action or event to occur. In the latter case, the barrier serves, for instance, to slow down uncontrolled releases of matter and energy, to limit the reach of the consequences, or to weaken them in other ways (p. 68).

Hollnagel's comments on barriers can be highly influential for safety professionals who seek more precise knowledge about how accidents happen. To paraphrase Hollnagel, an accident happens because of the failures of one or more barriers. Think about that premise.

When Hollnagel says that "failure of a barrier only rarely is a cause in itself" the implication is that the failure probably results from deficiencies in the respective management systems.

In *System Safety for the 21st Century,* Richard A. Stephans (2004) gave a rather broad definition of barriers. And his definition is considered appropriate.

> Barrier—Anything used to control, prevent, or impede energy flows. Types of barriers include physical, equipment design, warning devices, procedures and work processes, knowledge and skills, and supervision. Barriers may be control or safety barriers or act as both (p. 357).

In the *NRI MORT User's Manual: For Use with the Management Oversight & Risk Tree* (2009), definitions of barriers and controls are considered jointly. This is what is said.

> [In this Manual], the barriers and controls branch consider whether adequate barriers and controls were in place to prevent vulnerable persons and objects from being exposed to harmful energy flows and/or environmental conditions (p. 5).

Analyzing for the adequacy or inadequacy of barriers and controls is particularly significant in applying the MORT system. Coverage of the subject in the NRI publication is extensive, as is also the case in the 1992 version of the *MORT User's Manual: For Use with the Management Oversight and Risk Tree* issued by the U.S. Department of Energy (Chapter 22 in this book pertains to MORT).

It is important to note that barriers are not exclusively physical. They may include, as Stephans says, aspects such as having a qualified staff, training, supervision, appropriate procedures, maintenance, communications, etc.

Safety practitioners and decision-makers should appreciate the importance of barriers in the practice of safety.

ALL ENGINEERS ARE ALSO SAFETY ENGINEERS

Why encourage all engineers to recognize that they are also safety engineers and that they become adept in making risk assessments? For two reasons:

- over many years, as engineers and designers continued to improve facilities and work systems, their reductions of hazards and risks had a major impact on the decreases in fatalities and fatality rates and in serious injuries, although they may not be aware of it;
- to make them additionally cognizant that the decisions they make in the design and redesign processes have a major influence on the safety of an operation.

How significant can the decisions made by engineers and designers? Joe Stephenson (1991) wrote the following in *System Safety 2000* (p. 10). Richard Stephans duplicated Stephenson's statement in *System Safety for the 21st Century* (p. 13).

The safety of an operation is determined long before the people, procedures, and plant and hardware come together at the work site to perform a given task.

Thousands, literally thousands, of safety-related decisions are made by engineers in their day-to-day design and redesign work. Usually, those decisions meet (or exceed) applicable safety-related codes and standards with respect to such as (to name but a few): the contour of exterior grounds; sidewalks and parking lots; building foundations; facility layout and configuration; floor materials; roof supports; process selection and design; determination of the work methods; aisle spacing; traffic flow; hardware; equipment; tooling; materials to be used; energy choices and controls; lighting, heating, and ventilation; fire protection; and environmental concerns.

If designers recognize additionally that their decisions have a major influence on the existence or nonexistence of operational hazards, there will be fewer hazards for which retrofitting is necessary. Retrofitting is expensive and sometimes not done.

An ideal would be to have engineers become adept in making risk assessments and say, as they sign off on plans for new or altered facilities and systems, that they have given appropriate attention to avoiding serious injuries, illnesses and fatalities, and the occurrence of low probability/serious consequence events.

RISK ASSESSMENTS: VALUE AND TRENDING

Reviews of several texts on problem-solving techniques indicate that their authors agree on at least one vitally important premise—that the first step in problem-solving is to define the problem.

To prevent accidents, management must identify the potentials of hazards and their related risks. To know of those potentials, hazards identification and analyses and risk assessments must be made.

Thus, making risk assessments should also be considered an identification activity for potential problems.

A significant document that promotes making risk assessments is the previously mentioned final report published by the CSB on an explosion and fire that occurred at the Macondo Deepwater horizon rig in the Gulf of Mexico on April 20, 2010. CBS says that

> Companies need an effective, and realistic, risk reduction goal because they cannot eliminate every risk completely—absolute safety is not possible. The question then becomes, when are efforts to reduce the level of residual risk sufficient? This challenge led to reducing risk to a level as low as is reasonably practicable, or ALARP, an important concept to explore in risk reduction practices. (Vol. 3, p. 170)

This is strong stuff, especially since it comes from a prestigious governmental agency. What does all this mean? Safety practitioners should understand that organizations must have an effective and realistic risk assessment and reduction system.

Risk assessments made in the design process provide opportunities to avoid bringing hazards into the workplace. Risk assessments made in existing operations can identify the potentials of discovered hazards and their accompanying risks so that preventive action can be taken in the form of effective barriers and controls. Management would apply a hierarchy of controls (see Chapter 11) to determine what actions should be taken to achieve acceptable risk levels.

Two components must be addressed in developing a risk assessment—probability of occurrence and severity of outcome. Hazard identification and analysis establishes severity—the probable harm or damage that could result if an incident occurs. To convert a hazard analysis into a risk assessment, a probability of occurrence factor must be added. Then, risk levels can be established (e.g., Low, Moderate, Serious, High) and priorities can be set.

The author believes that risk assessment should be the core of an Operational Risk Management System. If employees at all levels had more knowledge and awareness of hazards and risks, there would be fewer serious injuries, illnesses, and fatalities. Getting the required knowledge embedded into the minds of all employees requires a major, long-term endeavor. Crafting specifically directed communication and training programs will be necessary to achieve the awareness and knowledge required—the culture change.

Literature on making risk assessments is abundant. For example, in the ANSI standard ANSI/ASSE Z690.3-2011—*Risk Assessment Techniques*—reviews are included of 31 techniques. Examples are such as Primary Hazard Analysis, Fault Tree Analysis, Hazard and Operably Studies, Bow-Tie Analysis, Markov Analysis, and Bayesian Statistics. In this book, use of The Five Why Problem-Solving Technique is promoted, as in Chapter 21.

With great emphasis, this author states that having employees become capable of making risk assessments and encouraging that they adopt a mindset whereby identifying and analyzing hazards and the risks that derive from them become integral in how they approach and think about the work they do would be a major step forward in reducing serious injuries and fatalities.

Having knowledge of hazard identification and analysis and risk assessments become rooted within an organization's culture is the type of innovative action needed to further reduce serious injury and fatality potential.

In August 2008, the European Union (n.d.) gave importance to risk assessment in an undated bulletin that is no longer available on the Internet. But it is available in "Prevention and Control Strategies" issued by OSHWiki, the Health & Safety Laboratory, UK. This is what the European Union said:

> Risk assessment is the cornerstone of the European approach to prevent occupational accidents and ill health. It is the start of the health and safety management approach. If it is not done well or not at all the appropriate preventative measures are unlikely to be identified or put in place (pages not numbered).

It is highly significant that the European Union declared that "Risk assessment is the cornerstone of the European approach to prevent occupational accidents and ill health." That statement is foundational and should be supported by safety professionals.

If an organization chooses to improve on serious injury, illness, and fatality prevention, an early step would be to define the potentials for such incidents to occur. Risk assessments will serve that purpose.

Some safety practitioners have used the term "precursors" to identify severity potentials. What is a precursor? A composite definition is—a harbinger that foreshadows what is to come. And seeking precursors is a valid and worthwhile venture.

But searching for and making modifications for precursors is not sufficient. If precursors exist, the reasons for their existence—deficiencies in management systems—must also be determined and acted upon. If the precursors exist because of design shortcomings, information about the corrective actions taken should also be communicated to design people so that they can eliminate similar precursors in future designs.

Trending—Risk Assessment Related Literature and Services

There has been a huge increase in risk assessment-related literature in the recent past as a journey into the Internet will reveal. Mention is made here of references that are of particular interest:

> Lyon, Bruce and Bruce Hollcroft. "Risk Assessments: Top 10 Pitfalls & Tips for Improvement." Park ridge, IL: Professional Safety, December 2012.
>
> Lyon, Bruce, Bryce Hollcroft and Georgi Povov. *Risk Assessment: A Practical Guide to Assessing Operational Risk*. Hoboken, NJ: John Wiley, 2016. (Fred Manuele wrote the first chapter.)
>
> Lyon Bruce and Georgi Popov. *Risk Management Tools for Safety Professionals*. Park Ridge, IL: American Society of Safety Professionals, 2018.

Main, Bruce W. *Risk Assessment: Basics and Benchmarks*. Ann Arbor, MI: Design Safety Engineering, Inc., 2004.

Main, Bruce W. *Risk Assessment: Challenges and Opportunities*. Ann Arbor, MI: Design Safety Engineering, Inc., 2012.

But that is only a beginning. Additional representative examples follow. They represent only a fraction of the literature and services now available on the Internet.

- In 1996, the European Commission, based in Luxembourg, issued a 64-page paper titled "Guidance on Risk Assessment at Work." (Downloadable).
- Based in Dublin, the Health and Safety Authority issued a "Guide on Manual Handling Risk Assessment in the Manufacturing Sector." (Downloadable).
- There is now a security standard that proposes that risk assessments be made (2012).
- The Centre for Policy Leadership issued "The Role of Risk Management in Data Protection." (2014, Downloadable).
- In 2016, the European Agency for Safety at Work (n.d.) issued "The OiRA sectoral tools (which are to) enable micro and small enterprises to carry out risk assessments. (17 pages and downloadable. Listed in References.)

Safety professionals—you will be expected to provide guidance on making risk assessments.

PREVENTION THROUGH DESIGN

In *Guidance on the Principles of Safe Design for Work*, comments are made about the results of a study on the "contribution that the design of machinery and equipment has on the incidence of fatalities and injuries in Australia." They say:

> Of the 210 identified workplace fatalities, 77 (37%) definitely or probably had design-related issues involved. Design contributes to at least 30% of work-related serious nonfatal injuries. (p. 6)

A review made by this author of a variety of incident investigation reports (not limited to machinery) concluded that there were implications of workplace and work methods design inadequacies in over 35%.

To reduce serious injury, illness, and fatality potential, the author proposes that Prevention through Design be established as a separately identified element within an Operational Risk Management System.

To provide the needed education for designers, safety professionals are encouraged to develop supportive data on incidents in which design shortcomings were identified and undertake a major effort to have the American National Standard ANSI/ASSE Z590.3-2011 be accepted as a design guide. The title of Z590.3-2011 is *Prevention through Design: Guidelines for Addressing Occupational Hazards and Risks in Design and Redesign Processes*. This is the **Scope** of Z590.3:

> This standard provides guidance on including prevention through design concepts within an occupational safety and health management system. Through the application of these concepts, decisions pertaining to occupational hazards and risks can be incorporated into the process of design and redesign of work premises, tools, equipment, machinery, substances, and work processes including their construction, manufacture, use, maintenance, and ultimate disposal or reuse. This standard provides guidance for a life-cycle assessment and design model that balances environmental and occupational safety and health goals over the life span of a facility, process, or product.

Z590.3 says that, in so far as is practicable, the goal shall be to assure that the design selected attains the following:

- An acceptable risk level is achieved, as defined in this standard.
- The probability of personnel making human errors because of design inadequacies is as low as reasonably practicable.
- The ability of personnel to defeat the work system and the work methods prescribed is as low as reasonably practicable.
- The work processes prescribed take into consideration human factors (ergonomics)—the capabilities and limitations of the work population.
- Hazards and risks with respect to access and the means for maintenance are at as low as reasonably practicable.
- The need for personal protective equipment is as low as reasonably practicable, and aid is provided for its use where is necessary (e.g., anchor points for fall protection).
- Applicable laws, codes, regulations, and standards have been met.
- Any recognized code of practice, internal or external has been considered.

Proposing that Prevention through Design be a specifically defined element in an Operational Risk Management System is also influenced by ongoing transitions observed in the methods to eliminate or reduce the occurrence of human error. Some specialists in human error prevention now say that the first action to be taken when human errors occur is to consider the design of the system and the work methods relative to the errors.

MANAGEMENT OF CHANGE/PRE-JOB PLANNING

Management of change/Pre-job planning is a process to be applied before modifications are made continuously throughout the modification activity to assure that

- Hazards are identified and analyzed and risks are assessed.
- Appropriate avoidance, elimination, or control decisions are made so that acceptable risk levels are achieved and maintained during the change process.
- New hazards are not knowingly created by the change.
- The change does not impact negatively on previously resolved hazards.
- The change does not make the potential for harm of an existing hazard more severe.

In the management of change (MOC) process, consideration as applicable would be given to

- The safety of employees making the changes.
- Employees in adjacent areas.
- Employees who will be engaged in operations after changes are made.
- Environmental, public, product safety, and quality.
- Avoiding property damage and business interruption.

A section that appeared earlier in this chapter is titled "Activities in Which Serious Injuries and Fatalities Occur." Having an effective MOC/pre-job planning system in place would have served well to reduce the probability of serious injuries and fatalities occurring in the operational categories listed in that section.

The data clearly establishes that the potential for serious injuries and fatalities occurring can be diminished by having a MOC/pre-job planning system in place as a separately identified element within an Operational Risk Management System.

INCIDENT INVESTIGATION AND ANALYSIS

Research has shown that the quality of incident investigations in many companies has been less than stellar.

One of the purposes of this chapter is to convince safety professionals that their learned advice when incident investigations are made is seriously needed at all levels of management.

This is an important subject for which inadequacy is often condoned by management. (See Manuele (2014). "Incident Investigation: Our Methods Are Flawed" for comments on the shortcomings found in many incident investigation systems and how they may be improved.)

Incident investigations should be a good source from which determinations can be made about the adequacy of barriers and controls and possible deficiencies in management systems. Opportunities to reduce risks and avoid injury or damage to people, property, and the environment are lost when incident investigations do not identify the reality of causal factors.

It is suggested that safety professionals analyze the incident investigation system in place for their effectiveness, asking—do they really identify causal factors.

Also, it should be recognized that incident investigation reports could also be a good reference from which to select leading indicators.

CONCLUSION

Reductions made in the occurrence of accidents and injuries of all types in recent years have been notable and significant. Safety practitioners who gave advice to management personnel who achieved those reductions should feel good about themselves.

But the record is clear. Reductions in injuries having lesser severity are greater than those resulting in more lost work days, and the rate for the more serious injuries may have plateaued. While the reductions in the number of fatalities and the fatality rates have been commendable, the data plainly shows that numbers and rates have not improved in recent years. They are close to stationary.

This provides opportunities for safety practitioners to do the analyses recommended and develop the advice to be given to management so that continued reductions can be made.

REFERENCES

ANSI/ASSE Z590.3. *Prevention Through Design: Guidelines for Addressing Occupational Hazards and Risks in Design and Redesign Processes.* Park Ridge, IL: American Society of Safety Professionals, 2011 (R 2016).

ANSI/ASSE Z690.3. *Risk Assessment Techniques.* Park Ridge, IL: American Society of Safety Professionals, 2011.

Antonello, K. "2105 State of the Line". Boca Raton, FL: National Council on Compensation Insurance, 2015.

Bureau of Labor Statistics. *Census of Fatal Occupational Injuries.* Washington, DC: Department of Labor, 2011–2017.

Bureau of Labor Statistics. *Lost-Worktime Injuries and Illnesses: Characteristics and Resulting Time Away from Work. Table 7 for 1995.* Washington, DC: Department of Labor, 1996.

Chapanis, A. "The Error-Provocative Situation" In *The Measurement of Safety Performance*, William E. Tarrants, ed. New York: Garland Publishing, 1980.

Dekker, S. *The Field Guide to Understanding Human Error.* Burlington, VT: Ashgate Publishing Company, 2006.

Deming, W.E. *Out of the Crisis.* Cambridge, MA: Massachusetts Institute of Technology for Advanced Engineering Study, 1982.

European Agency for Safety at Work. Brussels: "OiRA: Free and Simple Tools for A Straightforward Risk Assessment Process." Available at https://osha.europa.eu/en/tools-and-publications/oira. Accessed June 1, 2019 (n.d.).

European Union statement on risk assessment. "Prevention and Control Strategies," UK: OSHWiki, The Health & Safety Laboratory, 2017. Available at https://oshwiki.eu/wiki/Prevention_and_control_strategies#Principles_deriving_from_the_legislative_Framework. Accessed June 1, 2019 (n.d.).

Guidelines for Preventing Human Error in Process Safety. New York: Center for Chemical Process Safety of the American Institute of Chemical Engineers, 1994.

Hammer, W. *Handbook of System and Product Safety.* Englewood, NJ: Prentice Hall, 1972.

Heinrich, H.W. *Industrial Accident Prevention*, Fourth Edition. New York, NY: McGraw-Hill Book Company, 1959.

Hollnagel, E. *Barriers and Accident Prevention.* Burlington, VT: Ashgate Publishing, 2004.

Johnson, W. *MORT Safety Assurance Systems.* Itasca, IL: National Safety Council, 1980 (Also published by Marcel Dekker, NY).

Kletz, T. *An Engineer's View of Human Error*, Second Edition. Rugby, Warwickshire, UK: Institution of Chemical Engineers, 1991.

Lyon, B. and B. Hollcroft. *Risk Assessments: Top 10 Pitfalls & Tips for Improvement.* Park Ridge, IL: Professional Safety, December 2012.

Lyon, B., Hollcroft, B., and G Povov. *Risk Assessment: A Practical Guide to Assessing Operational Risk.* Hoboken, NJ: John Wiley, 2016 (Fred Manuele wrote the first chapter).

Lyon, B. and G. Popov. *Risk Management Tools for Safety Professionals.* Park Ridge, IL: American Society of Safety Professionals, 2018.

Main, B.W. *Risk Assessment: Basics and Benchmarks.* Ann Arbor, MI: Design Safety Engineering, Inc., 2004.

Main, B.W. *Risk Assessment: Challenges and Opportunities.* Ann Arbor, MI: Design Safety Engineering Inc., 2012.

Manuele, F.A. "Incident Investigation: Our Methods Are Flawed," Professional Safety, October 2014.

Manuele, F.A. "Reviewing Heinrich: Dislodging Two Myths from the Practice of Safety" (2011).

MORT, The Management Oversight and Risk Tree. Washington, DC: U.S. Department of Energy (DOE), 1992.

NCCI—Workers Compensation Claim Frequency, Exhibit 20 for 2011. Boca Raton, FL: National Council on Compensation Insurance, 2011.

NRI MORT User's Manual, NR-1: For Use with the Management Oversight & Risk Tree. Delft, The Netherlands: The *Noordwijk Risk* Initiative Foundation, 2009.

Petersen, D. Safety Management, *Second Edition.* Des Plaines, IL: American Society of Safety Engineers, 1998.

Reason, J. *Human Error.* New York, NY: Cambridge University Press, 1990.

Reason, J. *Managing the Risks of Organizational Accidents.* Burlington, VT: Ashgate Publishing Company, 1997.

Stephans, R.A. *System Safety in the 21st Century.* Hoboken, NJ: John Wiley & Sons, 2004

Stephenson, J. *System Safety 2000.* Hoboken, NJ: John Wiley & Sons, 1991.

U.S. Census Bureau. Number of Firms, Number of Establishments, Employment, and Annual Payroll by Enterprise Employment Size for the United States, All Industries: 2015.

U.S. Chemical Safety and Hazard Investigation Board (CSB) Investigation Report on the Drilling Rig Explosion and Fire at the Macondo Well, Washington, DC. Available at http://www.csb.gov/the-us-chemical-safety-boards-investigation-into-the-macondo-disaster-finds-offshore-risk-management-and-regulatory-oversight-still-inadequate-in-gulf-of-mexico/. Accessed June 1, 2019 (n.d.).

Whittingham, R.B. *The Blame Machine: Why Human Error Causes Accidents*. Burlington, MA: Elsevier Butterworth-Heinemann, 2004.

FURTHER READING

ANSI/ASSP Z10.0-2019. *Occupational Health and Safety Management Systems*. Park Ridge, IL: American Society of Safety Professionals, 2019.

ANSI/ASSP/ISO 45001-2018. *Occupational Health and Safety Management Systems—Requirements with Guidance for Use*. Park Ridge, IL: American Society of Safety Professionals, 2018.

Bureau of Labor Statistics. *Employer-Reported Workplace Injuries and Illnesses. Table 7*. Washington, DC: Department of Labor, 2015.

Bureau of Labor Statistics. *Nonfatal Injuries and Illnesses Requiring Days Away from Work, for 2014, Table 9*. Washington, DC: Department of Labor, 2015.

Manuele, F.A. "Serious injuries & fatalities—A call for a new focus on their prevention,". Professional Safety, December 2008.

Manuele, F.A. *On the Practice of Safety*, Fourth Edition. New York: John Wiley & Sons, 2013.

Manuele, F.A. "Root-cause analysis: uncovering the hows and whys of incidents," Professional Safety, May 2016.

CHAPTER 8

HUMAN ERROR AVOIDANCE

Many injuries, illnesses, and fatalities result from avoidable human errors and organizational, cultural, technical, and management systems deficiencies often lead to those errors. Although focusing on the management decision-making that may be the source of human errors was proposed many years ago, doing so has been infrequently prominent in the work of safety professionals.

Fortunately, a renewed interest has emerged on being able to explain why human errors occur in the occupational setting. It is now more common to see statements such as the following in safety-related literature.

- To prevent human error sometimes requires changes in the design of the workplace and the work methods.
- Managers may wish to address human error by "getting into the heads" of their employees with training being the default corrective action—which will not be effective if error potential is designed into the work.
- The effects of decisions made at a management level accumulate over time and become the contributing factors for incidents resulting in serious results.
- Workers may commit errors because of system defects that result from poor design, inadequate maintenance, and management decisions about how the work is to be done.
- It is management's responsibility to anticipate errors and to have work systems and work methods designed so as to reduce error potential.

Advanced Safety Management: Focusing on Z10.0, 45001, and Serious Injury Prevention, Third Edition. Fred A. Manuele.
© 2020 John Wiley & Sons, Inc. Published 2020 by John Wiley & Sons, Inc.

158 HUMAN ERROR AVOIDANCE

Safety professionals will be more effective as they give counsel on injury and illness prevention if they are aware of the reality of human error causal factors. This chapter

- Relates human error prevention to ergonomics
- Brings attention to human errors that derive from
  - Management decisions that result in unacceptable risk levels
  - Extremes of cost reduction
  - Safety management systems deficiencies
  - Design and engineering decisions
  - Overly stressing work methods
  - Error-provocative operations
- Provides a selected literature and resource review

This chapter is not a text on human error avoidance and reduction. Contents relate to the author's experience and to selections from publications that provide knowledge on the subject that safety professionals should have. Safety professionals need not be concerned over a lack of resources. Enter "human error reduction" into a search engine and that will show a huge number of resources that are available.

HUMAN ERROR AND ERGONOMICS

Several years ago, a group of safety directors were asked how they would view estimates of 50% of the number of workers compensation claims and 60% of total claim costs being ergonomics- and musculoskeletal-related. There was a general agreement that those estimates could be close on a macro basis. But they cautioned that within their own companies, there were differences and that variations by industry could be significant.

A similar inquiry was made for this chapter. Although the number of workers compensation claims has been substantially reduced, musculoskeletal and ergonomics-related incidents are still a major part of the whole.

In some organizations, it has been recognized that help from designers is necessary to reduce human errors—unsafe acts—in the workplace. As an example, a colleague who is employed in a Fortune 500 company spent nearly all of 2018 visiting with engineering units throughout our country to bring to their attention the error-provocative design shortcomings that were corrected on the shop floor. He said that a very large share of the design shortcomings was ergonomics-related.

ORGANIZATIONAL TRANSITION

A transition has taken place that recognizes the greater significance of the term "ergonomics" and resulted in ergonomics and human factors becoming close to synonymous terms. What was the organization known as The Human Factors Society has become the Human Factors and Ergonomics Society. This transition has resulted in a broadened perspective being taken in attempting to match the work to the human in a way that increases value and reduces costs.

There is no other field that is comparable to ergonomics/human factors in which safety professionals can show by their successes that they provide added value. Safety professionals who are involved in ergonomics/human factors tell stories about how they initiated studies because of injury potential and productivity improved and costs were reduced.

DEFINING HUMAN ERROR AND HUMAN ERROR REDUCTION

It seems that every author on human error has his or her own definition, and they vary somewhat. Some are obscure and esoteric. Nevertheless, the many definitions have some similarities. Two selected definitions of human error follow. James Reason's principal research area has been in human error and the way people and organizational processes contribute to the breakdown of complex, well-defended technologies. In *Human Error*, Reason (1990) offers this definition:

> Error will be taken as a generic term to encompass all those occasions in which a planned sequence of mental or physical activities fails to achieve its intended outcome, and when these failures cannot be attributed to the intervention of some chance agency. (p. 9)

Reason's book was written for cognitive psychologists, human factors professionals, safety managers, and reliability engineers. His definition covers all the bases, but is not quite as specific as is needed in the occupational setting.

Trevor Kletz (1991), in *An Engineer's View of Human Error*, gives a definition that relates more precisely to the places in which people work:

> I have tried to show that so-called human errors are events of different types (slips and lapses, mistakes, violations, errors of judgment, mismatches and ignorance of responsibilities), made by different people (managers, designers, operators, construction workers, maintenance workers and so on) and that different actions are required to prevent them happening again: in some cases better training or instructions, in other cases better enforcement of the rules, in most cases a change in the work situation. (p. 181)

Kletz's definition of human error fits well with the author's experience. For simplicity and to have a terse definition of human error that relates directly to the occupational setting in which the exposures to injuries and illnesses occur, this definition is presented:

> Human error: a decision, an oversight, or a personnel action or inaction out of which the potential arises for the occurrence of a harmful incident or exposure.

Then, the goal of human error avoidance and reduction is to have the probability be as low as reasonably practicable that decisions or oversights, made individually or accumulatively, and personnel actions or inactions, will bring about the occurrence of harmful incidents and exposures.

A REVIEW OF SELECTED LITERATURE

Chapanis

Dr. Alphonse Chapanis was exceptionally well known in ergonomics and human factors engineering circles. His work can be traced back to World War II and it is still occasionally cited. He is often quoted, particularly on the benefits of considering the capabilities and limitations of workers as systems are designed.

Chapanis coined the phrase error-provocative and was strong on designing to avoid work methods that were error-provocative. (Chapanis also coined the phrase—To err is human: To forgive, design).

Because of Chapanis' influence and additional research, the author could comfortably make this observation concerning management decision-making for every type of occupational injury:

> If the design of the workplace or the work methods is error-provocative, you can be sure that human errors will occur.

Chapanis was the author of a chapter titled "The Error-Provocative Situation" in *The Measurement of Safety Performance*, a 1980 publication. The following are very brief excerpts from that chapter. Note that they relate to decision-making above the worker level:

- The improvement in system performance that can be realized from the redesign of equipment is usually greater than the gains that can be realized from the selection and training of personnel.
- Design characteristics that increase the probability of error include a job, situation, or system which:
  a. Violates operator expectations
  b. Requires performance beyond what an operator can deliver
  c. Induces fatigue
  d. Provides inadequate facilities or information for the operator
  e. Is unnecessarily difficult or unpleasant
  f. Is unnecessarily dangerous

Improvement in system performance and the design of the operating system is principally a management responsibility, although it is wise to seek worker input in the improvement process. An appropriate goal for safety professionals is to educate decision-makers so that avoiding the creation of error-provocative work situations is ingrained into their thinking.

Hammer

Willie Hammer, in *Handbook of System and Product Safety*, published in 1972, recognized that a worker's human error could be the result of decisions made by personnel at a higher level. His premise is still applicable. Hammer wrote this:

> Almost every mishap can be traced ultimately to personnel error, although it may not have been error on the part of the person immediately involved in the mishap. It may have been committed by a designer, a worker manufacturing the equipment, a maintenance worker, or almost anyone other than the person present when the accident occurred.
>
> It often happens that the person involved is overwhelmed by failures due to causes beyond his or her control, failures that could have been forestalled by incorporation of suitable measures in the design stage. In many instances, due consideration was not given to human capabilities and limitations and to the factors that can and may affect a human being. (p. 68)

Hammer made plain that it would be advantageous in the practice of safety to look for causal factors that occur above the level of a worker who may have been involved in a mishap.

Reason

James Reason's *Human Error* is also a highly recommended resource. First published in 1990, it has since had 12 reprintings. Reason discusses: The Nature of error; Studies of human error; Performance levels and error types; Cognitive under-specification and error forms; A design for a fallible machine; The detection of errors; Latent errors and system disasters; and Assessing and reducing the human error risk.

Reason acknowledges that the bulk of his book favors theory rather than practice and that this final chapter—"Assessing and Reducing the Human Error Risk"—seeks to "redress the balance by focusing on remedial possibilities." He also says that this chapter was written with safety professionals and psychologists in mind.

Reason recognizes the major accomplishments in recent years in risk assessment and in the development of the knowledge necessary to design systems for error avoidance and resiliency.

Another of James Reason's books—*Managing the Risks of Organizational Accidents*—is a "must" read for safety professionals who want an education in human error reduction. It was published in 1997 and has been reprinted five times. Reason writes about how the effects of decisions *accumulate over time* and become the contributing factors for incidents resulting in serious injuries or damage when all the circumstances necessary for the occurrence of a major event come together. Reason writes this:

> Latent conditions arise from strategic and other top-level decisions made by governments, regulators, manufacturers, designers and organizational

managers. The impact of these decisions' spreads throughout the organization, shaping a distinctive corporate culture and creating error-producing factors within the individual workplaces. (p. 10)

Writers and researchers have reiterated that errors are made at an organizational, managerial, and design levels, that they form a distinctive corporate culture, and they create overly stressing and error-producing situations within the occupational setting. Reducing the probability of such human errors occurring is the new frontier for safety professionals.

Norman

Donald A. Norman's *The Psychology of Everyday Things* (1988) is recommended as an additional and important resource. Norman's background is in both engineering and the social sciences. He wrote this

> Most major accidents follow a series of breakdowns and errors, problem after problem, each making the next more likely. Seldom does a major accident occur without numerous failures: equipment malfunctions, unusual events, a series of apparently unrelated breakdowns and errors that culminate in major disaster; yet no single step has appeared to be serious. In many cases, the people noted the problem but explained it away, finding a logical explanation for the otherwise deviant observation. (p. 128)

It is highly recommended that the following comments by Reason and Norman be kept in mind as attempts are made to reduce human error:

> Reason: The impact of [top-level] decisions spreads throughout the organization, shaping a distinctive corporate culture and creating error-producing factors within individual workplaces.
>
> Norman: In many cases, the people noted the problem but explained it away, finding a logical explanation for the otherwise deviant observation.

CCPS

One of the best references on human error prevention is the book *Guidelines for Preventing Human Error in Process Safety* (1994), to which I often refer. Although written for the process industry, the following excerpts are generic. They advise on where human errors occur, who commits them, and at what levels, the influence of organizational culture, and where attention is needed to reduce the occurrence of human errors. The author suggests that safety practitioners seriously consider the following statements and relate them to the injuries and illnesses of which they are familiar:

- It is readily acknowledged that human errors at the operational level are a primary contributor to the failure of systems. It is often not recognized, however, that these errors frequently arise from failures at the management, design, or technical expert levels of the company (p. xiii).

- Almost all major accident investigations in recent years have shown that human error was a significant causal factor at the level of design, operations, maintenance, or the management process. (p. 5)
- One central principle presented in this book is the need to consider the organizational factors that create the preconditions for errors, as well as the immediate causes. (p. 5)

Note particularly that "failures [were] at the management, design, or technical expert levels of the company" and that "human error was a significant causal factor at the level of design, operations, maintenance, or the management process." Those failures and human errors affect the design of the workplace and the work methods—the operating system.

Thus, it is reasonable to suggest that the focus in the practice of safety should be on improving the operating system by stressing prevention through design and the need for adequate and effective barriers so that acceptable risk levels can be achieved and maintained.

Also, with respect to human error—the unsafe act—attention must be given specifically to "the organizational factors that create the preconditions for errors." Then, in the design process, attempts would be made to anticipate and avoid "preconditions for error." That calls for including the necessary barriers.

Although it has been known for quite some time that the foundations for human errors may be in "failures at the management, design, or technical expert levels of an organization," offering counsel to reduce human errors at those levels is not usually a significant element within safety management systems. As the interest in serious injury prevention becomes more prominent, safety professionals will be additionally challenged to become knowledgeable about reducing human errors above the worker level.

Whittingham

R.B. Whittingham's book *The Blame Machine: Why Human Error Causes Accidents* (2004) emphasizes that human errors and defective management systems can be causal factors for major accidents. From the Preface:

> *The Blame Machine* describes how disasters and serious accidents result from recurring, but potentially avoidable, human errors. It shows how such errors are preventable because they result from defective systems within a company.

Johnson

W. Johnson is the author of *Human Error, Safety and Systems Development*, a 2004 publication. It is another text indicating that a transition is taking place whereby additional emphasis is being given to human error reduction.

> Recent developments in a range of industries have increased concern over the design, development, management and control of safety-critical systems.

Attention has now been focused upon the role of human error both in the development and in the operation of complex systems. (p. 6)

Strater

Cognition and Safety: An Integrated Approach to Systems Design and Assessment was written by Oliver Strater and published in 2005. Strater's purpose is to promote having risk assessments made in the design process. This is a worthy goal.

Studies have shown that it is the exception when engineering students acquire knowledge about hazards, risks, and risk assessments. That results sometimes, as Strater says, in designers creating constraints at the sharp-end, which eventually lead to human errors. "Sharp-end" is a British term, meaning the point where the work or task is done. Strater says this in his Preface:

> Safety suffers from the variety of methods and models used to assess human performance. For example, operation is interested about human error while design is aligning the system to workload or situational awareness. This gap decouples safety assessment from design. As a result, design creates constraints at the sharp-end, which eventually leads to human errors.

Dekker

Sidney Dekker is the author of *The Field Guide to Understanding Human Error* (2006). Dekker's doctorate is in Cognitive System Engineering—acquired at Ohio State University. The author recommends Dekker's *Field Guide* highly to all safety professionals.

Only a few excerpts from Dekker's book follow. They direct a person who gives counsel on error avoidance and reduction to try to affect the decisions made above the worker level as a primary initiative. Achieving success in such an endeavor results in a culture change.

- Human error is not a cause of failure. Human error is the effect, or symptom, of deeper trouble. Human error is ... systematically connected to features of people's tools, tasks, and operating systems. (p. 15)
- Sources of error are structural, not personal. If you want to understand human error, you have to dig into the system in which people work. You have to stop looking for people's shortcomings. (p. 17)

Dekker quoted from James Reason's book *Human Error* as follows:

> Rather than being the main instigator of an accident, operators tend to be the inheritors of system defects created by poor design, incorrect installation, faulty maintenance and bad management decisions. Their part is usually that of adding the final garnish to a lethal brew whose ingredients have already been long in the cooking. (p. 88 in Dekker; p. 173 in Reason's *Human Error*)

THE CONCEPT OF DRIFT

Particularly in difficult economic times, management decision-making may result in drift and deterioration in safety-related systems. Jens Rasmussen (1997) was one of the first authors to comment on drift in the decision-making that results in the creation of an accumulation of adverse conditions and practices that could lead eventually to incidents resulting in serious injuries and fatalities. In *Risk Management in a Dynamic Society: A Modelling Problem*, he wrote that

> Companies today live in a very aggressive and competitive environment which will focus the incentives of decision makers on short term financial and survival criteria rather than long term criteria concerning welfare, safety, and environmental impact. (p. 186)

James Reason expands on the concept of drift. Reason wrote in *Managing the Risks of Organizational Accidents* about how the effects of decisions *accumulate over time* and become the contributing factors for incidents resulting in serious injuries or damage when all the circumstances necessary for the occurrence of a major event come together.

Whittingham also writes about errors made in the decision-making process in *The Blame Machine: Why Human Error Causes Accidents*. He writes:

> Failures in decision-making have been identified as a principal cause of a number of major accidents. Decision-making errors are always latent and their effects can be delayed for extremely long periods of time before the errors are detected. Another problem is the difficulty in predicting and understanding how decision making errors are propagated down through an organization to, eventually, lead to an accident. (p. 31)

Signs of the effects of drift become apparent when nonconformity from what had been considered appropriate practice occurs. Sidney Dekker (2011) is the author of *Drift into Failure*. A section of his book is devoted to "Normalizing Deviance, Structural Secrecy, and Practical Drift." He comments on decisions taken that made sense and were considered favorable at the time. He then observes

> Each little decision they made, each little step, was only a small step away from what had previously been seen as acceptable risk. (p. 103)
>
> It was easy to reach agreement on the notion that production pressure and economic constraints are a powerful combination that could lead to adverse events. (p. 105)

Dekker uses the term "normalization of deviance." That term was used for a time to refer to situations in which departing from what was considered good risk management practice and acceptance of deviations that resulted in an increase in the risk level occurred. Each such situation is an alert for observant safety professionals and requires a good analytic and risk management approach to determine the actions to be taken. The author does not say that reversing a trend of drift will be easy.

CONCLUSION

General observations can be drawn from the several sources cited here and the author's experience:

- Human errors, of commission or omission, are factors in the occurrence of nearly all hazards-related incidents.
- Typical safety management systems do not address human error reduction above the worker level, particularly on an anticipatory basis.
- You cannot change the human condition, but you can change the conditions under which people work.
- The solutions to most human performance problems are technical rather than psychological.
- Potentials for human error derive largely from top-level decisions, and the impact of those decisions' spreads throughout the organization, shaping a distinctive corporate culture and creating overly stressing and error-provocative situations.
- To avoid hazard-related incidents resulting in serious injuries, human error potentials must be addressed at the cultural, organizational, management systems, design, and engineering levels, and with respect to the work methods prescribed.

All this spells opportunity for safety professionals to acquire knowledge with respect to human error avoidance and reduction and to enhance their professional status. Human error reduction may very well become the frontier for the practice of safety.

Sidney Dekker said in *The Field Guide to Understanding Human Error* that some readers may believe that his intent is to let line workers "off the hook" and that his writing "looks like a ploy to excuse defective operators" (p. 196).

He assures readers—that is not his intent. And that is not the intent here. If a problem in a given situation is identified as an individual employee's performance, the solution should focus on altering that person's performance.

But if the employee was set up to err because of the design of the workplace or the work methods, trying to influence employee performance will not solve the problem.

REFERENCES

Chapanis, A. The Error-Provocative Situation" In *The Measurement of Safety Performance*, William E. Tarrants, ed. New York: Garland Publishing, 1980.

Dekker, S. *Drift into Failure*. Burlington, VT: Ashgate Publishing Company, 2011.

Dekker, S. *The Field Guide to Understanding Human Error*. Burlington, VT: Ashgate Publishing Company, 2006.

Guidelines for Preventing Human Error in Process Safety. New York: Center for Chemical Process Safety of the American Institute of Chemical Engineers, 1994.

Hammer, W. *Handbook of System and Product Safety*. Englewood, NJ: Prentice Hall, 1972.

Johnson, W. *Human Error, Safety and Systems Development*. Norwell, MA: Kluwer Academic Publishers, 2004.

Kletz, T. *An Engineer's View of Human Error*, Second Edition. Rugby, Warwickshire, UK: Institution of Chemical Engineers, 1991.

Norman, D.A. *The Psychology of Everyday Things*. New York: Basic Books, 1988.

Rasmussen, J. "Risk management in a dynamic society: a modelling problem." *Safety Science*, Vol. 27, 2/3, pp. 183–213, 1997. Elsevier Science Ltd. Printed in Great Britain.

Reason, J. *Human Error*. Cambridge, UK: Cambridge University Press, 1990.

Reason, J. *Managing the Risks of Organizational Accidents*. Burlington, VT: Ashgate Publishing Company, 1997.

Strater, O. *Cognition and Safety: An Integrated Approach to Systems Design and Assessment*. Burlington, VT: Ashgate Publishing Company, 2005.

Whittingham, RB. *The Blame Machine: Why Human Error Causes Accidents*. Burlington, MA: Elsevier Butterworth Heinemann, 2004.

FURTHER READING

Manuele, FA. *On the Practice of Safety*, Fourth Edition. New York Hoboken, NJ: John Wiley & Sons, 2013.

CHAPTER 9

ON HAZARDS ANALYSES AND RISK ASSESSMENTS

Both Z10.0 and 45001 have requirements for organizations to have systems in place for hazards identification and analyses, risk assessments and prioritization. Safety professionals can expect that being able to make documented risk assessments will be necessary for job retention and career enhancement. That premise has acquired weight because of the more frequent inclusion in safety standards and guidelines of provisions requiring that hazards be identified and analyzed and that risk assessments be made.

Since safety professionals are to analyze hazards and assess the risks that derive from them, the question logically follows—what do they need to know? This chapter provides a primer that will serve many of the hazard analysis and risk assessment needs that safety professionals will encounter. This chapter

- Defines the terms that must be understood in the hazard analysis and risk assessment processes
- Establishes the parameters for a hazard analysis
- Indicates how a hazard analysis is extended into a risk assessment
- Establishes that the goal is to achieve acceptable risk levels
- Includes *A Hazard Analysis and Risk Assessment Guide*

Advanced Safety Management: Focusing on Z10.0, 45001, and Serious Injury Prevention, Third Edition. Fred A. Manuele.
© 2020 John Wiley & Sons, Inc. Published 2020 by John Wiley & Sons, Inc.

- Gives examples of the terms used in risk assessment matrices and their variations
- Presents examples of basic, two-dimensional risk assessment matrices
- Describes several of the most commonly used hazard analysis and risk assessment techniques

DEFINING HAZARD, HAZARD ANALYSIS, RISK, AND RISK ASSESSMENT

When a safety professional identifies a hazard and its potential for harm or damage and decides on the probability that an injurious or damaging incident can occur, a risk assessment has been subjectively made. In doing so, for the simpler and less complex hazards and risks, the assessment may be based entirely on *a priori* knowledge and experience, without documentation. Making informal risk assessments has been an integral part of the practice of safety professionals from time immemorial.

Recent developments take the risk assessment subject to a higher level within the practice of safety. By formalizing the hazard analysis and risk assessment process, a better appreciation of the significance of individual risks is achieved. As risk levels are categorized and prioritized, more intelligent decisions can be made with respect to their elimination or reduction. For the hazard analysis and risk assessment process, it is necessary that agreement be reached on the definitions of hazards, hazards analyses, risks, and risk assessments.

Z10.0 and 45001are occupational health and safety management system standards and their terminologies are, understandably, worker injury and illness-related. They do not include considerations for injury to the public, possible damage to the environment, or damage to property or business down time. This chapter has a broader purpose than Z10.0 and 45001, as will be seen.

- A hazard is defined as the potential for harm—to people, property, or the environment. If there is no potential for harm, injury or damage cannot occur. Hazards encompass all aspects of technology or activity that produce risk. Hazards include the characteristics of things and the actions or inactions of people.
- A hazard analysis is made to estimate the severity of harm or damage that could result from a hazards-related incident or exposure. In the hazard analysis process, it is not necessary to include an estimate of incident or exposure probability. Examples of hazards analyses that do not include probability indicators are the estimates made by fire protection engineers of Maximum Foreseeable Loss or Maximum Probable Loss for insurance purposes, and the hazard analysis requirements of *OSHA's Rule for Process Safety Management of Highly Hazardous Chemicals, 1910.119* (1992).
- Whether hazardous situations are simple or complex, the process for making a hazard analysis will address the following questions:

DEFINING HAZARD, HAZARD ANALYSIS, RISK, AND RISK ASSESSMENT   **171**

1. Is there a potential for harm, deriving from aspects of the technology or activity, the characteristics of things, or the actions or inactions of people?
2. Can the potential be realized?
3. Who and what are exposed to harm or damage?
4. What is the frequency of endangerment?
5. What will the consequences be as represented by the severity of harm or damage if the potential of a hazard is realized?

- Making a hazard analysis is necessary to and precedes making a risk assessment. William Johnson (1980, p 245) said this about hazard analysis in *MORT Safety Assurance Systems*: "Hazard analysis is the most important safety process in that, if that fails, all other processes are likely to be ineffective." Johnson's premise is sound.
- Risk is defined as an estimate of the probability of a hazards-related incident or exposure occurring and the severity of harm or damage that could result.
- Probability is defined as the likelihood of a hazard's potential being realized and initiating an incident or exposure that could result in harm or damage—for a selected unit of time, events, population, items, or activity being considered.
- Severity is defined as the extent of harm or damage that could result from a hazard-related incident or exposures.
- Risk assessment is a process that commences with hazard identification and analysis, through which an estimate is made of the probable extent of severity of harm or damage if an incident or exposure occurs, and concludes with an estimate of the probability of the incident or exposure occurring.

Excerpts that follow are responsive to the question—What is "Risk." They are taken from the paper titled *Framework for Environmental Health Risk Management* (1997) which was issued by The Presidential/Congressional Commission on Risk Assessment and Risk Management. They are an indication of the widespread adoption of the foregoing definitions.

What Is "Risk"

Risk is defined as the probability that a substance or situation will produce harm under specified conditions. Risk is a combination of two factors:

- The **probability** that an adverse event will occur;
- The **consequences** of the adverse event.
  Risk encompasses impacts on public health and on the environment and arises from **exposure** and **hazard**. Risk does not exist if exposure to a harmful substance or situation does not or will not occur. Hazard is determined by whether a particular substance or situation has the potential to cause harmful effects.

MAKING A HAZARD ANALYSIS/RISK ASSESSMENT

For many hazards and the risks that derive from them, knowledge gained by safety professionals through education and experience will lead to proper conclusions on how to attain an acceptable risk level without bringing teams of people together for discussion.

For the more complex situations, it is vital to seek the counsel of experienced personnel who are familiar with the work or process. Reaching group consensus is a highly desirable goal. Sometimes, for what a safety professional considers obvious, achieving consensus is still desirable so that buy-in is obtained for the actions to be taken.

A general guide follows on how to make a hazard analysis and how to extend the process into a risk assessment. This guide is applicable to every phase of the life cycle of facilities, equipment, processes, and materials—for example, design, construction, operation, maintenance, and disposal.

> *Whatever the simplicity or complexity of the hazard/risk situation, and whatever analysis method is used, the following thought and action process is applicable.*

A Hazard Analysis and Risk Assessment Guide

1. Select a Risk Assessment Matrix A risk assessment matrix provides a method to categorize combinations of probability and severity, thus establishing risk levels. A matrix helps in communicating with decision-makers and influencing their decisions on risks and the actions to be taken to ameliorate them. Also, risk assessment matrices can be used to compare and prioritize risks and to effectively allocate mitigation resources.

Definitions of the levels of probability and severity used in risk assessment matrices vary greatly. That reflects the differences in the perceptions people have of risk. Since a risk assessment matrix is a management decision tool, management personnel at appropriate levels must agree on the definitions of the terms to be used. In so doing, management establishes the levels of risk that require reduction and those that are acceptable.

Repeating for emphasis: Safety professionals must understand that definitions of terms for incident probability and severity and for risk levels vary greatly. Thus, they should tailor a risk assessment matrix to suit the hazards and risks and the management tolerance for risk with which they deal.

Examples of definitions used for incident probability and severity are presented here as well as for risk categories and risk assessment matrices. They are to provide safety professionals with a broad base of information from which choices can be made in developing the matrix considered appropriate for their clients' needs.

The breadth of possibilities in drafting a risk assessment matrix is extensive. Matrices have been developed that display only one or a combination of several of

MAKING A HAZARD ANALYSIS/RISK ASSESSMENT   **173**

the following injury or damage classes: employees, members of the public, facilities, equipment, product, operation down time, and the environment.

2. Establish the Analysis Parameters   Select a manageable task, system, process, or product to be analyzed, establish its boundaries and operating phase (standard operation, maintenance, startup), and define its interface with other tasks or systems, if appropriate. Determine the scope of the analysis in terms of what can be harmed or damaged: people (the public, employees); property; equipment; productivity; the environment.

For this primer, two-dimensional risk assessment matrices are discussed. And they should be adequate for most all risks with which safety professionals are involved, but not all. Comments are made on variations of two categories of terms: the **severity** of harm or damage that could result from a hazards-related incident or exposure; and the **probability** that the incident or exposure could occur. They also show the **risk levels** that derive from the various combinations of severity and probability.

Some organizations have adopted three- and four-dimensional risk assessment systems for a variety of risk categories, but particularly for the safety of products. A review of three- and four-dimensional risk assessment systems is given in Chapter 10, "Three- and Four-Dimensional Risk Scoring Systems."

3. Identify the Hazards   A frame of thinking should be adopted that gets to the bases of causal factors, which are hazards. These questions would be asked: What are the aspects of technology or activity that produces risk? What are the characteristics of things or the actions or inactions of people that present a potential for harm? Depending on the complexity of the hazardous situation, some or all of the following may apply:

1. Use intuitive engineering and operational sense: This is paramount throughout.
2. Examine system specifications and expectations.
3. Review codes, regulations, and consensus standards.
4. Interview current or intended system users or operators.
5. Consult checklists.
6. Review studies from other similar systems.
7. Consider the potential for unwanted energy releases.
8. Take into account possible exposures to hazardous environments.
9. Review historical data—industry experience, incident investigation reports, OSHA, and National Safety Council data, manufacturer's literature.
10. Brainstorm.

4. Consider the Failure Modes   Define the possible failure modes that would result in realization of the potentials of hazards. Ask: What circumstances can arise that would result in the occurrence of an undesirable event? What controls are in place that mitigates against the occurrence of such an event or exposure? How effective are the controls?

5. Determine the Frequency and Duration of Exposure For each harm or damage category selected for the analysis (people, property, business interruption, etc.), estimate the frequency and duration of exposure to the hazard. This is a very important part of this exercise. For instance, in a workplace situation, ask—how often is a task performed, how long is the exposure period, and how many people are exposed? More judgments than one might realize will be made in this process.

6. Assess the Severity of Consequences The purpose is to determine the magnitude of harm or damage that could result. Informed speculations are made to establish the consequences of an incident or exposure: the number of injuries or illnesses and their severity, and fatalities; the value of property or equipment damaged; the time for which productivity will be lost; and the extent of environmental damage. Historical data can be of value as a baseline but should not be treated as the primary or sole determinant. Judgment of knowledgeable and competent personnel should prevail.

On a subjective basis, the goal is to decide on the worst credible consequences should an incident or exposure occur, not the worst conceivable consequence.

When the severity of the outcome of a hazard-related incident or exposure is determined, a hazard analysis has been completed.

7. Determine Occurrence Probability Extending the hazard analysis into a risk assessment requires the one additional step of estimating the likelihood, the probability, of a hazardous event or exposure occurring. Unless empirical data are available, and that would be a rarity, the process of selecting incident or exposure probability is subjective. As is the case when determining severity of consequences, historical data can be of value as a baseline but should not be the sole determinant. Judgment of knowledgeable and competent personnel, controls in place and their capability, and lessons learned with respect to similar systems should prevail.

For the more complex hazardous situations, it should be considered a necessity that brainstorming sessions be held with knowledgeable people and that consensus be reached in determining occurrence probability and severity.

To be meaningful, probability has to be related to an interval base of some sort, such as a unit of time or activity, events, units produced, or the life cycle of a facility, equipment, process, or a product.

8. Define the Initial Risk Conclude with a statement that addresses the probability of a hazards-related incident or exposure occurring; the expected severity of adverse results; and a risk category. (For example—High, Serious, Moderate, or Low risk.) Using a risk assessment matrix for that purpose assists in communicating on the risk level.

9. Risk Prioritization A risk ranking system should be adopted so that priorities can be established. Since the risk assessment exercise is subjective, the risk ranking system would also be subjective. Prioritizing risks gives management the knowledge needed for intelligent allocation of resources with respect to risk avoidance, elimination, or control.

10. Select and Implement Risk Reduction and Control Methods When the initial risk assessment indicates that elimination or reduction measures are to be taken to achieve acceptable risk levels:

- Alternate proposals for the design and operational changes necessary would be recommended.
- In their order of effectiveness, the actions outlined in Chapter 11, "Hierarchy of Controls," would be the base upon which remedial proposals are made.
- If informal judgments on costs are considered inadequate for a risk situation, formal action would be taken to establish remediation cost for each proposal and an estimate would be given of its effectiveness in achieving risk reduction to an acceptable risk level.
- Risk avoidance, elimination, or reduction methods would be selected and implemented.
- A tracking system should be put in place to assure completion of the actions undertaken.

11. Assess the Residual Risk Residual risk is defined as the risk remaining after preventive measures have been taken. No matter how effective the preventive actions, there will always be residual risk if a facility exists or an activity continues. Attaining zero risk is not possible.

- If the residual risk is not acceptable, the action outline set forth in this hazard analysis and risk assessment process would be applied again.
- The risk assessment process is to continue until an acceptable risk level is attained.
- If an acceptable risk level cannot be achieved, operations should not be continued, except in unusual and emergency circumstances or as a closely monitored and limited exception and with approval of the person having authority to accept the risk.
- Information on residual risks is to be communicated to relevant stakeholders.

Even though the residual risk may be acceptable, management should consider taking additional risk reduction measures if the cost is reasonable, particularly if doing so resolves concerns of employees. While some employee concerns may seem excessive, management should understand that employee perceptions are their reality.

12. Risk Acceptance Decision-Making Risk acceptance decisions shall be made at the appropriate management levels:

- Temporary acceptance of high and serious risks shall be made at the top management level.
- Responsibility for moderate or low risks may be assigned to lower management levels.

13. Documentation To the extent appropriate for a given risk situation, documentation should include the

- The names and job titles/qualifications of the persons who did the risk assessment.
- Risk assessment method(s) used.
- Hazards identified.
- Risks deriving from the hazards.
- Avoidance, elimination, reduction, and control measures taken to attain acceptable risk levels.

Consideration would be given to additional documentation in certain cases that includes relevant information about the use, effectiveness of design, problems during startup and continued use, and changes to designs over the life cycle of the design. The foregoing applies whether the documentation is done directly under the direction of management or by a supplier of services.

14. Reassess the Risk: Follow-Up on Actions Taken This step is a part of all effective problem-solving exercises. Follow-up activity would determine that the

- Problem was resolved, only partially resolved, or not resolved and acceptable risk levels were or were not achieved.
- Actions taken did or did not create new hazards.

If acceptable risk levels were not achieved or new hazards were introduced, the risk is to be re-evaluated and other countermeasures are to be proposed and taken. Repeating for emphasis: acceptable risk levels are to be obtained.

DESCRIPTIONS: PROBABILITY AND SEVERITY

Examples follow in Tables 9.1–9.6 to show variations in the terms used and their descriptions. They have been adopted in a variety of applied risk assessment processes for probability of occurrence and severity of consequence. There is no one right method in selecting probability and severity categories and their descriptions.

Tables 9.4–9.6 show how severity of harm or damage categories can be related to several types of adverse consequences and levels of harm or damage.

TABLE 9.1 Example A: Probability Descriptions

Descriptive Word	Probability Descriptions
Frequent	Likely to occur repeatedly
Probable	Likely to occur several times
Occasional	Likely to occur sometime
Remote	Not likely to occur
Improbable	So unlikely that one can assume occurrence will not be experienced

TABLE 9.2 Example B: Probability Descriptions

Descriptive Word	Probability Descriptions
Frequent	Occurs often, continuously experienced
Probable	Occurs several times
Occasional	Occurs sporadically, occurs sometimes
Seldom	Remote chance of occurrence; unlikely but could occur sometime
Unlikely	Can assume incident will not occur

TABLE 9.3 Example C: Probability Descriptions

Descriptive Word	Probability Descriptions
Frequent	Could occur annually
Likely	Could occur once in 2 years
Possible	Not more than once in 5 years
Rare	Not more than once in 10 years
Unlikely	Not more than once in 20 years

TABLE 9.4 Example A: Severity Descriptions for Multiple Harm and Damage Categories

Catastrophic	Death or permanent total disability, system loss, major property damage, and major business down time
Critical	Permanent, partial, or temporary disability in excess of three months, major system damage, significant property damage and down time
Marginal	Minor injury, lost workday accident, minor system damage, minor property damage and little down time
Negligible	First aid or minor medical treatment, minor system impairment

TABLE 9.5 Example B: Severity Descriptions for Multiple Harm and Damage Categories

Catastrophic	One or more fatalities, total system loss, chemical release with lasting environmental or public health impact
Critical	Disabling injury or illness, major property damage and business down time, chemical release with temporary environmental or public health impact
Marginal	Medical treatment or restricted work, minor subsystem loss or damage, chemical release triggering external reporting requirements
Negligible	First aid only, nonserious equipment or facility damage, chemical release requiring only routine cleanup without reporting

TABLE 9.6 Example C: Severity Descriptions for Multiple Harm and Damage Categories

Category: Descriptive Word	People: Employees, Public	Facilities, Product or Equipment Loss (in US$)	Operations Down Time	Environmental Damage
Catastrophic	Fatality	Exceeds 3 million	Exceeds six months	Major event, requires more than 2 years for full recovery
Critical	Disabling injury or illness	500,000 to 3 million	Four weeks to six months	Significant event, injury or illness. Requires 1 to 2 years for full recovery
Marginal	Minor injury or illness	50,000 to 500,000	Two days to four weeks	Recovery time is less than 1 year
Negligible	Injury requires only first aid	Less than 50,000	Less than two days	Minor damage, easily repaired. Little time for recovery

EXAMPLES OF RISK ASSESSMENT MATRICES

Three examples of risk assessment matrices follow. An adaptation is shown in Table 9.7 of the "Risk Assessment Matrix" in MIL-STD-882E, the *Department of Defense Standard Practice for System Safety* (2000, p. 11).

MIL-STD-882, first issued in 1969, is the grandfather of risk assessment matrices. All of the over 30 variations of matrices the author has collected include the basics found in the 882 standards. They include event probability categories, severity of harm or damage ranges, and risk gradings.

Table 9.8 shows a risk assessment matrix that combines types of severity categories and uses alpha risk gradings.

EXAMPLES OF RISK ASSESSMENT MATRICES   179

TABLE 9.7 Risk Assessment Matrix

Occurrence Probability	Severity of Consequence			
	Catastrophic	Critical	Marginal	Negligible
Frequent	High	High	Serious	Medium
Occasional	High	Serious	Medium	Low
Remote	Serious	Medium	Medium	Low
Improbable	Medium	Medium	Medium	Low
Eliminated	This category is used only for identified hazards that are totally removed			

TABLE 9.8 Risk Assessment Matrix: Alpha Risk Level Indicators: Probability That Something Will Go Wrong

Severity Categories	Frequent: Likely to Occur Immediately or Soon	Likely: Quite Likely to Occur Often	Occasional: May Occur in Time	Seldom: Not Likely to Occur But Possible	Unlikely: Unlikely to Occur
Catastrophic, death, multiple injuries, severe property, or environmental damage	E	E	E	H	M
Critical, serious injuries, significant property, or environmental damage	E	H	H	M	L
Marginal, may cause minor injuries, financial loss, or negative publicity	H	M	M	L	L
Negligible, minimum threat to persons or damage to property	M	L	L	L	L

E: Extremely High-Risk; H: High Risk; M: Moderate Risk; L: Low Risk.

The risk assessment matrix shown in Figure 9.1 is a composite of matrices that include numerical values for probability and severity levels that are transposed into qualitative risk gradings. It was originally developed for people who prefer to deal with numbers rather than qualitative indicators. To provide a one-page tool that could be used by safety professionals who wanted to get shop floor personnel involved in risk assessments, extensions were made to include severity and probability descriptions and action levels.

In two instances, shop floor personnel said to safety professionals that relating numbers to each other first—such as 6 to 12—was a big help in understanding whether the risk category was moderate or serious. If using a risk assessment matrix in which numbers are used to begin with to establish risk levels makes the

Risk Assessment Matrix

Severity Levels and Values	Occurrence Probabilities and Values				
	Unlikely (1)	Seldom (2)	Occasional (3)	Likely (4)	Frequent (5)
Catastrophic (5)	5	10	15	20	25
Critical (4)	4	8	12	16	20
Marginal (3)	3	6	9	12	15
Negligible (2)	2	4	6	8	10
Insignificant (1)	1	2	3	4	5

Numbers were intuitively derived. They are qualitative, not quantitative. They have meaning only in relation to each other.

Incident or Exposure Severity Descriptions

Catastrophic: One or more fatalities, total system loss and major business down time, environmental release with lasting impact on others with respect to health, property damage or business interruption.

Critical: Disabling injury or illness, major property damage and business down time, environmental release with temporary impact on others with respect to health, property damage or business interruption.

Marginal: Medical treatment or restricted work, minor subsystem loss or property damage, environmental release triggering external reporting requirements.

Negligible: First aid or minor medical treatment only, non-serious equipment or facility damage, environmental release requiring routine cleanup without reporting.

Insignificant: Inconsequential with respect to injuries or illnesses, system loss or down time, or environmental release.

Incident or Exposure Probability Descriptions

Unlikely: Improbable, unrealistically perceivable.
Seldom: Could occur but hardly ever.
Occasional: Could occur intermittently.
Likely: Probably will occur several times.
Frequent: Likely to occur repeatedly.

Risk Levels

Combining the Severity and Occurrence Probability values yields a risk score in the matrix. The risks and the action levels are categorized below.

Risk Categories, Scoring, and Action Levels

	Risk Score	Action Level
Low risk	1 to 5	Remedial action discretionary.
Moderate risk	6 to 9	Remedial action to be taken at appropriate time.
Serious risk	10 to 14	Remedial action to be given high priority.
High risk	15 or greater	Immediate action necessary. Operation not permissible except in an unusual circumstance and as a closely monitored and limited exception with approval of the person having authority to accept the risk.

FIGURE 9.1 Risk assessment matrix: numerical gradings.

process more understandable and acceptable for operating personnel that should be encouraged.

It is important to understand that the numbers in Figure 9.1 were intuitively derived: they are qualitative—not quantitative. Thus, the numbers have value only in relation to each other and that is the case for all risk scoring systems that are not based on hard probability and severity numbers, which are rarely available.

RISK ASSESSMENT

Risk assessment is both an art and a science. Since establishing risk levels is largely a matter of judgment, people will come to different conclusions in a given situation. Although there are no rules with respect to the terms used to establish qualitative risk levels, a matrix, as a minimum, should show probability and severity categories and risk gradings. However, safety professionals should draft matrices with which they are comfortable. Since risk assessment matrices are valuable communication tools, the terms used in them must be agreed upon and the education time necessary to achieve an understanding of them must be allocated.

ON ACCEPTABLE RISK

Always, the theoretical goal would be to eliminate all hazards, but that is not feasible. When a hazard cannot be eliminated, the associated risk should be reduced to a level as low as reasonably practicable and acceptable, taking into consideration the cost constraints in relation to the extent of risk reduction to be obtained, and performance requirements.

Applying the concept of ALARP—as low as reasonably practicable—is recognized as a valuable tool to assist in determining acceptable risk levels. Thus far, this chapter has dealt with hazards, risks, probability, and severity. In applying the ALARP concept, economics is brought into the decision-making. ALARP is defined as follows: ALARP is that level of risk which can be further lowered only by an increment in resource expenditure that is disproportionate in relation to resulting decrement of risk. However, when risk reduction measures are concluded, the risks must be acceptable.

MANAGEMENT DECISION LEVELS

Going through the exercise of creating and reaching agreement on a risk assessment matrix and the management decision levels adds to a safety professional's effectiveness in communicating about risks and obtaining consideration of the remedial actions recommended. For the following comments on acceptable risk levels and the management actions to be taken to achieve them, keep in mind that

- an acceptable risk level must be tolerable in the situation being considered
- while economic considerations are a part of the decision-making, the risk level is to be as low as reasonably practicable and acceptable
- special consideration should be given to preventing incidents resulting in serious injuries, illnesses, and fatalities

The author's opinions follow on management actions to be taken for various categories of risk. Others may have different views:

- If the risk category for worker injury or illness or environmental damage or other property damage is High, operation is not permissible except in an unusual circumstance and as a closely monitored and limited exception, only with approval of the person having authority to accept the risk. If it is determined that the cost to reduce the risk to an acceptable level is excessive in relation to the risk reduction benefit to be achieved, the operation should cease in all but rare situations (e.g., society accepts the risks of deep-sea fishing, a high hazard occupation).
- If the risk category is Serious, the risk is not acceptable and action should be undertaken on a high priority basis, meaning very soon, to lower the risk to an acceptable level. While arrangements are made to reduce the risk, an extra heavy application of the lower levels in the hierarchy of controls (warning systems, blocking off work areas, administrative controls, training, personal protective equipment) is in order. If it is determined that the cost to reduce the risk to an acceptable level is excessive in relation to the risk reduction benefit to be achieved, the operation should cease in all but rare situations.
- Where the risk category is Medium, even though the probability ratings for severe injury or illness are "Improbable" or "Remote," and the probability rating for minor injury is "Occasional," and the probability ratings for negligible injury are "Frequent" or "Probable," remedial action should be taken, in good time, to reduce the risk in accord with good economics. This is the risk category where the lower levels in the hierarchy of controls, if more extensively and effectively applied, may be sufficient to achieve acceptable risk levels.
- Where the risk category is Low, the risk is considered acceptable. Nevertheless, there will be times when it is good business management and employee relations to reduce Low risks further if they are perceived by employees to be more serious. Remember, employee perceptions are their reality.

Some of the risk assessment matrices shown in this chapter combine elements pertaining to personal injury with the financial impact of an incident represented by an amount of property damage, business interruption time, and time to recover from an incident.

Safety professionals who have made such combinations in their risk assessment matrices say that they get better management response to their proposals for risk reduction if they tie the severity of injury potential to avoiding operational property damage, down time, and business interruption and to avoiding environmental damage. That has been the author's experience.

DESCRIPTIONS OF HAZARDS ANALYSES AND RISK ASSESSMENT TECHNIQUES

Over the past 60 years, a large and unwieldy number of hazard analysis and risk assessment techniques have been developed. For example: Pat Clemens (1982) gave brief descriptions of 25 techniques in "A Compendium of Hazard Identification & Evaluation Techniques for System Safety Applications."

In the *System Safety Analysis Handbook*, 101 methods are described. In *ANSI/ASSE Z690.3*-2011, the title of which is *Risk Assessment Techniques*, comparisons and descriptions are given of 31 risk assessment techniques.

Brief descriptions will be given here of purposely selected hazard analysis techniques. If a safety professional understands all of them and is capable of bringing them to bear in resolving hazard and risk situations, she or he will be exceptionally well qualified to meet risk assessment requirements in Z10.0 and 45001.

As a practical matter, having knowledge of three risk assessment concepts will be sufficient to address most occupational safety and health risk situations. They are Preliminary Hazard Analysis; the What–If Checklist Analysis Method; and Failure Mode and Effects and Analysis.

It is important to understand that each of those techniques complement, rather than supplant, each other. Selecting the technique or a combination of techniques to be used to analyze a hazardous situation requires good judgment based on knowledge and experience. Qualitative rather than quantitative judgments will prevail. For all but the complex risks, qualitative judgments will be sufficient.

Sound quantitative data on incident probability of occurrence and severity of outcome is seldom available. Colleagues who insist on having quantitative risk assessments are not pleased when I say that most quantitative risk assessments are really qualitative risk assessments because so many judgments have to be made in the process to decide on probability and severity.

Preliminary Hazard Analysis (PHA): Hazard Analysis and Risk Assessment

Originally, the preliminary hazards analysis (PHA) technique was used to identify and evaluate hazards in the early stages of the design process. However, in actual practice, the technique has attained much broader use. Principles on which preliminary hazards analyses are based have been adopted for use not only in the initial design process but also in assessing the risks of existing products or operations.

For example, a standard adopted by the International Organization for Standardization is applied in the European Community and the standard requires that risk assessments be made for all machinery that is to go into a workplace in member countries. That standard is EN ISO 12100-2010: *Safety of Machinery. General principles for design. Risk assessment and risk reduction.* Those risk assessment requirements have been met in some companies by applying an adaptation of the PHA technique.

In reality, the PHA technique needs a new name, reflecting its broader usage. In one organization, the process is called Hazard Analysis and Risk Assessment, a designation that is coming into greater usage since it is more descriptive of its purpose.

(Also, take note to avoid confusion: for *OSHA's Rule for Process Safety Management of Highly Hazardous Chemicals* and for *EPA's Risk Management Program for Chemical Accidental Release Prevention* (1996), PHA stands for Process Hazard Analysis.)

Headings on preliminary hazard analysis forms will include the typical identification data: date, names of evaluators, department, and location. The following information is usually included in a preliminary hazard analysis form:

- A hazard description, sometimes called a hazard scenario.
- A description of the task, operation, system, or product being analyzed.
- The exposures that are to be analyzed: people (employees, the public); facility, product or equipment loss; operation down time; environmental damage.
- The probability interval to be considered: unit of time or activity; events; units produced; life cycle.
- A numerical or alpha indicator for the severity of harm or damage that might result if the hazard's potential is realized.
- A numerical or alpha indicator for the occurrence probability.
- A risk assessment code, using the agreed-upon Risk Assessment Matrix.
- Remedial action to be taken, if risk reduction is needed.

A communication accompanies the analysis, explaining the assumptions made and the rationale for them. Comment would be made on the assignment of responsibilities for the remedial actions to be taken, and when. A Hazard Analysis and Risk Assessment Worksheet (formerly called a Preliminary Hazard Analysis Worksheet) appears as Addendum A in this chapter. That form, and other similar forms, requires entry of severity, probability, and risk codes before and after countermeasures are taken.

Figure 9.1 in this chapter is an add-on to Addendum A.

On Developing a Risk Coding System

Assume that there are to be four severity categories, having the numerical codes: Catastrophic—1; Critical—2; Marginal—3; Negligible—4.

Assume that there are to be five categories of occurrence probability, having alpha codes: Frequent—A; Probable—B; Occasional—C; Remote—D; Improbable—E.

Table 9.9 displays combinations of numerical and alpha indicators and the risk codes that derive from them.

TABLE 9.9   Risk Assessment Matrix, Including Probability and Severity Codes

Occurrence Probability	Severity Categories			
	Catastrophic—1	Critical—2	Marginal—3	Negligible—4
Frequent—A	High	High	Serious	Medium
Probable—B	High	High	Serious	Medium
Occasional—C	High	Serious	Medium	Low
Remote—D	Serious	Medium	Medium	Low
Improbable—E	Medium	Medium	Low	Low

Risk codes would then be as follows, taking into account the combinations of the severity and probability codes.

Combinations	Risk Category	Risk Code
A-1, A-2, B-1, B-2, C-1	High	H
A-3, B-3, C-2, D-1	Serious	S
A-4, B-4, C-3, D-2, D-3, E-1, E-2, E-3	Medium	M
C-4, D-4, E-4	Low	L

The foregoing is intended as an illustration from which suitable adaptations can be made by safety professionals.

What–If Analysis

For a What–If Analysis, a group of people (as little as two, but often several more) use a brainstorming approach to identify hazards, hazard scenarios, how incidents can occur, and what their probable consequences might be. Questions posed during the brainstorming session may commence with What–If, as in "What if the air conditioning system fails in the computer room?" or may be expressions of more general concern, as in "I worry about the possibly of spillage and chemical contamination during truck offloading."

All of the questions are recorded and assigned for investigation. Each subject of concern is then addressed by one or more team members. They would consider the potential of the hazardous situation and the adequacy or inadequacy of risk controls in effect, suggesting additional risk reduction measures if appropriate.

Checklist Analysis

Checklists are primarily adaptations from published standards, codes, and industry practices. There are many such checklists. They consist of lists of questions pertaining to the applicable standards and practices—usually with a yes or no or not applicable response. Their purpose is to identify deviations from the expected, and thereby possible hazards. A checklist analysis requires a walk-through of the area to be surveyed.

Checklists are easy to use and provide a cost-effective way to identify customarily recognized hazards. Nevertheless, the quality of checklists is dependent on the experience of the people who develop them. Further, they must be crafted to suit particular needs. If a checklist is not complete, the analysis may not identify some hazardous situations.

A General Design Safety Checklist appears in Chapter 12, "Safety Design Reviews." It can serve as a resource for those who choose to build their own checklists.

What–If/Checklist Analysis

A What–If/Checklist hazard analysis technique combines the creative, brainstorming aspects of the What–If method with the systematic approach of a Checklist. Combining the techniques can compensate for the weaknesses of each. The What–If part of the process, using a brainstorming method, can help the team identify hazards that have the potential to be the causal factors for incidents, even though no such incidents have yet occurred.

The checklist provides a systematic approach for review that can serve as an idea generator during the brainstorming process. Usually, a team experienced in the design, operation, and maintenance of the operation performs the analysis. The number of people required depends, of course, on the operation's complexity.

Failure Modes and Effects Analysis (FMEA)

In several industries, failure modes and effects analyses have been the techniques of choice by design engineers for reliability and safety considerations. They are used to evaluate the ways in which equipment fails and the response of the system to those failures. Although an FMEA is typically made early in the design process, the technique can also serve well as an analysis tool throughout the life of equipment or a process.

An FMEA produces qualitative, systematic lists that include the failure modes, the effects of each failure, safeguards that exist, and additional actions that may be necessary. For example, for a pump, the failure modes would include such as fails to stop when required; stops when required to run; seal leaks or ruptures; and pump case leaks or ruptures.

Both the immediate effects and the impact on other equipment would be recorded. Generally, when analyzing impacts, the probable worst case is assumed and analysts would conclude that existing safeguards do or do not work.

Although an FMEA can be made by one person, it is typical for a team to be appointed when there is complexity. In either case, the traditional process is similar:

1. Identify the item or function to be analyzed
2. Define the failure modes
3. Record the failure causes
4. Determine the failure effects
5. Enter a severity code and a probability code for each effect
6. Enter a risk code
7. Record the actions required to reduce the risk to an acceptable level

Note that the FMEA process described here requires entry of probability, severity, and risk codes. The example of a Risk Assessment Matrix shown in Table 9.9 and the Risk Category codes that follow it fulfill the needs for traditional FMEA purposes.

A Failure Modes and Effects Analysis form on which those codes would be entered is provided as Addendum B.

Hazard and Operability (HAZOP) Analysis

The Hazard and Operability Analysis technique was developed to identify both hazards and operability problems in chemical process plants. An interdisciplinary team and an experienced team leader are required. In a HAZOP application, a process or operation is systematically reviewed to identify deviations from desired practices that could lead to adverse consequences. HAZOPs can be used at any stage in the life of a process.

HAZOPs usually require a series of meetings in which the team, using process drawings, systematically evaluates the impact of deviations from the desired practices. The team leader uses a set of guide words to develop discussions. As the team reviews each step in a process, they record any deviations, along with their causes, consequences, safeguards, and required actions, or the need for more information to evaluate the deviation.

Fault Tree Analysis (FTA)

A fault tree analysis (FTA) is a top-down, deductive logic model that traces the failure pathways for a predetermined, undesirable condition or event, called the TOP Event. An FTA can be carried out either quantitatively or subjectively.

An FTA generates a fault tree (a symbolic logic model) entering failure probabilities for the combinations of equipment failures and human errors that can result in the accident. Each immediate causal factor is examined to determine its subordinate causal factors until the root causal factors are identified.

The strength of an FTA is its ability to identify combinations of basic equipment and human failures that can lead to an accident, allowing the analyst to focus preventive measures on significant basic causes. An FTA has particular value when analyzing highly redundant systems and high-energy systems in which high severity events can occur.

For systems vulnerable to single failures that can lead to accidents, the FMEA and HAZOP techniques are better suited. FTA is often used when another technique has identified a hazardous situation that requires more detailed analysis. Making an FTA of other than the simplest systems requires the talent of experienced analysts.

Management Oversight and Risk Tree—MORT

As Pat Clemens wrote in a previously cited paper, the Management Oversight and Risk Tree (MORT) technique applies "a pre-designed, systematized logic tree to the identification of total system risks, both those inherent in physical equipment and processes and those which arise from operational/management inadequacies." MORT is an incident investigation and analysis technique. It is discussed here for a particular purpose. There are four major stages in operational risk management, as follows:

1. *Preoperational stage*: in the initial planning, design, specification, prototyping, and construction processes, where the opportunities are greatest and the costs are lowest for hazard and risk avoidance, elimination, reduction, or control.
2. *Operational stage*: where hazards and risks are identified and evaluated and mitigation actions are taken through redesign initiatives or changes in work methods before incidents or exposures occur.
3. *Postincident stage*: where investigations are made of incidents and exposures to determine the causal factors which will lead to appropriate interventions and acceptable risk levels.
4. *Postoperational stage*: when demolition, decommissioning, or reusing/rebuilding operations are undertaken.

All of the hazard analysis and risk assessment techniques previously discussed relate *principally* to the design process or achieving risk reduction in the operational mode *before hazards-related incidents occur*. MORT was developed for use *principally* when incident investigations are made—the postincident stage. In the Introduction for the NRI MORT User's Manual (2009), the following comments are made.

The Management Oversight and Risk Tree (MORT) method is an analytical procedure for inquiring into causes and contributing factors of accidents and incidents. The MORT method is a logical expression of the functions needed by an organization to manage its risks effectively. MORT reflects a philosophy which holds that the most effective way of managing safety is to make it an integral part of business management and operational control. (p. viii)

MORT is a comprehensive analytical procedure that provides a disciplined method for determining the systemic causes and contributing factors of accidents. MORT directs the user to the hazards and risks deriving from both system design and procedural shortcomings. When properly used in the postincident stage of the practice of safety, MORT provides an excellent resource from which decisions can be made to redesign technical and procedural aspects of operations (Chapter 22 in this book is on MORT).

Avoiding Unrealistic Expectations

As has been said previously, making hazards analyses and risk assessments is both an art and a science. Whatever methodology is chosen—the simplest or the most complex—many judgments will be made in determining the severity potentials of hazards and the probably of incidents and exposures occurring.

Even though appliers of the risk assessment methodologies make informed judgments, they may disagree on which hazards are most important because of their severity potential and which risks deserve the highest priority. One way to resolve those differences is to have qualified teams participate when the hazards and risks are considered significant, the intent being to reach consensus.

Some who oppose the use of qualitative risk assessment techniques do so because the outcomes are not stated in absolutely assured, precise numbers. Such accuracy is not attainable because incident probability data is lacking and severity of event outcomes is a best estimate. Expecting such results is unrealistic.

Fortunately, recognition continues to grow that hazard analysis and risk assessment methods, although largely qualitative, add value to operational risk management decision-making.

ADDITIONAL RESOURCES

A Risk Assessment Tool (n.d.), made available by the European Agency for Safety and Health at Work, can be found at http://hwi.osha.europa.eu/ra_tools_generic/.

Parts I and II provide basic information on risk assessment. The subject matter consists of but eight pages. Nevertheless, it takes a reader through a basic hazard identification and risk assessment process.

Part III provides checklists for hazard identification and selection of preventive measures for 10 subjects—e.g., moving machinery electrical installations and equipment, fire, etc.

Part IV consists of checklists for seven occupation settings—e.g., office work, food processing, small-scale surface mining, etc.

In the United Kingdom, the Health and Safety Executive (2008) recently updated "Five Steps to Risk Assessment." It can be accessed at http://www.hse.gov.uk/risk/fivesteps.htm. It is also a basic, uncomplicated system.

If a safety professional undertakes to inform supervisors and workers on the fundamentals of hazard identification and risk assessment, say at safety meetings, the two foregoing resources will serve well in developing the presentations and the written material.

A particularly valuable reference is the *Guidelines for Hazard Evaluation Procedures, Second Edition with Worked Examples* issued by the Center for Chemical Process Safety (1992). Although this text is issued by a chemical industry organization, it is largely generic.

The Basics of FMEA is a 75-page 5 × 7-inch paperback written by McDermott, Mikulak, and Beauregard (1996). As the title indicates, it is a primer on the failure mode and effect analysis process.

In Chapter 10, "Three- and Four-Dimensional Numerical Risk Scoring Systems," comment is made on the FMEA publication issued for the semiconductor industry by International SEMATECH. It is well done. Its uniqueness is that it contains ratings for environmental, safety, and health considerations, all of which are vital in the process described.

OSHA has issued a paper on Job Hazards Analysis which has enough guidance points to warrant inclusion of a modified version of it as Addendum C in this chapter. Its definition of a hazard is the same as the author's—A hazard is the potential for harm. The paper also says that "Ideally, after you identify uncontrolled hazards, you will take steps to eliminate or reduce them to an acceptable risk level." But the data on eliminating or reducing hazards is a bit shallow.

Should safety professionals want to acquire extensive and valuable texts devoted entirely to applications in risk assessment, it is suggested that they consider two books written by Bruce Main, president of design safety engineering.

1. *Risk Assessment: Basics and Benchmarks* is a 485-page treatise published in 2004.
2. *Risk Assessment: Challenges and Opportunities* is a 364-page text published in 2012.
   With respect to risk assessment, Main has been a researcher, writer, consultant, software developer, instructor, and a leader in standards development.
   They could also consider:
3. *Risk Assessment: A Practical Guide to Assessing Operational Risks*, by Popov, Lyon, and Hollcroft (2016).
4. *Risk Management Tools for Safety Professionals* by Lyon and Popov (2018).

CONCLUSION

This chapter is but a primer on hazard analysis and risk assessment. Its purpose is to provide a foundation for those who perceive that having additional knowledge in this aspect of safety and health risk management provides opportunity for professional growth, accomplishment, and recognition. Having that knowledge adds to one's ability to evaluate hazardous situations and make more convincing presentations to management for resource allocation to accomplish the risk reduction measures proposed.

Safety professionals can expect that having knowledge about hazard analysis and risk assessment techniques will be required for job retention and career enhancement. Fortunately, it is not difficult to acquire the knowledge and skill required to fulfill almost all of their needs.

As safety and health professionals become more involved in risk assessments, they will come to understand that professional safety practice requires attention to the two distinct aspects of risk:

- Avoiding, eliminating, or reducing the **probability** of a hazard-related incident or exposure occurring
- Having the **severity** of harm or damage potential be as low as reasonably practicable

REFERENCES

"A Risk Assessment Tool". European Agency for Safety and Health at Work. Available at http://hwi.osha.europa.eu/ra_tools_generic/. Accessed June 1, 2019. n.d.

ANSI/ASSE Z690.3 – *Risk Assessment Techniques*. Des Plaines, IL: American Society of Safety Professionals, 2011.

Clemens, P. "A compendium of hazard identification and evaluation techniques for system safety application." *Hazard Prevention*, March/April, 1982.

EN ISO 12100-2010. *Safety of Machinery. General principles for Design. Risk assessment and Risk reduction*. Geneva, Switzerland: International Organization for Standardization, 2010.

EPA's *Risk Management Program for Chemical Accidental Release Prevention*. Washington, DC: U.S. Environmental Agency, 1996. Available at https://www.epa.gov/rmp/risk-management-plan-rmp-rule-overview. Accessed June 1, 2019.

"Five Steps to Risk Assessment." London, UK: The Health and Safety Executive, 2008.

Framework for Environmental Health Risk Management, Final Report, Volume 1. Washington, DC: The Presidential/Congressional Commission on Risk Assessment and Risk Management, 1997.

Guidelines for Hazard Evaluation Procedures, Second Edition. With Worked Examples. New York: Center for Chemical Process Safety of the American Institute of Chemical Engineers, 1992.

Johnson, W.G. *MORT Safety Assurance Systems*. New York: Marcel Dekker, 1980.

Lyon, B. and G. Popov. *Risk Management Tools for Safety Professionals*. Park Ridge, IL: American Society of Safety Professionals, 2018.

Main, B. *Risk Assessment: Basics and Benchmarks*. Ann Arbor, Michigan: Design Safety Engineering, 2004. Available at www.designsafe.com.

Main, B. *Risk Assessment: Challenges and Opportunities*. Ann Arbor, Michigan: Design Safety Engineering, 2012. Available at www.designsafe.com.

McDermott, R.E., R.J. Mikulak, and M.R. Beauregard. *The Basics of FMEA*. New York, NY: Productivity Press, 1996.

Mil-STD-882E. *Standard Practice for System Safety*. Washington, DC: Department of Defense, 2000. Available at http://www.systemsafetyskeptic.com/yahoo_site_admin/assets/docs/MIL-STD-882E_final.135152939.pdf. Accessed June 1, 2019.

NRI MORT User's Manual. Delft, The Netherlands: Noordwijk Risk Initiative Foundation, 2009. Enter "NRI MORT User's Manual" into a search engine for a free download.

OSHA's Rule for Process Safety Management of Highly Hazardous Chemicals, 1910.119 of 29 CFR 1910. Washington, DC: Department of Labor, Occupational Safety and Health Administration, 1992.

Popov, G., B. Lyon, and B. Hollcroft. *Risk Assessment: A Practical Guide to Assessing Operational Risks*. Hoboken, NJ: John Wiley & Sons, 2016.

FURTHER READING

ANSI/ASSP Z10.0-2019. *Occupational Health and Safety Management Systems*. Park Ridge, IL: American Society of Safety Professionals, 2019.

ANSI/ASSP/ISO 45001-2018. *Occupational Health and Safety Management Systems—Requirements with Guidance for Use*. Park Ridge, IL: American Society of Safety Professionals, 2018.

Failure Mode and Effects Analysis (FMEA): A Guide for Continuous Improvement for the Semiconductor Equipment Industry. Technology Transfer #92020963A-ENG. Austin, TX: International SEMATECH, 1992. Available at http://dewihardiningtyas.lecture.ub.ac.id/files/2012/09/FMEA-Guideline.pdf. Accessed June 1, 2019.

Job Hazard Analysis. U.S. Department of Labor, Occupational Safety and Health Administration, OSHA 3071, 2002. (Revised). Available at http://www.osha.gov/Publications/osha3071.html. Accessed June 20, 2019.

Stephans, R.A. and W.W. Talso. *System Safety Analysis Handbook*, Second Edition. Albuquerque, NM: New Mexico Chapter of the System Safety Society, 1997.

ADDENDUM A-1

Preliminary Hazard Analysis and Risk Assessment
Codes are in Addendum A–2

Prepared by _____ Date_____ Location_____ Project No._____ Page No.____of Pages___
Description of task/operation/process _____

Hazards Identified	Risk Before			Risk Code	Risk Reduction Measures	Risk After		Risk Code
	E	S	P			S	P	

Advanced Safety Management: Focusing on Z10.0, 45001, and Serious Injury Prevention, Third Edition. Fred A. Manuele.
© 2020 John Wiley & Sons, Inc. Published 2020 by John Wiley & Sons, Inc.

ADDENDUM A-2

Risk Assessment Matrix

Severity Levels and Values	Occurrence Probabilities and Values				
	Unlikely (1)	Seldom (2)	Occasional (3)	Likely (4)	Frequent (5)
Catastrophic (5)	5	10	15	20	25
Critical (4)	4	8	12	16	20
Marginal (3)	3	6	9	12	15
Negligible (2)	2	4	6	8	10
Insignificant (1)	1	2	3	4	5

Numbers were intuitively derived. They are qualitative, not quantitative and have meaning only in relation to each other.

Exposure Codes: P – personnel; E – environment; D – damage – facility, equipment, business interruption; M – material, product

Severity Descriptions

C – Catastrophic: One or more fatalities, total system loss and major business down time, environmental release with lasting impact on others with respect to health, property damage or business interruption.

Cr – Critical: Disabling injury or illness, major property damage and business down time, environmental release with temporary impact on others with respect to health, property damage or business interruption.

M – Marginal: Medical treatment or restricted work, minor subsystem loss or property damage, environmental release triggering external reporting requirements.

N – Negligible: First aid or minor medical treatment only, non-serious equipment or facility damage, environmental release requiring routine cleanup without reporting.

I – Insignificant: Inconsequential with respect to injuries or illnesses, system loss or down time, or environmental release.

Probability Descriptions

U – Unlikely: Could occur but hardly ever.
O – Occasional: Could occur intermittently.
L – Likely: Probably will occur several times.
F – Frequent: Likely to occur repeatedly.

Risk Levels

Combining the Severity and Occurrence Probability values yields a risk score in the matrix. The risks and the action levels are categorized below.

Risk Categories, Scoring, and Action Levels

Category	Risk Score	Action Level
L – Low risk	1 to 5	Remedial action discretionary.
M – Moderate risk	6 to 9	Remedial action to be taken at appropriate time.
S – Serious risk	10 to 14	Remedial action to be given high priority.
H – High risk	15 or greater	Immediate action necessary. Operation not permissible except in an unusual circumstance and as a closely monitored and limited exception only with approval of the person having authority to accept the risk.

ADDENDUM B

FMEA FORM

SYSTEM:
SUBSYSTEM:
REFERENCE DRAWING:

FAILURE MODE AND EFFECTS ANALYSIS (FMEA)
FAULT CODE#

DATE:
SHEET:
PREPARED BY:

Subsystem/ Module & Function	Potential Failure Mode	Potential Local Effect(s) of Failure	Potential End Effect(s) of Failure	SEV	Cr	Potential Cause(s) of Failure	OCC	Current Controls/ Fault Detection	DET	RPN	Recommended Action(s)	Area/ Individual Responsible & Completion Date(s)	Action Taken	SEV	OCC	DET	RPN
Subsystem name and function	"How can this subsystem fail to perform its function?" "What will an operator see?"	The local effect of the subsystem or end user	Downtime # of hours at the system level. 2. Safety 3. Environmental 4. Scrap loss	7	1	"How can this failure occur?" Describe in terms of something that can be corrected or controlled. 2. Refer to specific errors or malfunctions	5	"What mechanisms are in place that could detect, prevent, or minimize the impact of this cause?"	6	210	"How can we change the design to eliminate the problem?" "How can we detect (fault isolate) this cause for this failure?" "How should we test to ensure the failure has been eliminated?" "What PM procedures should we recommend?" 1. Begin with highest RPN. 2. Could say "no action" or "Further study is required." 3. An idea written here does not imply corrective action.	"Who is going to take responsibility?" "When will it be done?"	"What was done to correct the problem?" Examples: Engineering Change, Software revision, no recommended action at this time due to obsolescence, etc.	6	1	1	6

Critical Failure symbol (Cr). Used to identify critical failures that must be addressed (i.e., whenever safety is an issue).

Severity Ranking (1–10) (see Severity Table)

Occurrence Ranking (1–10) (see Occurrence Table)

Detection Ranking (1–10) (see Detection Table)

Risk Priority Number
RPN = Severity*Occurrence*Detection

How did the "Action Taken" change the RPN?

196

ADDENDUM C

This presentation is an adaption from *Job Hazard Analysis*, U.S. Department of Labor, Occupational Safety and Health Administration, OSHA 3071, 2002 (revised).
Available at http://www.osha.gov/Publications/osha3071.html .

What is a Hazard?

A hazard is the potential for harm. In practical terms, a hazard often is associated with a condition or activity that, if left uncontrolled, can result in an injury or illness.

Identifying hazards and eliminating or controlling them as early as possible will help prevent injuries and illnesses.

What is a job hazard analysis?

A job hazard analysis is a technique that focuses on job tasks as a way to identify hazards before they occur. It focuses on the relationship between the worker, the task, the tools, and the work environment. Ideally, after you identify uncontrolled hazards, you will take steps to eliminate or reduce them to an acceptable risk level.

Why is Job Hazard Analysis Important?

Many workers are injured and killed at the workplace every day in the United States. Safety and health can add value to your business, your job, and your life.

You can help prevent workplace injuries and illnesses by looking at your workplace operations, establishing proper job procedures, and ensuring that all employees are trained properly.

One of the best ways to determine and establish proper work procedures is to conduct a job hazard analysis. A job hazard analysis is one component of the larger commitment of a safety and health management system.

Advanced Safety Management: Focusing on Z10.0, 45001, and Serious Injury Prevention, Third Edition. Fred A. Manuele.
© 2020 John Wiley & Sons, Inc. Published 2020 by John Wiley & Sons, Inc.

What is the Value of a Job Hazard Analysis?

Supervisors can use the findings of a job hazard analysis to eliminate and prevent hazards in their workplaces. This is likely to result in fewer worker injuries and illnesses; safer, more effective work methods; reduced workers' compensation costs; and increased worker productivity. The analysis also can be a valuable tool for training new employees in the steps required to perform their jobs safely.

For a job hazard analysis to be effective, management must demonstrate its commitment to safety and health and follow through to correct any uncontrolled hazards identified. Otherwise, management will lose credibility and employees may hesitate to go to management when dangerous conditions threaten them.

What Jobs are Appropriate for a Job Hazard Analysis?

A job hazard analysis can be conducted on many jobs in your workplace. Priority should go to the following types of jobs:

- Jobs with the highest injury or illness rates;
- Jobs with the potential to cause severe or disabling injuries or illness, even if there is no history of previous accidents;
- Jobs in which one simple human error could lead to a severe accident or injury;
- Jobs that are new to your operation or have undergone changes in processes and procedures; and
- Jobs complex enough to require written instructions.

Where do I Begin?

Involve your employees. It is very important to involve your employees in the hazard analysis process. They have a unique understanding of the job, and this knowledge is invaluable for finding hazards. Involving employees will help minimize oversights, ensure a quality analysis, and get workers to "buy in" to the solutions because they will share ownership in their safety and health program.

Review your accident history. Review with your employees your worksite's history of accidents and occupational illnesses that needed treatment, losses that required repair or replacement, and any "near misses" — events in which an accident or loss did not occur, but could have. These events are indicators that the existing hazard controls (if any) may not be adequate and deserve more scrutiny.

Conduct a preliminary job review. Discuss with your employees the hazards they know exist in their current work and surroundings. Brainstorm with them for ideas to eliminate or control those hazards.

If any hazards exist that pose an immediate danger to an employee's life or health, take immediate action to protect the worker. Any problems that can be corrected easily should be corrected as soon as possible. Do not wait to complete your job hazard analysis.

This will demonstrate your commitment to safety and health and enable you to focus on the hazards and jobs that need more study because of their complexity. For those hazards determined to present unacceptable risks, evaluate types of hazard controls.

List, rank, and set priorities for hazardous jobs. List jobs with hazards that present unacceptable risks, based on those most likely to occur and with the most severe consequences. These jobs should be your first priority for analysis.

Outline the steps or tasks. Nearly every job can be broken down into job tasks or steps. When beginning a job hazard analysis, watch the employee perform the job and list each step as the worker takes it. Be sure to record enough information to describe each job action without getting overly detailed. Avoid making the breakdown of steps so detailed that it becomes unnecessarily long or so broad that it does not include basic steps. You may find it valuable to get input from other workers who have performed the same job.

Later, review the job steps with the employee to make sure that you have not omitted something. Point out that you are evaluating the job itself, not the employee's job performance. Include the employee in all phases of the analysis—from reviewing the job steps and procedures to discussing uncontrolled hazards and recommended solutions.

Sometimes, in conducting a job hazard analysis, it may be helpful to photograph or videotape the worker performing the job. These visual records can be handy references when doing a more detailed analysis of the work.

How do I Identify Workplace Hazards?

A job hazard analysis is an exercise in detective work. Your goal is to discover the following:

- What can go wrong?
- What are the consequences?
- How could it arise?
- What are other contributing factors?
- How likely is it that the hazard will occur?

To make your job hazard analysis useful, document the answers to these questions in a consistent manner. Describing a hazard in this way helps to ensure that your efforts to eliminate the hazard and implement hazard controls help target the most important contributors to the hazard. Good hazard scenarios describe:

- Where it is happening (environment),
- Who or what it is happening to (exposure),
- What precipitates the hazard (trigger),
- The outcome that would occur should it happen (consequence), and
- Any other contributing factors.

Rarely is a hazard a simple case of one singular cause resulting in one singular effect. More frequently, many contributing factors tend to line up in a certain way to create the hazard. Here is an example of a hazard scenario:

In the metal shop (environment), while clearing a snag (trigger), a worker's hand (exposure) comes into contact with a rotating pulley. It pulls his hand into the machine and severs his fingers (consequences) quickly.

To perform a job hazard analysis, you would ask:

- What can go wrong? The worker's hand could come into contact with a rotating object that "catches" it and pulls it into the machine.
- What are the consequences? The worker could receive a severe injury and lose fingers and hands.
- How could it happen? The accident could happen as a result of the worker trying to clear a snag during operations or as part of a maintenance activity while the pulley is operating. Obviously, this hazard scenario could not occur if the pulley is not rotating.
- What are other contributing factors? This hazard occurs very quickly. It does not give the worker much opportunity to recover or prevent it once his hand comes into contact with the pulley. This is an important factor, because it helps you determine the severity and likelihood of an accident when selecting appropriate hazard controls.

  Unfortunately, experience has shown that training is not very effective in hazard control when triggering events happen quickly because humans can react only so quickly.
- How likely is it that the hazard will occur? This determination requires some judgment. If there have been "near-misses" or actual cases, then the likelihood of a recurrence would be considered high. If the pulley is exposed and easily accessible, that also is a consideration.

In the example, the likelihood that the hazard will occur is high because there is no guard preventing contact, and the operation is performed while the machine is running.

By following the steps in this example, you can organize your hazard analysis activities.

HOW DO I CORRECT OR PREVENT HAZARDS?

After reviewing your list of hazards with the employee, consider what control methods will eliminate or reduce them. The most effective controls are engineering controls that physically change a machine or work environment to prevent employee exposure to the hazard. The more reliable or less likely a hazard control can be circumvented, the better. If this is not feasible, administrative controls may be appropriate.

This may involve changing how employees do their jobs.

Discuss your recommendations with all employees who perform the job and consider their responses carefully. If you plan to introduce new or modified job procedures, be sure they understand what they are required to do and the reasons for the changes.

CHAPTER 10

THREE- AND FOUR-DIMENSIONAL RISK SCORING SYSTEMS

In all the risk assessment matrices discussed in Chapter 9, "On Hazards Analyses and Risk Assessments," risk levels are based on the *probability* of an event or exposure occurring and the *severity* of harm or damage that could result. Therefore, they are two-dimensional risk scoring systems. For the hazards and risks that most safety professionals encounter, two-dimensional systems will be sufficient. Some risk scoring systems now in use are three or four dimensional.

Safety professionals can expect that variations of numerical risk scoring systems will proliferate. For the knowledge that safety professionals should have, this chapter informs on

- Transitions in numerical risk assessment methods
- Cautions to be considered in using three- and four-dimensional systems, particularly those that diminish the significance of severity potential
- Three- and four-dimensional numerical risk scorings systems
- An extended three-dimensional numerical risk scoring system that includes a method to justify the risk amelioration costs in relation to the amount of risk reduction to be attained
- A numerical risk scoring system developed by this author

Advanced Safety Management: Focusing on Z10.0, 45001, and Serious Injury Prevention, Third Edition. Fred A. Manuele.
© 2020 John Wiley & Sons, Inc. Published 2020 by John Wiley & Sons, Inc.

TRANSITIONS IN RISK ASSESSMENT

It is typical for engineers to have a passion for quantification and the application of numbers. They become comfortable with statistical measurements. Quantification is stressed in their education. Engineering texts still quote Lord Kelvin who wrote the following over 100 years ago:

> When you can measure what you are speaking about, and express it in numbers, you know something about it: but when you cannot measure it, when you cannot express it in numbers, your knowledge is of a meager and unsatisfactory kind. Accessed June 20, 2019 (http://www.cromwell-intl.com/3d/Index.html)

Some practitioners in risk assessment disagree with Lord Kelvin's absolutist statement, such as Vernon L. Grose. The quote above from Lord Kelvin can also be found in his book *Managing Risk: Systematic Loss Prevention for Executives* (1987, p. 259). Grose expressed concern over the use of "numerology" in making risk assessments, and cautions about the

> Increasingly common abuse of mathematical statistics by "numericalizing without data" to obtain risk probabilities, and the deception-by-numbers that lulls executives into a false sense of security. (p. 259)

Nevertheless, the influence that engineers have had in developing risk assessment methods is obvious. Their passion for numerical precision encourages the use of quantitative methods. As an example, in at least two industries quantitative Failure Modes and Effects Analyses (FMEAs), rather than qualitative FMEAs, are now required to meet quality assurance requirements. Mostly, engineers make those FMEAs.

Similarly, some engineers are using numerical risk scoring systems to meet the risk assessment requirements placed on manufacturers who sell machinery that is to go into workplaces in European Community countries. One such system in use for that purpose is four dimensional. Its negative aspects will be discussed later.

Also, a three-dimensional numerical risk scoring system is in use in a segment of the heavy machinery manufacturing industry. It was introduced by engineering personnel to meet the demands for product safety.

A NEED FOR CAUTION AND PERCEPTIVE EVALUATION

Although three- and four-dimensional risk scoring systems are now more commonly used, there are several reasons to be inquisitive and cautious concerning: their content and the results in their application; the meanings of the terms used in them; the numerical values applied to the gradations within elements to be scored; and how they are applied in determining risk levels. The author's observations follow:

- Creating numerical risk scorings begins with subjective judgments on the values to be given to the elements in the system, and those judgments are then translated into numbers. Thus, judgmental observations become finite numbers which then leads to an image of preciseness. Further, those numbers are multiplied or totaled to produce a risk score, giving the risk assessment process a scientific appearance. For this observer, that risk assessment process is still as much art as science.
- There are no universally applied rules to assign value numbers to elements to be scored. Value numbers in all numerical risk scoring systems are judgmental and reflect the experience and views of those who create a system.
- Historically, frequency of exposure has been one of the elements considered to determine event probability. However, giving frequency of exposure its own multiplier, separate from and in addition to a probability multiplier, diminishes the needed emphasis on the severity of harm or damage that could result from an event or exposure.
- Meanings of terms such as probability, likelihood, frequency of exposure or endangerment, and severity used in risk scoring systems are not consistent. Also, it may be that within a system, a term may appear more than once and have different usages. Thus, consistency in the meanings of the terms used in a system may not be maintained.
- If risk scoring systems are applied rigidly whereby a higher score indicates a greater risk than a lower score, and the scores are not considered in light of employee, community, social, and financial concerns, the best interests of an organization may not be well served. This applies especially to low probability incidents that may have serious consequences, for which risk scores may be low.

THREE- AND FOUR-DIMENSIONAL NUMERICAL RISK SCORING SYSTEMS

There are substantial variations in the elements to be scored in the three-dimensional and four-dimensional numerical risk-scoring systems reviewed in this chapter. To commence the discussion, excerpts are taken from a National Safety Council (2009) publication—*Accident Prevention Manual: Administration & Programs, 13th Edition*—and appear here with permission.

NATIONAL SAFETY COUNCIL

In the Accident Prevention Manual, the Council outlines a system for "Ranking Hazards by Risk (Severity, Probability, and Exposure)." It is duplicated here with permission. The text says that "Ranking provides a consistent guide for corrective action, specifying which hazards warrant immediate action, which have secondary priority, and which can be addressed in the future" (p. 155).

204   THREE- AND FOUR-DIMENSIONAL RISK SCORING SYSTEMS

In the process, for each identified hazard, numerical scorings as follows are given to Severity, Exposure, and Probability. It is significant that they are totaled to arrive at a final rating, not multiplied in sequence as is the case in other three- and four-dimensional risk scoring systems. Thus, the status of the severity rating is not diminished as much.

Severity: Consider the potential losses or destructive and disruptive consequences that are most likely to occur if the job/task is performed improperly. Assumptions would be made as to the worst credible outcome of an incident in selecting a numerical rating for severity. Use the point values listed in Table 10.1.

TABLE 10.1   NSC Severity Descriptions and Point Values

1.	Negligible	Probably no injury or illness; no production loss; no lost work days
2.	Marginal	Minor injury or illness; minor property damage
3.	Critical	Severe injury or occupational illness with lost time; major property damage; no permanent disability or fatality
4.	Catastrophic	Permanent disability; loss of life; loss of facility or major process

Probability: Consider the probability of loss that occurs each time the job is performed. The key question is, how likely is it that things will go wrong when this job is performed? Probability is influenced by a number of factors, such as the hazards associated with the task, difficulty in performing the job, the complexity of the job, and whether the work methods are error-provocative. Use the point values Table 10.2.

TABLE 10.2   NSC Probability Descriptions and Point Values

1.	Low probability of loss occurrence
2.	Moderate probability of loss occurrence
3.	High probability of loss occurrence

Exposure: Consider the number of employees that perform the job, and how often. Use the following point values Table 10.3.

After numerical point values are given to each job/task for severity, probability, and exposure, the point values are added to produce a total risk assessment code (RAC). A total score can be as low as 3 or as high as 10. So each task is given a risk ranking

TABLE 10.3   NSC Exposure Descriptions and Point Values

1.	A few employees perform the task up to a few times a day
2.	A few employees perform the task frequently
3.	Many employees perform the task frequently

from which judgments can be made in setting priorities, and management can use these values to select particular jobs for further analysis.

But the Council's text clearly states that "The RAC rating scale is a guideline and is not intended to be used as an absolute measurement system" (p. 156).

In the Council's system, there are four severity ratings, while probability and exposure each have three. Choosing to set up such a system and deciding on the numerical values to be assigned is entirely at the discretion of the creator of the system.

However, a variation of this system, or any other system, should keep severity, frequency of exposure, and probability considerations in balance so that disproportionate weightings are not given to any one of the categories.

FAILURE MODE AND EFFECTS ANALYSIS (FMEA)

SEMATECH is a consortium of semiconductor equipment manufacturing companies. Its *Failure Mode and Effects Analysis (FMEA): A Guide for Continuous Improvement for the Semiconductor Equipment Industry* (n.d.) describes its recommended FMEA technique. This guide was chosen to illustrate a three-dimensional risk assessment technique because of its uniqueness and special character:

- consideration of environmental, safety, and health needs is prominent throughout the analysis.
- there is a separate table for severity levels that encompasses several forms of harm, damage, or loss—injury to personnel, exposure to harmful chemicals or radiation, fire, or a release of chemicals to the environment (p. 14).
- the value of counsel available from safety engineering personnel is recognized by including that designation— safety engineering—among the titles of people who would form a team to conduct FMEAs (p. 4).

With the permission of SEMATECH, excerpts from that publication appear here. In the SEMATECH FMEA, three criteria are to be numerically ranked: Severity, Occurrence, and Detection. This is how they are defined in the *Guide*. As was stated earlier in this chapter, definitions of the terms used in risk scoring systems vary and the definitions in this system have their own distinct characteristics.

Severity ranking criteria: Customer satisfaction is key in determining the effect of a failure mode. Safety criticality is also determined at this time related to Environmental, Safety, and Health (ES&H) levels. Based on this information, a severity ranking is used to determine criticality of the failure mode on the subassembly to the end effect. The end (global) effect of the failure mode is the one to be used for determining the severity ranking. Calculating the severity levels provides for a classification ranking that encompasses safety, production continuity, scrap loss, etc. (p. 8).

This system also provides for the use of an Environmental, Safety, and Health Severity Code which is a qualitative means of representing the worst-case incident that could result from an equipment or process failure or for lack of a contingency plan for such an incident (p. 14).

Occurrence ranking criteria: The probability that a failure will occur during the expected life of the system can be described in potential occurrences per unit of time (p. 15).

Detection ranking criteria: this section provides a ranking based on an assessment of the probability that the failure mode will be detected given the controls in place (p. 16).

Two scoring tables are given for the severity ranking: one has five possible scoring levels and relates to the effect a failure may have on customer relations; the other pertains to environmental, safety, and health levels and has four possible scoring levels. The following are abbreviated versions of the severity criteria (Tables 10.4 and 10.5).

Note that for most of the scoring levels for the Severity, Occurrence, and Detection criteria, two numerical scoring possibilities are available for each narrative description.

TABLE 10.4 SEMATACH Scoring Table for Severity Ranking—Customer Related

Rank	Description
1–2	Failure is of such a minor nature that the customer (internal or external) will probably not detect the failure
3–5	Failure will result in slight customer annoyance and/or slight deterioration of part or system performance
6–7	Failure will result in customer dissatisfaction and/or slight deterioration of part or system performance
8–9	Failure will result in high degree of customer or cause nonfunctionality of system
10	Failure will result in major customer dissatisfaction and cause nonsystem operation or noncompliance with government regulations (p. 14)

TABLE 10.5 SEMATECH Scoring Table for Severity Ranking—ES&H Definitions

Rank	Severity Level	Description
10	Catastrophic I	A failure results in major injury or death of personnel
7–9	Critical II	A failure results in minor injury to personnel, personnel exposure to harmful chemicals or radiation, a fire or a release of chemicals to the environment
4–6	Major III	A failure results in a low-level exposure to personnel, or activates facility alarm system
1–3	Minor IV	A failure results in minor system damage but does not cause injury to personnel, allow any kind of exposure to operational or service personnel, or allow any release of chemicals into environment (p. 14)

A statement, as follows, defines the risk level to be achieved: "ES&H severity levels are patterned after the industry standard, *SEMI S2-91- Product Safety Guidelines*. All equipment should be designed to level IV severity. Types I, II, III are considered unacceptable risks" (p. 16).

Thus, the guide clearly establishes that Type IV severity is the acceptable risk level. It is well known that this industry has established high environmental, safety, and health standards for itself. A slightly reduced display of the Occurrence Ranking Criteria follows. It interestingly relates probability to a time operating interval and expresses probability as a failure rate (Table 10.6).

TABLE 10.6   SEMATECT Scoring Table for Occurrence Ranking Criteria

Rank	Description
1	An unlikely probability of occurrence during the item operating time interval. Unlikely is defined as a single failure mode (FM) probability <0.001 of the overall probability of failure during the time operating interval
2–3	A remote probability of occurrence during the item operating time interval (i.e., once every two months), defined as a single FM probability >0.001 but <0.01
4–6	An occasional probability of occurrence during the item operating time interval (i.e., once a month), defined as a single FM probability >0.01 but <0.10
7–9	A moderate probability of occurrence during the item time operating interval (i.e., once every two weeks), defined as a single FM probability >0.10 but <0.20
10	A high probability of occurrence during the item time operating interval (i.e. once a week), defined as a single FM probability >0.20 (p. 15)

A condensed version of the Detection ranking scale is given in Table 10.7. It pertains to what the verification system and/or controls in place are expected to accomplish.

TABLE 10.7   SEMATECH Scoring Table for Detection Ranking Criteria

Rank	Probability That The Defect Will Be Detected
1–2	Very high—almost certain
3–4	High—good chance
5–7	Moderate—likely
8–9	Low—not likely
10	Very low (or zero)

At this point, a Risk Priority Number (RPN) is calculated:

$$RPN = \text{Severity} \times \text{Occurrence} \times \text{Detection}$$

$$RPN = S \times O \times D$$

For each failure mode, codes for severity, occurrence probability, and detectability and the RPN are entered into the FMEA form. Recommended actions are listed to resolve situations that need attention. After the recommended actions are taken, revised codes and a priority number are entered, reflecting the effect of the work done to reduce the risk.

There is a major difference in this system that should be of particular interest to safety professionals. In the semiconductor industry FMEA form, "Cr" is at the top of a column. This is a Critical Failure Symbol. A "Y" is to be entered for "yes" if the failure potential is considered safety-critical.

That gives importance to the Environmental, Safety, and Health Severity Level Definitions previously shown. On the form, the instruction is that Critical Failures must be addressed when safety is an issue.

Although the SEMATECH FMEA publication was drafted for the semiconductor equipment industry, it is highly recommended as a reference if three-dimensional risk scoring systems are to be considered. Its inclusion of environmental, safety, and health considerations makes it unique and exceptionally valuable. It is downloadable, free, at http://dewihardiningtyas.lecture.ub.ac.id/files/2012/09/FMEA-Guideline.pdf.

Addendum A is an example of an FMEA form that relates to the SEMATECH provisions. It includes instruction on its use.

A THREE-DIMENSIONAL RISK ASSESSMENT SYSTEM—HEAVY EQUIPMENT BUILDERS

Hazard analysis and risk assessment systems have been used for many years by certain heavy equipment manufacturers in their design processes to build inherently safer products. In the use of those systems, the principle concern is product safety and the avoidance of injury to equipment users or to bystanders.

There are variations in the terms used in the several versions of this industry's Hazard Analysis System. In one version, the risk assessment variables are severity of the injury, frequency of exposure, and the probability that injury is not avoided and injury occurs.

In another version, the parameters to be scored are severity, frequency, and vulnerability. This latter version and its definitions will be discussed here. Definitions of terms follow:

- *Severity*: The most probable injury which would be expected from an accident.
- *Frequency:* An estimate of how often a product user or bystander may be exposed to a hazard. The hazard may result from a machine part or system failure or from a man–machine interface failure.
- *Vulnerability:* The degree of user or bystander susceptibility to injury when exposed to a hazard. The likelihood that personal injury will occur once exposure to a hazard has occurred, taking into consideration the detectability of

a hazard, risk assumptions, presence of environmental or stress conditions, skills, attitudes, etc.

For each of these three variables, there are five possible ratings, all using identical numerical values: 1, 3, 5, 7, and 9. Other numerical selections are equally valid as long as they are proportional and giving excessive weight to one of the factors being scored is avoided. The ratings are in Tables 10.8–10.10.

TABLE 10.8   Severity Scale—S

1.	Minor first aid. Immediate return to work/activity
3.	Doctor's office or emergency room treatment. Up to one week lost time
5.	Hospitalization. Up to one month lost time. No loss in work capacity
7.	Permanent partial loss in work capacity. Increased work difficulty
9.	Death or complete disability

TABLE 10.9   Frequency Scale—F

1.	Theoretically can occur, but highly unlikely during the life of the machine population
3.	Only once in the life of a small percentage (10%) of the product
5.	Once per use season or once annually
7.	Once daily
9.	Continuous exposure

TABLE 10.10   Vulnerability Scale—V

1.	Practically impossible to complete injury sequence
3.	Remotely possible, but unlikely
5.	Some conditions favorable to completing the injury sequence
7.	Very possible, but not assured
9.	Almost certain to complete injury sequence

The literature on this system indicates the following: Hazards analyses and risk assessments are done by a team; Consensus is to be reached on risk scores; Risk scores are listed and ranked; High and low values would be examined, and actions for improvement are to be developed when required. This is the risk scoring system:

$$\text{Risk Score} = \text{Severity} \times \text{Frequency} \times \text{Vulnerability}$$

$$\text{Risk Score} = S \times F \times V$$

This risk scoring system gives equal weight to all variables. Thus, the needed emphasis on severity of injury is diminished when the risk score is produced. In a discussion with a user of this system, it was acknowledged that, with all variables being given equal weight, the significance of severity was subordinated. However, it was also said that, in practice, the team gives the potential for severe injury due consideration when deciding on product improvement recommendations.

THE WILLIAM T. FINE SYSTEM: A THREE-DIMENSIONAL NUMERICAL RISK SCORING MODEL

While at the Naval Ordnance Laboratory in White Oak, Maryland, in 1971, William T. Fine submitted a report titled "Mathematical Evaluations for Controlling Hazards." Fine was an early proponent of determining whether the expenditure needed to reduce risk could be justified after considering the amount of reduction to be attained.

In a sense, Fine's work was a precursor of the concept on which ALARP is based. ALARP is defined as that level of risk which can be further lowered by an increment in resource expenditure that cannot be justified by the resulting decrement in risk. Fine made this comment about his work in the paper's Abstract.

> A formula has been devised which weighs the controlling factors and "calculates the risk" of a hazardous situation, giving a numerical evaluation to the urgency for remedial attention to the hazard. Calculated Risk Scores are then used to establish priorities for corrective action. An additional formula weighs the estimated cost and effectiveness of any contemplated corrective action against the Risk Score and gives an indication on whether the cost is justified.

Fine's paper is thought-provoking, valuable, a good resource, and somewhat scarce. A reduced version of Fine's paper can be found in Addendum B to this chapter. Although reduced, the substance of the paper has been maintained. For the entirety of the paper, go to https://apps.dtic.mil/dtic/tr/fulltext/u2/722011.pdf.

A FOUR-DIMENSIONAL NUMERICAL RISK SCORING SYSTEM

Although reference is made here to a four-dimensional risk scoring system as it appears in *Pilz—Guide to Machinery Safety, Sixth Edition* (1999), it should be understood that this system has appeared elsewhere and is in the public domain. For example, it has been used with minor modification by at least one U.S. Company to meet the European Community risk assessment requirements as set forth in the International Organization for Standardization standard *EN ISO 12100-2010: Safety of Machinery. General principles for design. Risk assessment and risk reduction.* As will be illustrated later, this risk scoring system has major shortcomings.

In the Pilz text, the following statement is made early in Chapter 4.0—"Background to risk assessment."

In simple terms, there are only two real factors to consider:

- The severity of foreseeable injuries (ranging from a bruise to a fatality)
- The probability of their occurrence. (p. 75)

As the text proceeds, however, the system becomes more complex. There are four elements to be scored:

- Likelihood of occurrence/contact with hazard (LO)
- Frequency of exposure to the hazard (FE)
- Degree of possible harm (DPH), taking into account the worst-possible case
- Number of persons exposed to the hazard (NP)—(p. 85)

This system has a particular focus. It applies only to personal injury, particularly to employees, that could derive from machinery operation. All of the listings for "Degree of possible harm" involve personal injuries. There are no entries for possible damage to property or the environment. Terms used to establish gradations and scores in the Likelihood of occurrence and Frequency of exposure categories are comparable to those in previously cited risk assessment systems. They follow in the next few tables.

Note the position within the listings of "Even chance—could happen" in the Likelihood category, and "Annually" in the Frequency of exposure category. A computation appears later in which these ratings are used (Tables 10.11–10.13).

The fourth dimension in this system is Number of persons exposed to the hazard (Table 10.14).

TABLE 10.11 Likelihood of Occurrence (LO) and the Scores

Category	Scores
Almost impossible—possible only under extreme circumstances	0.033
Highly unlikely—though conceivable	1
Unlikely—but could occur	1.5
Possible—but unusual	2
Even chance—could happen	5
Probable—not surprising	8
Likely—only to be expected	10
Certain—no doubt	15

TABLE 10.12 Frequency of Exposures to the Hazards (FE) and the Scores

Frequency of Exposure	Scores
Annually	0.5
Monthly	1
Weekly	1.5
Daily	2.5
Hourly	4
Constantly	5

THREE- AND FOUR-DIMENSIONAL RISK SCORING SYSTEMS

TABLE 10.13 Degree of Possible Harm and the Scores

Degree of Harm	Scores
Scratch/bruise	0.1
Laceration/mild ill-effect	0.5
Break minor bone or minor illness (temporary)	2.0
Break major bone or major illness (temporary)	4.0
Loss of one limb, eye, hearing loss (permanent)	6.0
Loss of two limbs, eyes (permanent)	10.0
Fatality	50.0

TABLE 10.14 Number of Persons Exposed and the Scores

Number of Persons	Scores
1–2 persons	1
3–7 persons	2
8–15 persons	4
16–50 persons	8
50+ persons	12

Using the numerical ratings given to each element, the formula produces a Risk level by simple multiplication. The formula and the Risk levels follow (Table 10.15):

$$LO \times FE \times DPH \times NP = \text{Risk level}$$

TABLE 10.15 Risk Level

Risk Level	Scoring Range
Negligible: Presenting very little risk to health and safety	0–5
Low but significant: Containing hazards that require control measures	5–50
High: Having potentially dangerous hazards, which require control measures to be implemented urgently	50–500
Unacceptable: Continued operation in this state is unacceptable	500+

Assume that in a company's annual shut down for retooling and maintenance, a task is to be performed for which the likelihood of occurrence of a hazardous event was rated as "Even chance—could happen" and the outcome would be Fatalities. In the illustration in Table 10.16, the ratings are purposely kept the same for likelihood of occurrence, frequency of exposure, and degree of possible harm. Ratings vary for the number of people exposed (Table 10.16).

For such a hazard scenario, if there is an even chance that two people could be killed, the risk level is Low. That is not good risk management. Computations using this four-dimensional method never fall within the Unacceptable risk level (500+)

TABLE 10.16 Usage of a Four-Dimensional Risk Scoring System

Ratings	Number of Persons Exposed				
	2	7	15	50	51
Likelihood: even chance	5	5	5	5	5
Frequency of exposure: annual	0.5	0.5	0.5	0.5	0.5
Degree of possible harm: fatality	15	15	15	15	15
For: number of persons exposed	1	2	4	8	12
Risk levels	37.5	75	150	300	450
	Low but significant	High	High	High	High

category, even when there is an even chance that 51 people will be killed. Score levels in this system are ill-conceived. They greatly diminish the value of life.

Using the number of persons exposed as a category in risk assessment requires careful consideration. In most risk assessment matrices, a single death or a permanent total disability is in the Catastrophic category. For example, in the FMEA system published by SEMATECH, a single fatality gets a Catastrophe rating. In the Fine system, a fatality that has an even chance of occurring and an annual exposure falls in the unacceptable risk category.

In the scenario just previously described, an occurrence resulting in 51 fatalities does not receive the severity grading it deserves. This risk scoring system is not acceptable.

A MODEL OF THREE-DIMENSIONAL NUMERICAL RISK SCORING SYSTEM

One of the aims in this author's study of multidimensional numerical risk scoring systems was to determine whether a model could be proposed that

1. Serves the needs of those who are more comfortable with statistics.
2. Addresses the strong beliefs of those who want frequency of exposure given separate consideration in the risk assessment process.
3. Maintains credibility and efficacy.

As the work proceeded in crafting the numerical risk-scoring model presented here, the following guidelines emerged and were adopted.

- In a statistical risk scoring system, all scores are to meet a plausibility test. High-risk scores must be produced for high risks: a lower level risk is not to fall in a high-risk category.
- To create a focal point for the relative placing of other risk scores, it was assumed that an incident having a severity outcome of one or more fatalities,

with an even chance of occurring (50/50), and an annual frequency of exposure, the risk score must be High.
- Frequency of exposure can be separately evaluated without negatively affecting the validity of risk determinations, provided that adequate weighting is given to severity of outcome in the scoring system.
- Adaptations can be made of the risk assessment matrices shown in Chapter 9, "On Hazards Analyses and Risk Assessment," to develop a single risk-scoring model that addresses injury to people (employees and the public); facilities, product, or equipment loss; operations down time; and chemical releases and environmental damage.
- If the number of gradations for probability, frequency of exposure, or severity is excessive, distinctions between them are difficult to make: Five gradations were chosen for probability and for frequency of exposure; for severity, there are four gradations.
- Although an attempt was made not to use a descriptive word more than once in the elements to be scored, it was not successful.

DEFINITIONS

Definitions follow for the terms relevant to the numerical risk scoring system presented here:

- *Hazard*: the potential for harm to people, property, and the environment. The dual nature of hazards must be understood. Hazards encompass all aspects of technology or activity that produce risk. Hazards include the characteristics of things and the actions or inactions of people.
- *Risk*: an estimate of the probability (likelihood) of a hazards-related incident or exposure occurring and the severity of harm or damage that could result.
- *Probability*: an estimate of the likelihood an incident or exposure occurring that could result in harm or damage—for the selected unit of time, events, population, items, or activity being considered.
- *Frequency of exposure*: the frequency and duration of exposure to the hazard, over time.
- *Severity*: an estimate of the magnitude of the harm or damage that could result from a hazard-related incident or exposure, taking into consideration the number of people exposed, occupational health and environmental exposures, the potential for property damage, and production loss.

THE RISK SCORE FORMULA

Having decided to include frequency of exposure as a separate element to be scored, the question became how to assure that scoring computations, particularly for severity, are not adversely skewed. Thus, for this Risk Score Formula, the rating for frequency of exposure is not an equal multiplier—rather, it is added to the rating for occurrence probability to produce a mid-level score that is then multiplied by the severity rating.

Risk Score = (Probability Rating + Frequency of Exposure Rating)
× Severity Rating

$$RS = (PR + FER) \times SR$$

GRADATION AND SCORING DEVELOPMENT

To achieve plausibility, a variety of scorings were tested until credible results were obtained. The gradations and ratings selected are shown in Table 10.17.

TABLE 10.17 Descriptive Words and Ratings

Probability		Frequency of Exposure		Severity	
Frequent (Fre)	15	Often (Of)	13	Catastrophic (Cat)	50
Likely (Lik)	9	Occasional (Oc)	10	Critical (Cri)	40
Occasional (Occ)	4	Infrequent (In)	7	Medium (Med)	25
Remote (Rem)	1	Seldom (Se)	4	Minimal (Min)	10
Improbable (Imp)	0.5				

WHAT THE DESCRIPTIVE WORDS MEAN

To give substance to the words used to establish gradations within the probability, severity, and frequency of exposure categories, their meanings as they are used in this risk scoring system are presented in Tables 10.18–10.20.

TABLE 10.18 Incident Probability

Category: Descriptive Word	Definition: Applies for the Selected Unit of Time, Events, Population, Items, or Activity
Frequent	Likely to occur repeatedly, to even chance
Likely	Likely to occur several times
Occasional	Occurs sporadically, likely to occur sometime
Remote	Not likely to occur, but could possibly occur
Improbable	So unlikely, can assume occurrence will not be experienced

TABLE 10.19 Severity of Consequences

Category: Descriptive Word	People: Employees, Public	Facilities, Product, or Equipment Loss (in US$)	Operations Down Time	Environmental Damage
Catastrophic	Fatality	Exceeds 2 million	Exceeds six months	Major event, requiring several years for recovery
Critical	Disabling injury or illness	500,000 to 2 million	Four weeks to six months	Event requires one to two years for recovery
Marginal	Minor injury or illness	50,000 to 500,000	Two days to four weeks	Recovery time is less than one year
Negligible	No injury or illness	Less than 50,000	Less than two days	Minor damage, easily repaired

TABLE 10.20 Frequency of Exposure

Category: Descriptive Word	Definition
Often	Continues to daily
Occasional	Daily to monthly
Infrequent	Monthly to yearly
Seldom	Less than yearly

TABLE 10.21 Risk Categories, Score Levels, and Action or Risk Acceptance Levels

Risk Category	Score Levels	Remedial Action or Acceptance
Low	199 and below	Risk is acceptable: remedial action discretionary
Moderate	200 to 499	Remedial action to be taken at appropriate time
Serious	500 to 799	Remedial action to be given high priority
High	800 and above	Immediate action necessary. Operation not permissible except in an unusual circumstance and as a closely monitored and limited exception with approval of the person having authority to accept the risk

Four risk categories were considered adequate—Low, Moderate, Serious, and High. Risk scores for each category were assigned, as well as management decision indicators with respect to risk reduction actions to be taken or for risk acceptance. They are shown in Table 10.21.

It must be understood that the score levels provided in Table 10.22 were established through subjective judgments. They should be considered as advisory

TABLE 10.22 The Risk Scoring System

Probability of Category and Rating		Frequency of Exposure and Rating		Mid-Score	Catastrophic	Final Score	Critical	Final Score	Medium	Final Score	Minimal	Final Score
Fre	15	Of	13	28	50	1,400	40	1,120	25	700	10	280
		Oc	10	25	50	1,250	40	1,000	25	625	10	250
		In	7	22	50	1,100	40	880	25	550	10	220
		Se	4	19	50	950	40	760	25	475	10	190
Lik	9	Of	13	22	50	1,100	40	880	25	550	10	220
		Oc	10	19	50	950	40	760	25	475	10	190
		In	7	16	50	800	40	640	25	400	10	160
		Se	4	13	50	650	40	520	25	325	10	130
Occ	4	Of	13	17	50	850	40	680	25	425	10	170
		Oc	10	14	50	700	40	560	25	350	10	140
		In	7	11	50	550	40	440	25	275	10	110
		Se	4	8	50	400	40	320	25	200	10	80
Rem	1	Of	13	14	50	700	40	560	25	350	10	140
		Oc	10	11	50	550	40	440	25	275	10	110
		In	7	8	50	400	40	320	25	200	10	80
		Se	4	5	50	250	40	200	25	125	10	50
Imp	0.5	Of	13	13.5	50	675	40	540	25	338	10	135
		Oc	10	10.5	50	525	40	420	25	263	10	105
		In	7	7.5	50	375	40	300	25	188	10	75
		Se	4	4.5	50	225	40	180	25	113	10	45

TABLE 10.23   Example of How the Risk Scoring System is Applied

Probability of Occurrence	Score	Frequency of Exposure	Score	Severity	Score	Risk Score
Frequent	15	Often	13	Critical	40	$(15 + 13) \times 40 = 1120$
Likely	9	Occasional	10	Critical	40	$(9 + 10) \times 50 = 950$
Occasional	4	Infrequent	7	Medium	25	$(4 + 7) \times 25 = 275$
Remote	1	Seldom	4	Minimal	10	$(1 + 4) \times 10 = 50$

indicators and not as absolutes in management decision-making. For example, in a real-world case having high severity potential, remedial actions brought the risk score down to 225, which is very close to 199, the level where the risk would be considered acceptable and not requiring further remedial action. Work proceeded, with close observation.

Table 10.22 displays the Risk Scoring System. It shows the various combinations with respect to incident probability, the frequency of exposure, and the severity of consequences; how probability and frequency ratings are totaled to become mid-scores; and the final score—arrived at by multiplying the severity score by the midscore.

The goal was to create a three-dimensional numerical risk scoring system that serves the needs of those who are more comfortable with statistics in their risk assessments; addresses the strong beliefs of those who want frequency of exposure given separate consideration in the risk assessment process; and maintains credibility and efficacy. An example follows in Table 10.23, demonstrating how the risk scoring system is applied.

CONCLUSION

Three-dimensional numerical risk scoring systems can be crafted and have credibility. But how are such systems to be used? Numerical risk scores carry an image of preciseness and that can influence decision-making and priority setting. But in reality, they should not be the sole or absolute determinant.

In the research for this chapter, one of the most interesting discussions of the real-world use of a numerical risk scoring system took place with a person whose principle interest is the safety of the products her company produces. The risk determination system in that company requires that an independent facilitator serve as a discussion leader and that the risk review committee consist of at least five knowledgeable people. Consensus on risk scores and recommendations for any necessary risk reduction actions must be reached.

In these deliberations, it is common that severity scores for injuries to employees, users, or bystanders are moved up a level or two from what is computed. For example, an injury requiring but a visit to a doctor and immediate return to work receives the lowest severity rating in the scoring system.

However, when arriving at the recommendations to be made affecting product or operations design or revising the standard operating procedures or the procedures in instruction manuals for users, the review group regularly gives more weight to the injury than the scoring system indicates is necessary. Recommendations to achieve risk reduction are often influenced by the possible damage to corporate image, customer relations, employee relations, and an assumed societal responsibility.

Numerical risk-scoring systems can serve a need. But it should be remembered that they consist of numerics arrived at through subjective judgments. Risk assessment is still as much art as science.

REFERENCES

Accident Prevention Manual, Administration & Programs, 13th Edition. Itasca, IL: National Safety Council, 2009.

EN ISO 12100-2010: Safety of Machinery. General Principles for Design. Risk Assessment and Risk Reduction. Geneva, Switzerland: International Organization for Standardization, 2010.

Failure Mode and Effects Analysis (FMEA): A Guide for Continuous Improvement for the Semiconductor Equipment Industry. Technology. Available at http://dewihardiningtyas.lecture.ub.ac.id/files/2012/09/FMEA-Guideline.pdf. Accessed June 1, 2019. n.d.

Fine, W.T. *Mathematical Evaluations for Controlling Hazards.* White Oak, MD: Naval Ordinance Laboratory, 1971. Available at https://apps.dtic.mil/dtic/tr/fulltext/u2/722011.pdf. Accessed June 1, 2019.

Grose, V.L. *Managing Risk: Systematic Loss Prevention for Executives.* Originally published by Prentice-Hall, 1987. Now available through Omega Systems Group, Incorporated, Arlington, VA.

Pilz: Guide to Machinery Safety, Sixth Edition. Farmington, MI: Pilz Automation Technology, 1999.

ADDENDUM A

FMEA FORM

Advanced Safety Management: Focusing on Z10.0, 45001, and Serious Injury Prevention,
Third Edition. Fred A. Manuele.
© 2020 John Wiley & Sons, Inc. Published 2020 by John Wiley & Sons, Inc.

FMEA FORM
SYSTEM: DATE:
SUBSYSTEM: FAILURE MODE AND EFFECTS ANALYSIS (FMEA)   SHEET:
REFERENCE DRAWING: FAULT CODE#   PREPARED BY:

Subsystem/ Module & Function	Potential Failure Mode	Potential Local Effect(s) of Failure	Potential End Effect(s) of Failure	SEV	Cr	Potential Cause(s) of Failure	OCC	Current Controls/ Fault Detection	DET	RPN	Recommended Action(s)	Area/ Individual Responsible & Completion Date(s)	Action Taken	SEV	OCC	DET	RPN
Subsystem name and function	"How can this subsystem fail to perform its function?" "What will an operator see?"	The local effect of the subsystem or end user	Downtime # of hours at the system level. 2. Safety 3. Environmental 4. Scrap loss	7	1	"How can this failure occur?" Describe in terms of something that can be corrected or controlled. 2. Refer to specific errors or malfunctions	5	"What mechanisms are in place that could detect, prevent, or minimize the impact of this cause?"	6	210	"How can we change the design to eliminate the problem?" "How can we detect (fault isolate) this cause for this failure?" "How should we test to ensure the failure has been eliminated?" "What PM procedures should we recommend?" 1. Begin with highest RPN. 2. Could say "no action" or "Further study is required." 3. An idea written here does not imply corrective action.	"Who is going to take responsibility?" "When will it be done?"	"What was done to correct the problem?" Examples: Engineering Change, Software revision, no recommended action at this time due to obsolescence, etc.	6	1	1	6

Critical Failure symbol (Cr). Used to identify critical failures that must be addressed (i.e., whenever safety is an issue).

Severity Ranking (1–10) (see Severity Table)

Occurrence Ranking (1–10) (see Occurrence Table)

Detection Ranking (1–10) (see Detection Table)

Risk Priority Number RPN = Severity*Occurrence*Detection

How did the "Action Taken" change the RPN?

ADDENDUM B

MATHEMATICAL EVALUATIONS FOR CONTROLLING HAZARDS

William T. Fine
Naval Ordinance Laboratory
White Oak, Silver springs, MD
1971

Abstract: To facilitate expeditious control of hazards for accident prevention purposes, two great needs have been recognized. These are for:

1. a method to determine the relative seriousness of all hazards for guidance in assigning priorities for preventive effort; and
2. a method to give a definite determination as to whether the estimated cost of the contemplated corrective action to eliminate a hazard is justified.

To supply these needs, a formula has been devised which weighs the controlling factors and "calculates the risk" of a hazardous situation, giving a numerical evaluation to the urgency for remedial attention to the hazard. Calculated Risk Scores are then used to establish priorities for corrective effort. An additional formula weighs the estimated cost and effectiveness of any contemplated corrective action against the Risk Score and gives a determination as to whether the cost is justified.

CONTENTS

Chapter 1. Introduction
Section

   1. General
   2. Definitions

Chapter 2. Formula for Evaluating the Seriousness of a Risk Due to a Hazard

   1. General
   2. The Formula
   3. Examples
   4. Summarizing Risk Scores
   5. Results and Uses of Summary Risk Scores

Chapter 3. Formula to determine Justification for Recommended Corrective Action

   1. General
   2. The Formula
   3. Criteria for Justification
   4. Examples
   5. Recommended Procedure for Using the "J" Formula
   6. Exception to Reliance on the "J" Formula

Chapter 1: Introduction

General. The purpose of this chapter is to illustrate the need for quantitative evaluations to aid in the control of hazards and to explain the general plan of this report.

A problem frequently facing the head of any (field type) safety organization is to determine just how serious each known hazard is, and to decide to what extent he should concentrate his resources and strive to get each situation corrected. Normal safety routines such as inspections and investigations usually produce varying lists of hazards which cannot all be corrected at once. Decisions must be made as to which ones are the most urgent. On costly projects, management often asks whether the risk due to the hazard justifies the cost of the work required to eliminate it. Since budgets are limited, there is necessity to assign priorities for costly projects to eliminate hazards.

The question of whether a costly engineering project is justified is usually answered by a general opinion which may be little better than guesswork. Unfortunately, in many cases, the decision to undertake any costly correction of a hazard depends to a great extent on the salesmanship of safety personnel. As a result, due to insufficient information, the cost of correcting a very serious hazard may be considered prohibitive by management, and the project postponed; or due to excellent selling jobs by Safety, highly expensive engineering or construction jobs may be approved when the risks involved really do not justify them.

In Chapter 2 of this report, a formula is presented to "calculate the risk" due to a hazard, or to quantitatively evaluate the potential severity of a hazardous situation. Use of this formula will provide a logical system for safety and management to determine priorities for attention to hazardous situations, and guidance for safety personnel in determining the areas where their efforts should be concentrated.

In Chapter 3 of this report, a formula is presented for determining whether or not the cost of eliminating a hazard is justified. Use of the formula will provide a solid foundation upon which safety personnel may base their recommendations for engineering-type corrective action. It will assure that projects which are not justified will not be recommended.

This report deals with justification of costs to eliminate hazards. This does not imply in any way that a cost, no matter how great, is not worthwhile if it will prevent an accident and save a human life. However, we must also consider accident prevention with reason and judgment. Budgets are not unlimited. Therefore, the maximum possible benefit for safety must be derived from any expenditure for safety. When an analysis results in a decision that the cost of certain measures to eliminate a hazard "is not justified," we do not say or suggest that the hazard is not serious and may be ignored.

We do say that, based on evaluation of the controlling factors, the return on the investment, or in other words, the amount of accident prevention benefit, is below the standards we have established. The amount of money involved will no doubt provide greater safety benefit if used to alleviate other higher-risk hazards which this system will identify. As for the hazard in question, less costly preventive measures should be sought.

Definitions: For the purpose of this presentation, three factors are defined as follows.

a. Hazard: Any unsafe condition or potential source of an accident. Examples are: an unguarded hole in the ground; defective brakes on a vehicle; a deteriorated wood ladder; a slippery road.

b. Hazard-event: An undesirable occurrence; the combination of a hazard with some activity or person which could start a sequence of events to end in an accident. Examples of hazard events are: a person walking through a field which contains a hazard such as an unguarded well opening; a person not wearing eye protection while in an eye hazardous area; a person driving a vehicle that has defective brakes; a man climbing up a defective ladder; a vehicle being driven on a slippery road.

c. Accident sequence: This chain of events or occurrences which take place starting with a "hazard-event" and ending with the consequences of an accident.

d. Additional definitions will be provided in later pages as needed.

Chapter 2: Formula for Evaluating the Seriousness of the Risk Due to a Hazard

General. The purpose of this chapter is to present a complete explanation of the method for quantitatively evaluating the seriousness of hazards, and some of the benefits that may be derived from such analyses.

The expression "a calculated risk" is often used as a catchall for any case when work is to be done without proper safety measures being taken. But usually such work is done without any actual calculation. By means of this formula, the risk is calculated. The seriousness of the risk due to a hazard is evaluated by considering the potential consequences of an accident, the exposure or frequency of occurrence of the hazard-event that could lead to the accident, and the probability that the hazard-event will result in the accident and consequences.

The Formula is as follows:

$$\text{Risk Score} = \text{Consequences} \times \text{Exposure} \times \text{Probability}$$

$$\text{Abbreviated: } R = C \times E \times P$$

Definitions of the elements of the formula and numerical ratings for the varying degrees of the elements are given below.

a. **Consequences C**: The most probable result of a potential accident, including injuries and property damage.

This is based on an appraisal of the entire situation surrounding the hazard, and accident experience.

Classifications and ratings are:

Description	Rating
(1) Catastrophic: numerous fatalities; extensive damage (over $1,000,000) major disruption of activities of national significance	100
(2) Multiple fatalities; damage $500,000 to $1,000,000	50
(3) Fatality, damage $100,000 to $500,000	25
(4) Extremely serious injury (amputation, permanent disability); damage $1,000 to $100,000	15
(5) Disabling injuries; damage up to $1,000	5
(6) Minor cuts, bruises, bumps, minor damage	1

b. **Exposure E**: *Frequency of occurrence of the hazard-event* – the undesired event which could star the accident-sequence. Classifications are below. Selection is based on observation, experience and knowledge of the activity concerned.

Description	Rating
The hazard event occurs:	
(1) Continuously, (or many times daily)	10
(2) Frequently (approximately once daily)	6
(3) Occasionally (from once per week to once per month)	3
(4) Unusually (from once per month to once per year)	2
(5) Rarely (it has been known to occur)	1
(6) Very rarely (not known to have occurred but considered remotely possible	0.5

c. **Probability P**: This is the likelihood that, once the hazard-event occurs, the complete accident-sequence of events will follow with the necessary timing and coincidence to result in the accident and consequences. This is determined by careful consideration of each step in the accident sequence all the way to the consequences, and based upon experience and knowledge of the activity, plus personal observations. Classifications and ratings follow:

Description	Rating
The accident-sequence, including the consequences:	
(1) Is the most likely and expected result if the hazard-event takes place	10
(2) Is quite possible, would not be unusual, has an even 50/50 chance	6
(3) Would be an unusual sequence or coincidence	3
(4) Would be a remotely possible coincidence. (It has happened here.)	1
(5) Extremely remote but conceivably possible. (Has never happened after many years of exposure.)	0.5
(6) Practically impossible sequence or coincidence; a "one in a million" possibility. (Has never happened in spite of exposure of many years.)	0.1

Examples The use of this formula is demonstrated by actual examples. Six widely different types of situations have been selected to illustrate the broad applicability of the formula. (*In this condensation of Fine's paper, three examples are given.*)

a. Example No. 1

1. *Problem.* There is a quarter-mile stretch of two-lane road used frequently by both vehicles and pedestrians departing or entering the grounds. There is no sidewalk, so pedestrians frequently walk in the road, especially when he grass is wet orsnow covered.

   There is little hazard to pedestrians when all the traffic is going in one direction only; but when vehicles are going in both directions and passing by each other, the vehicles require the entire width of the road, and pedestrians must then walk on the grass alongside the road: It is considered that an accidental fatality could occur if a pedestrian steps into the road, or remains in the road at a point where two vehicles are passing.

2. *Steps to Use the Risk Score Formula*:

   Step 1. List the accident-sequence of events that could result in the undesired consequence
   1. It is a wet or snowy day, making the grass along the road wet and uninviting to walk on.
   2. At quitting time, a line of vehicles, and some pedestrians are leaving the grounds, using this road.
   3. One pedestrian walks on the right side of this road, and he has an attitude which makes him oblivious to the traffic. (This is the hazard-event.)
   4. Although traffic is "one way" out at this time, one vehicle comes from the opposing direction causing the outgoing traffic line to move to the right edge of the road.
   5. The pedestrian on the right side of the road fails to observe the vehicles, and he remains in the road.
   6. The driver of one vehicle fails to notice the pedestrian and strikes him from the rear.
   7. Pedestrian is killed.

   Step 2. Determine values for elements of formula: *Consequences*: A fatality. Therefore $C = 25$.
   - *Exposure*: The hazard-event is event 3 above, the pedestrian remaining in road and refusing to-notice the line of traffic. It is considered that this type individual appears or is "created" by conditions occasionally. Therefore $E = 3$.
   - *Probability* of all events of the accident sequence following the hazard-event is: "conceivably possible, although it has never happened in many years." Reasoning is as follows: events 4, 5, 6 and 7 are individually unlikely, so the combination of their occurring simultaneously is extremely remote.

   Event 4 is unlikely because traffic is "one way" at quitting time.

Event 5 is unlikely because a number of drivers would undoubtedly sound their horns and force the pedestrian's attention.

Event 6 is unlikely because most drivers are not deliberately reckless.

Event 7, a fatality, is unlikely because vehicle speeds are not great on the road, and the most likely case would be a glancing blow and minor injury. Not even a minor injury has ever been reported here. In view of the above Probability $P = 0.5$.

Step 3. Substitute into formula and determine the Risk Score.

$$R = C \times E \times P = 25 \times 3 \times 0.5 = 37.5$$

(NOTE: The Risk Score or one case alone is meaningless. Additional hazardous situations must also be calculated for comparative purposes and a definite pattern. Additional cases are similarly calculated below.)

b. Example No.2

1. *Problem.* A 12,000 gallon propane storage tank is subject to two hazards. One hazard is the fact that the tank is located alongside a well-traveled road. The road slopes, and is occasionally slippery due to rain, snow or ice. It is considered possible that a vehicle (particularly a truck) could go out of control, leave the road, strike and rupture the tank, and cause a propane gas explosion and fire that could destroy several buildings, with consequences amounting to damage costing $200,000, plus a fatality.

    The second hazard is the tank's location close to ultra-high compressed air lines and equipment. A high pressure pipeline explosion could result from a malfunctioning safety valve, a human error in operating the equipment, damage to a pipeline, or from other causes. Blast or flying debris could conceivably strike the propane tank, rupture it and cause it to explode with the same consequences as for a run-away vehicle.

2. *Using the Risk Score Formula* (Note: In this case there are two hazards, so the evaluation is done in two parts, one for each of the hazards, and the total scores are added.)

Step 1. Consider—just the first hazard, that due to a vehicle. List the sequence of events that would result in an accident:
  1. Many vehicles are driven down the hill alongside the storage tank
  2. The road has suddenly become slippery due to an unexpected freezing rain.
  3. One truck starts to slide on the slippery road as it goes down this hill. NOTE: This is the "hazard-event" that starts the accident sequence.)
  4. The driver loses his steering control at a point when he is uphill from and approaching the tank.

5. Brakes fail to stop the vehicle from sliding.
6. Vehicle heads out of control toward the tank.
7. Vehicle strikes the tank with enough force to rupture it and permit the propane gas to leak out.
8. A spark ignites the propane
9. Explosion and conflagration occur.
10. Building and equipment damage is $200,000, and one man is killed.

Step 2. Substitute numerical values into formula:

- *Consequences:* One fatality and damage loss of $200,000. Therefore $C = 25$.

- *Exposure:* The hazard-event that would start the accident sequence is—the truck starting to slide on this road. This happened "rarely." Therefore $E = 1$.

- *Probability.* To decide on the likelihood that the complete accident-sequence will follow the occurrence of the hazard-event, we consider the probability of each event:

a. Loss of steering control to occur at the precise point in the road approaching the tank is possible but would be a coincidence.

b. Once the vehicle started to slide, if the road was ice covered, it would be expected that the brakes would fail to stop the slide.

c. The vehicle heading toward the tank is highly unlikely. Momentum would cause the vehicle to continue straight down the road.

d. The vehicle striking the tank with great force is extremely unlikely.

If a vehicle were sliding on an ice covered surface toward the tank, it would be easily diverted from its direction of travel by a number of obstructions between the road and the tank. When roads are slippery, travel is curtailed and drivers are cautioned to drive slowly. A slow rate of speed would be unlikely to produce enough force to damage the tank. The shape and position of the tank are such that a vehicle would tend to glance off it.

In summary, because of the highly unlikely nature of most of the events, this sequence has a one-in-a-million probability. It has never happened. But it is conceivable. Therefore, $P = 0.5$.

Step 3. Substitute in the formula.

$$\text{Risk Score} = 25 \times 1 \times 0.5 = 12.5$$

The entire process is to be repeated for the second hazard, which is the location near the high pressure air lines and equipment.

Step 1. List the sequence of events.
1. Normal daily activities involve operation of equipment and pressuring of pipelines, some of which are in the vicinity of the propane storage tank.
2. A pipeline containing air compressed to 3,000 pounds per square inch, approximately 50 feet away from the storage tank has become deteriorated or damaged. (This is the hazard-event.)
3. The pipeline bursts.
4. Metal debris is thrown by the blast in all directions, several pieces flying and striking the propane tank with such force that the tank is ruptured.
5. Propane starts to leak out of the tank.
6. A spark ignites the propane fumes.
7. The propane and air mixture explodes.
8. Building damage is $200,000, and one man is killed.

Step 2. Determine values and substitute in the formula.
- *Consequences:* One fatality and damage loss of $200,000. $C = 25$.
- *Exposure:* High pressure air lines have been known to have been neglected or damaged. Frequency of such occurrences is considered "unusual." Therefore $E = 2$.
- *Probability:* Now we estimate the likelihood that a damaged pipeline will explode and the explosion will occur close enough and with enough blast to throw debris and strike the propane tank with such force as to complete the accident sequence. Several bursts have occurred in the past few years, but none have damaged the propane tank. Few of the pipelines are close enough to endanger the tank. After careful consideration, the accident sequence is considered "very remotely possible." $P = 0.5$.

Step 3. Substituting into the formula.

$$\text{Risk Score} = 25 \times 2 \times 0.5 = 25$$

$$\text{Totaling Risk Score} = 12.5 + 25 = 37.5$$

Chapter 3. Formula to Determine the Justification for Recommended Corrective Actions

General. The purpose of this chapter is to describe the method of determining whether the cost of corrective action to alleviate a hazard is justified. Once a hazard has been recognized, appropriate corrective action must be tentatively decided upon and its cost estimated. Now the "Justification" formula can be used to determine whether the estimated cost is justified.

The Formula is as follows:

$$\text{Justification} = \frac{\text{Consequences} \times \text{Exposure} \times \text{Probability}}{\text{Cost Factor} \times \text{Degree of Correction}}$$

Elements are abbreviated

$$J = \frac{C \times E \times P}{CF \times DC}$$

It should be noted that the elements of the numerator of this formula are the same as the Risk Score formula described in Chapter 2. We have simply added a denominator made up of two additional elements which are as follows:

a. *Cost Factor CF:* A measure of the estimated dollar cost of the proposed corrective action. Classifications and ratings are:

COST	RATINGS
(1) Over $50,000	10
(2) $25,000 to #50,000	6
(3) $10,000 to $25,000	4
(4) $1,000 to $10,000	3
(5) $100 to $1,000	2
(6) $25 to $100	1
(7) Under $25	0.5

b. *Degree of Correction DC*: An estimate of the degree to which the proposed corrective action will eliminate or alleviate the hazard, forestall the hazard-event, or interrupt the accident sequence. This will be an opinion based on experience and knowledge of the activity concerned. Classifications and ratings are:

DESCRIPTION	RATING
(1) Hazard positively eliminated, 100%	1
(2) Hazard reduced at least 75%, but not completely	2
(3) Hazard reduced by 50 to 75%	3
(4) Hazard reduced by 25 to 50%	4
(5) Slight effect on hazard, less than 25%	6

Criteria for Justification

Values are substituted into the formula to determine the numerical value for Justification. The Critical Justification Rating is 10. For any rating over 10, the expenditure will be considered justified. For a score less than 10, the cost of the contemplated corrective action is not justified.

Note: The Critical Justification Rating has been arbitrarily set at 10, based on experience, judgment and the current budgetary situation. After extended experience at an individual organization, based on accident experience, budgetary situations, and appraisals of the safety status, it may be found desirable to raise or lower the critical score.

Examples The use of the Justification formula will be illustrated by the use of the same six examples discussed in Chapter 2. (Three will be shown here.)

a. Example No. 1. The hazard of pedestrians and vehicles using the same road. To reduce this risk, the corrective action being considered is to construct a sidewalk alongside the road, at an estimated cost of $1,500. The "J" formula is now used to determine whether this contemplated expenditure is justified.

1. *Substitute values in the "J" formula*

$$J = \frac{C \times E \times P}{CP \times DC}$$

   a. *C, E, and P,* for this situation were discussed as Example No. 1 in Chapter 2 of this report and determined to be 25, 3, and 0.5 respectively.
   b. *Cost Factor:* The estimated cost factor is $1,500. Therefore CF is 3.
   c. *Degree of Correction:* The probability of the hazard-event occurring is considered to be reduced at least 75 percent, but 100 percent, by the construction of the sidewalk: Therefore DC = 2.
   d. *Justification Rating*

$$J = \frac{25 \times 3 \times 0.5}{3 \times 2} = \frac{37.5}{6} = 6.25$$

2. *Conclusion.* "J" is less than 10. Therefore the cost of construction of the sidewalk is not justified.
   Note: This lack of sufficient justification evaluates the situation from the safety viewpoint only. Management could feel there is added justification for morale or other purposes.
3. *Additional consideration.* Since the Risk Score is still a substantial 37.5, other less costly corrective measures should be sought. This includes improved administrative controls to enforce one-way traffic, reduce speed, and encourage pedestrians to use another exit gate. This will reduce the Risk Score by reducing both Exposure and Probability.

MATHEMATICAL EVALUATIONS FOR CONTROLLING HAZARDS 233

b. Example No. 2. The hazard due to compressed air being used in a shop without proper pressure reduction nozzles.

The proposed corrective action is installation of proper pressure reducing nozzles on the 50 air hoses, at a cost of $8 each, or $400. To determine justification for the expenditure:

1. Determine values for the elements of the "J" formula:
    a. *C, E, and P,* were discussed in Example No. 2 of Chapter 2 and evaluated at 5, 10, and 6 respectively.
    b. *Cost Factor.* The cost of the corrective action is $400, so CF = 2.
    c. *Degree of Correction.* The corrective action will reduce the hazard by at least 50 percent, so DC = 3.
    d. Substituting in the formula:

$$J = \frac{5 \times 10 \times 6}{2 \times 3} = \frac{300}{6} = 50$$

2. *Conclusion.* "J" is well above 10. The cost of installing pressure reduction nozzles is strongly justified.

c. Example No. 3. The hazardous location of the 12,000 gallon propane storage tank.

The proposed corrective action is to relocate the tank to a place where it will be less likely to be damaged by any external source, at an estimated cost of $16,000.

1. Determine values for elements of the formula.
    a. *C, E, and P,* were determined in Example No. 3 of Chapter 2 to be 25, 1, and 1.5 (the two hazards combined.)
    b. *Cost Factor:* Cost of relocation is $16,000. CF = 4.
    c. *Degree of Correction.* In the very best location available, there still remains a remote possibility of damage to the tank, so DC = 2.
    d. Substituting in the formula:

$$J = \frac{25 \times 1 \times 1.5}{4 \times 2} = \frac{37.5}{8} = 4.7$$

2. *Conclusion.* Based on the established criteria, the cost of relocation of the tank is not justified.
3. It is emphasized that the conclusion in this case that the proposed corrective action is not justified does not mean that the hazard is of little or no significance. The Risk Score is still 37.5, and this remains of appreciable concern.

Since the potential consequences of an accident are quite severe, effort should be expended to reduce the risk, by reducing either the Exposure of the Probability, or devising other less costly corrective action. In this case, it is considered that an additional steel plate barrier could be erected to protect the tank from the compressed air activities, and one or two strong posts in the ground could minimize danger from the road. Thus, the Probability of serious damage to the truck, and the Risk Score, would be considerably lessened at a very nominal cost.

Recommended Procedure for Using the "J" Formula

A convenient **"J"** Formula worksheet is furnished for undertaking a hazard analysis to determine the Justification Rating. Once a hazard has been recognized, the following procedure is recommended:

a. State the problem briefly.
b. Decide on the most likely consequences of an accident due to the hazard.
c. Review all factors carefully, on the scene. List the actual step-by-step sequence of events that is most likely to result in the consequences chosen. You must be specific.
d. Decide on the most appropriate corrective action and obtain or make a rough estimate of its cost.
e. Consider carefully the effect of the proposed corrective action on the hazard, and estimate roughly the degree to which the dangerous situation will be alleviated.
f. If alternative corrective measures are possible, repeat steps (d) and (e) for them.
g. Select the hazard-event: the first undesirable occurrence that could start the accident sequence.
h. Consider the existing situation carefully to determine the frequency of the occurrence of the hazard-event: by on the scene observation, and then decide on the Exposure Rating. If in doubt between two ratings, interpolate.
i. For the Probability Rating, consider the likelihood of the occurrence of each event of the accident sequence, including the resulting injury and/or damage, and form an opinion based on the descriptive words. For example, if two unusual coincidences are required, this could be considered "remotely possible"; two "remotely possible" occurrences could be "conceivably possible" etc. If in doubt between two ratings, interpolate.

Endeavor to be consistent. Consider the occurrence of only the same consequences which were decided on in step (b) above. For example, if you decided on consequences of a fatality, then in this step you may only consider the probability of a fatality. If you also wish to consider lesser injuries, a separate and additional computation must be made, since both the Consequences and Probability evaluations would be different. Scores should be added.

j. You have now obtained ratings for all the elements of the "J" formula. Substitute in the formula and compute the Justification Score.
k. If alternative corrective measures are being considered to alleviate the hazard, compute their Justification Scores also.
l. If there are alternative corrective measures which have acceptable Justification Scores, the most desirable from the Safety standpoint is the one which would make the greatest reduction in the Risk Score. Therefore, for each alternative, assume that the corrective measures are in effect and re-compute the Risk Score. Of course this selection may also be affected by external (non-safety) considerations such as the size of investment required, the relative effects on morale, esthetics, efficiency, convenience, ease of implementation, etc.

Exception to Reliance on the "J" Formula

A highly hazardous situation may exist for which no corrective action that can be devised will give an acceptable Justification Score. Obviously in such a case, whatever corrective action is necessary to reduce the Risk Score should be taken, regardless of the Justification Score.

236 MATHEMATICAL EVALUATIONS FOR CONTROLLING HAZARDS

"J" FORMULA WORKSHEET

Problem:

Sequence of events or factors necessary for accident

1.
2.
3.
4.
5.
6.
7.
8.

Formula Factors	Rating
C Consequences_____	_____
E Exposure_____	_____
P Probability_____	_____
CF Cost Factor_____	_____
DC Degree of Correction_____	_____
J Justification: $J = \frac{C \times E \times P}{CF \times DC} = \frac{\times \ \times}{\times}$	

CHAPTER 11

HIERARCHIES OF CONTROL

A hierarchy is a system of persons or things ranked one above the other. Hierarchies of control in Z10.0 and 45001 provide

- A systematic way of thinking, considering steps in a ranked and sequential order;
- An effective way for decision makers to eliminate or reduce hazards and the risks that derive from them.

Acknowledging the premise—that risk reduction measures should be considered and taken in a prescribed order—represents an important step in the evolution of the practice of safety. A major premise in applying a hierarchy of controls is that the outcome of the actions taken is to be an acceptable risk level, defined as follows:

Acceptable risk is that risk for which the probability of a hazard-related incident or exposure occurring and the severity of harm or damage that could result are as low as reasonably practicable in the situation being considered.

That definition requires taking into consideration each of the two distinct aspects of risk as decisions are made to reduce risks:

- Avoiding, eliminating, or reducing the *probability* of a hazard-related incident or exposure occurring

---

Advanced Safety Management: Focusing on Z10.0, 45001, and Serious Injury Prevention, Third Edition. Fred A. Manuele.
© 2020 John Wiley & Sons, Inc. Published 2020 by John Wiley & Sons, Inc.

- Reducing the *severity* of harm or damage that may result, if an incident or exposure occurs

HIERARCHIES OF CONTROL IN 45001 AND Z10.0

Requirements in the hierarchies of control in the 45001 and Z10.0 standards are close to identical, but not quite. In 45001, under the subcaption "Eliminating Hazards and Reducing OH&S Risks" (Section 8.1.2), this is the opening sentence and the standard's hierarchy of controls:

The organization shall establish, implement and maintain process(es) for the elimination of hazards and reduction of OH&S risks using the following hierarchy of controls:

- Eliminate the hazard;
- Substitute with less hazardous processes, operations, material or equipment;
- Use engineering controls and reorganization of work;
- Use administrative controls, including training; and
- Use adequate personal protective equipment.

There is one Note for this hierarchy of controls. It says that in some countries, organizations are required to provide personal protective equipment without cost to workers.

In Z10.0, under the sub-caption "Hierarchy of Controls" (Section 8.4), this is its opening sentence and its hierarchy of controls.

The organization shall establish a process for achieving an acceptable level of risk reduction based upon the following preferred order of controls:

A. Elimination;
B. Substitution of less hazardous materials, processes, operations, or equipment;
C. Engineering controls;
D. Warnings;
E. Administrative controls, and
F. Personal protective equipment.

There are three lengthy notes following the Z10.0 hierarchy. Paraphrasing, the notes say that in the risk reduction process, organizations should consider:

- reliability, effectiveness, worker acceptance and the possible need for multiple controls and feedback loops to monitor control effectiveness;
- the possibility of reorganizing the work in the decision making process;
- evaluating the effectiveness of controls periodically since most controls erode over time.

Then Z10.0 provides guidance on the subjects that should be considered when the hierarchy of controls is to be applied, as follows:

Application of this hierarchy of controls to achieve an acceptable level of risk shall take into account:

- The nature and extent of the risks being controlled;
- The degree of risk reduction desired;
- The requirements of applicable local, federal, and state statutes, standards and regulations;
- Recognized best practices in industry;
- The effectiveness, reliability, and durability of the control being considered;
- Human factors (includes ergonomics);
- Available technology;
- Cost-effectiveness;
- Internal organization standards;
- Strategies to eliminate or mitigate potential health exposures including those not originating from work activities.

Although the hierarchies of control in the two standards are close to each other, note that the hierarchy in 45001 does not contain an element for Warnings.

Whatever standard or guideline safety professionals use, the intent in applying a hierarchy of controls is to reduce risks to an acceptable level so that the potential for injuries, illnesses, and fatalities occurring is also reduced.

THE LOGIC OF TAKING ACTION IN THE DESCENDING ORDER GIVEN

Comments follow on each of the action elements listed in Z10.0's "Hierarchy of Control" and 45001's "Eliminating Hazards and Reducing OH&S Risks." Also, the rationale is given for the order in which the action elements are placed. Taking actions as *feasible and practicable* in the prescribed order is the most effective means to achieve risk reduction.

A. Elimination

Use of the term "elimination" as the first step assumes that there is something in place to eliminate. In Chapter 13, "Prevention through Design," Avoidance precedes Elimination. The theory is plainly stated. If the introduction of hazards is avoided in the design and redesign processes, risks that derive from those hazards are also avoided.

But avoidance of hazards completely by modifying the design may not always be practicable. Then, the goal is elimination—to modify the design of the workplace or work methods, within practicable limits, so as to limit the:

- Probability of personnel making human errors because of design inadequacies
- Ability of personnel to defeat the work system and the work methods prescribed

Examples: Redesigning to eliminate or reduce the risk from;

- Fall hazards
- Ergonomic hazards
- Confined space hazards
- Noise hazards
- Chemical hazards.

Obviously, hazard elimination or reduction is the most effective way to remove or reduce existing risks. If a hazard is eliminated or reduced, the need to rely on worker behavior to avoid a risk is diminished.

B. Substitute with Less Hazardous Processes, Operations, Materials, or Equipment

Methods that illustrate substituting less hazardous methods, materials, for that which is more hazardous include:

- Using automated material handling equipment rather than manual material handling
- Providing an automatic feed system to reduce machine hazards
- Using a less hazardous cleaning material
- Reduce speed, force, amperage
- Reduce pressure, temperature
- Replacing an ancient steam heating system and its boiler explosion hazards with a hot air system.

Substitution of a less hazardous method or material may or may not result in equivalent risk reduction in relation to what might be the case if the hazards and risks were reduced to an acceptable level through system design or redesign.

Consider this example. Considerable manual material handling is often necessary in a mixing process for chemicals. A chemical reaction takes place and an employee sustains serious chemical burns. There are identical operations at two of the company's locations. At one location, a decision is made to redesign the operation so that it is completely enclosed, automatically fed, and operated by a computer from a control panel, thus greatly reducing operator exposure.

At the other location, funds for doing the same were not available. To reduce the risk, a substitution took place in this manner:

- It was arranged for the supplier to premix the chemicals before shipment
- Some mechanical feed equipment for the chemicals was also installed

The risk reduction achieved by substitution was not equivalent to that attained by redesigning the operation. But in the given situation, substitution is the action to take and the action could reduce risk substantially.

C. Engineering Controls

Substantial risk reduction can be achieved when safety devices are incorporated into the system in the form of engineering controls. Engineered safety devices are to prevent access to hazards by workers and they reduce the potential for worker error. They include devices such as

- Machine guards
- Interlock systems
- Circuit breakers
- Start-up alarms
- Presence sensing devices
- Safety nets
- Ventilation systems
- Sound enclosures
- Fall prevention systems
- Lift tables, conveyors, and balancers

D. Warnings

Warning system effectiveness, and the effectiveness of instructions, signs, and warning labels, rely considerably on administrative controls, such as training, drills, the quality of maintenance, and the reaction capabilities of people. Further, although vital in many situations, warning systems may be reactionary in that they alert persons only after a hazard's potential is in the process of being realized (e.g., a smoke alarm). Examples are

- Smoke detectors
- Alarm systems
- Backup alarms
- Chemical detection systems
- Signs

Comment is necessary on this author's preferred use of the term "warning systems" over warnings or warning signs. The terms "warnings" and "warning signs" appear in some published hierarchies of control—as is the case in Z10.0. The entirety of the needs of a warning system must be considered, for which warning signs or warning devices alone may be inadequate.

For example, the NFPA Life Safety Code 101 (2018) may require, among other things: detectors for smoke and products of combustion; automatic and manual audible and visible alarms; lighted exit signs; designated, alternate, properly lit exit paths; adequate spacing for personnel at the end of the exit path; proper hardware for doors; and emergency power systems. Obviously, much more than merely "Warnings" is needed.

E. Administrative Controls

Administrative controls rely on the methods chosen being appropriate in relation to the needs, the capabilities of people responsible for their delivery and application, the quality of supervision, and the expected performance of the workers. Some administrative controls are

- Personnel selection
- Developing appropriate work methods and procedures
- Training
- Supervision
- Motivation, behavior modification
- Work scheduling
- Job rotation
- Scheduled rest periods
- Maintenance
- Management of change
- Investigations
- Inspections

Achieving a superior level of effectiveness in all of these administrative methods is difficult, and not often attained.

F. Personal Protective Equipment

Proper use of personal protective equipment relies on an extensive series of supervisory and personnel actions, such as the identification of the type of equipment needed, its selection, fitting, training, inspection, maintenance, etc. Examples include

- Safety glasses
- Face shields
- Respirators
- Welding screens
- Safety shoes

- Gloves
- Hearing protection

Although the use of personal protective equipment is common and necessary in many occupational situations, it is the least effective method to deal with hazards and risks. Systems put in place for their use can easily be defeated. In the design process, one of the goals should be to reduce reliance on personal protective equipment to as low as reasonably practicable (ALARP).

APPLICATION OF THE HIERARCHY: EFFECTIVENESS OF ELEMENTS

For many risk situations, a combination of the risk management methods shown in the hierarchy of controls is necessary to achieve acceptable risk levels. But the expectation is that consideration will be given to each of the steps in a descending order, and that reasonable attempts will be made to eliminate or reduce hazards and their associated risks through steps higher in the hierarchy before lower steps are considered. A lower step in the hierarchy of controls is not to be chosen until practical applications of the preceding level or levels are exhausted.

Decision-makers should understand that, with respect to the levels of action shown in the hierarchies of control for Z10.0 and 45001, ameliorations to be taken in the first, second, and third action levels are more effective because they:

- Are *preventive* actions that eliminate or reduce risk by design, substitution, and engineering measures.
- Rely the least on personnel performance.
- Are not as easily defeated by supervisors or other employees.

Decision-makers should also understand that actions taken in the remaining levels rely greatly on the performance of personnel.

ON VARIATIONS IN HIERARCHIES OF CONTROL

In Z10.0, the hierarchy of controls has six elements. In 45001, there are five. Hierarchies in other published standards and guidelines have three to nine elements and their contents can vary, as will be seen in the following examples.

AT THE NATIONAL SAFETY COUNCIL

The third edition of the National Safety Council's *Accident Prevention Manual* was published in 1955. Section 4 is titled "Removing the Hazard from the Job," It provides a three step "order of effectiveness and preference." This is taken from the *Accident Prevention Manual.*

The engineer should include in his planning and follow through such measures as will attain one of the accident prevention goals listed as follows (in the order of effectiveness and preference):

- Elimination of the hazard from the machine, method, material, or plant structure.
- Guarding or otherwise minimizing the hazard at its source if the hazard cannot be eliminated.
- Guarding the person of the operator through the use of personal protective equipment if the hazard cannot be eliminated or guarded at its source.

Company policies should be such that safety can be designed and built into the job, rather than added after the job has been put into operation. (p. 4-1).

Please note that the foregoing was written in 1955. Establishing the concept that risk reduction actions should be taken in an order of effectiveness and preference was an important step in the evolution of the practice of safety. It implies that some steps in the process are preferable since they achieve greater risk reduction than others.

Declaring that safety policies should require that safety be designed and built into the job rather than be dealt with as an add-on is also a premise that influenced later versions of hierarchies of control.

DIRECTIVE: THE COUNCIL OF THE EUROPEAN COMMUNITIES

Although the excerpts taken from the Directive issued in 1989 on workplace safety by The Council of the European Communities are not labeled a hierarchy of controls, that's what they are—a nine element listing of "principles of prevention." It is intriguing because it:

- Commences with risk avoidance;
- Stresses adopting the work to the capabilities of individuals;
- Gives prominence to the design of the workplace and the choice of work equipment; and
- Advises employers to "to take account of changing circumstances and aim to improve existing situations."

The full title of the bulletin is: "Council Directive 89/391/EEC of 12 June 1989 on the introduction of measures to encourage improvements in the safety and health of workers at work." This bulletin is a very good read. It is available at http://www.legislation.gov.uk/eudr/1989/391/contents

In Article 6 of this bulletin, these are the General obligations on employers:

1. Within the context of his responsibilities, the employer shall take the measures necessary for the safety and health protection of workers, including prevention

of occupational risks and provision of information and training, as well as provision of the necessary organization and means. The employer shall be alert to the need to adjust these measures to take account of changing circumstances and aim to improve existing situations.

2. The employer shall implement the measures referred to in the first subparagraph of paragraph 1 on the basis of the following general principles of prevention:

   a. avoiding risks;
   b. evaluating the risks which cannot be avoided:
   c. combating the risks at source;
   d. adapting the work to the individual, especially as regards the design of work places, the choice of work equipment and the choice of working and production methods, with a view, in particular, to alleviating monotonous work and work at a predetermined work-rate and to reducing their effect on health.
   e. adapting to technical progress;
   f. replacing the dangerous by the non-dangerous or the less dangerous;
   g. developing a coherent overall prevention policy which covers technology, organization of work, working conditions, social relationships and the influence of factors related to the working environment;
   h. giving collective protective measures priority over individual protective measures;
   i. giving appropriate instructions to the workers.

MIL-STD-882E-2012

The Department of Defense *Standard Practice For System Safety, MIL-STD-882*, was first issued in 1969. It was a seminal document at that time. Four revisions of 882 have been issued over a span of 43 years. This standard has had considerable influence on the development of risk assessment and risk avoidance, elimination, and amelioration concepts.

Much of the wording on risk assessments and hierarchies of control in safety standards and guidelines issued throughout the world is comparable to that in the several versions of 882.

The fifth edition, issued in May 2012, is designated MIL-STD-882E. It is available on the Internet at https://www.system-safety.org/Documents/MIL-STD-882E.pdf. It can be downloaded, free.

A "System safety design order of precedence" is outlined in 882E. Precedence means: priority in order, rank, or importance. In 882D, the design order of preference had four elements. This latest version has five, as follows. (p. 13)

The system safety design order of precedence identifies alternative mitigation approaches and lists them in order of decreasing effectiveness.

   a. Eliminate hazards through design selection.

b. Reduce risk through design alteration.
c. Incorporate engineered features or devices.
d. Provide detection and warning devices.
e. Incorporate signage, procedures, training, and PPE.

Instruction is given in 882E to assure that:

- mitigation measures, if necessary, are selected and implemented to achieve an acceptable risk level
- implementation measures are evaluated to verify that they have achieved the risk reduction expected
- documentation comments on the validation steps taken and the logic in support of risk acceptance

In the foregoing, note particularly that hazards are to be *eliminated* in the design process; hazards are to be *reduced* through design alteration; an *acceptable level of risk* is to be achieved; and there is to be an evaluation system to assure that acceptable risk levels have been achieved.

ADDITIONAL GOALS TO BE CONSIDERED

As decisions are made in applying each step in a hierarchy of controls, the following should be considered as goals:

1. Avoiding work methods that are overly stressful, taking into consideration worker capabilities and limitations
2. Keeping the probability of human error ALARP by designing workplaces and work methods that are not error-provocative, meaning that they do not:
   - Violate operator expectations.
   - Require performance beyond what an operator can deliver.
   - Induce fatigue.
   - Provide adequate facilities or information for the operator.
   - Present unnecessarily difficult or unpleasant requirements.
   - Include unnecessarily dangerous methods. (as in Chapanis, 1980 p. 119)
3. Designing systems so that human interaction with equipment and processes is ALARP
4. Designing systems so that use of personal protective equipment is ALARP

ATTACHING THE HIERARCHY OF CONTROLS TO A PROBLEM-SOLVING TECHNIQUE

The following observations are a reflection of this author's experience—encompassing the design and engineering, the operational, the postincident, and the post-operational aspects of the practice of safety:

ATTACHING THE HIERARCHY OF CONTROLS TO A PROBLEM-SOLVING TECHNIQUE

- Safety practitioners often recommend solutions to resolve hazard/risk situations before they define the problem—that is, before they identify the specifics of the hazards and assess the associated risks
- Seldom do safety management systems have provisions to determine whether the preventive actions taken achieve the intended risk reduction over a long term.

Those observations led to research into the feasibility of encompassing the hierarchy of controls within a sound problem-solving technique which:

- Commences with problem identification and analysis.
- Requires measurement of results of the actions taken to determine their effectiveness and their permanence.
- Infers that taking further preventive measures if the residual risk is not acceptable.

The initial step in this research was to review what is proposed about the basics of problem-solving in several texts. Problem-solving methods outlined by authors have great similarity. A composite of those techniques is given in Table 11.1.

In every problem-solving method reviewed, the first steps are to identify and analyze the problem. Also, they end with a provision requiring that evaluations be made of the effects of the actions taken.

Figure 11.1 presents a logical sequence of actions that safety professionals should consider in resolving safety issues: identify and analyze the problem; consider the possible solutions; decide on and implement an action plan; and determine whether the actions taken achieved the intended risk reduction results.

The Safety Decision Hierarchy depicts a way of thinking about hazards and risks and establishes an effective order for risk avoidance, elimination, or amelioration. Why propose that safety professionals adopt a safety decision hierarchy? This quote from *The New Rational Manager*, reflecting the real-world observations of Kepner and Tregoe (1981) in dealing with many clients, makes the case:

> The most effective managers, from the announcement of a problem until its resolution, appeared to follow a clear formula in both the orderly sequence and the quality of their questions and actions. (p. vii)

It makes sense to apply a safety decision hierarchy encompassing methods in an orderly sequence of effectiveness to resolve safety issues.

TABLE 11.1 Problem-Solving Methodology

1. Identify the problem
2. Analyze the problem
3. Explore alternative solutions
4. Select a plan and take action
5. Examine the effects of the actions taken

248 HIERARCHIES OF CONTROL

A. **Problem identification and analysis**
    1. Identify and analyze hazards
    2. Assess the risks

B. **Consider these actions in their order of effectiveness**

    1. Eliminate risks
    2. Reduce risks by substituting less hazardous methods or materials
    3. Use engineering controls
    4. Provide warning systems
    5. Apply administrative controls (work methods, training, etc.)
    6. Provide personal protective equipment

C. **Select risk reduction measures and implement them**

D. **Measure for effectiveness**

E. **Accept the residual risk, or start over if it is unacceptable**

FIGURE 11.1 The Safety Decision Hierarchy.

ON PROBLEM IDENTIFICATION AND ANALYSIS

In applying The Safety Decision Hierarchy, the goal in the problem identification and analysis phase is to identify and analyze the hazards in place and assess their accompanying risks. If this Hierarchy included the design processes, there would be another step—avoidance in the design phase.

Hazard and risk situations cannot be intelligently dealt with until the hazards are analyzed and assessments are made of the probability of incidents or exposures occurring and the severity of their consequences are estimated.

Chapter 9, "On Hazards Analyses and Risk Assessments," is a resource for this problem identification and analysis phase.

EXPLORING ALTERNATIVE SOLUTIONS

Action steps shown in the section of The Safety Decision Hierarchy under the caption "Consider these actions, in their order of effectiveness" provide a base for considering risk elimination or reduction measures. Logic in support of those steps and the order in which they are listed was given previously in this chapter.

DECIDING AND TAKING ACTION

All facets of The Safety Decision Hierarchy apply when considering the hazards and risks in a specific facility, process, system, piece of equipment, or a tool in its simplest form. Also, they are broadly applicable in all four of the major aspects of the practice of safety:

- In the design processes, preoperational—where the opportunities are greatest and the costs are lower for hazard avoidance, elimination, or control
- In the operational mode—where, integrated within a continual improvement process, hazards are eliminated or controlled, before their potentials are realized and hazards-related incidents or exposures occur
- Postincident—through investigation of hazards-related incidents and exposures to determine and eliminate or control their causal factors
- Post operational—when demolition, decommissioning, or reusing/rebuilding operation are undertaken

MEASURING FOR EFFECTIVENESS

Provisions in Z10.0 and 45001 require that systems be in place to measure the effectiveness of the risk reduction measures taken. Those provisions are relative to the requirements to "Measure for effectiveness" in The Safety Decision Hierarchy.

Assuring that the actions taken accomplish what was intended is an integral step in the PDCA process. Follow-up activity would determine that the

- Problem was resolved, or only partially resolved, or not resolved
- Actions taken did or did not create new hazards

If the follow-up activity indicates that the residual risk is not acceptable, the thought process set forth in The Safety Decision Hierarchy would again be applied, commencing with hazard identification and analysis.

CONCLUSION

Hierarchies of control in 45001 and Z10.0 derive from work that has evolved over many years. As management implements and maintains processes to achieve acceptable risk levels, the hierarchies present the actions to be considered in a logical order. Applying a prescribed hierarchy of controls sequentially is a very important element in a health and safety management system.

Encompassing a hierarchy of controls within a sound problem-solving technique furthers the ability of management and safety professionals to achieve effective risk avoidance, elimination, or reduction, and to meet the requirements of certain provisions in Z10.0 and 45001. Adopting established problem-solving techniques to address hazard and risk situations is a fundamentally sound approach. That is the purpose of The Safety Decision Hierarchy.

REFERENCES

Accident Prevention Manual, Third Edition. Itasca, IL: National Safety Council, 1955.

ANSI/ASSP Z10.0-2019. *Occupational Health and Safety Management Systems*. Park Ridge, IL: American Society of Safety Professionals, 2019.

ANSI/ASSP/ISO 45001-2018. *Occupational Health and Safety Management Systems—Requirements with Guidance for Use*. Park Ridge, IL: American Society of Safety Professionals, 2018.

Chapanis, A. "The error-provocative situation." In *The Measurement of Safety Performance*. William E. Tarrants, ed. New York, NY: Garland Publishing, 1980.

Council Directive 89/391/EEC of 12 June 1989 on the introduction of measures to encourage improvements in the safety and health of workers at work. Available at http://www.legislation.gov.uk/eudr/1989/391/contents. Accessed June 1, 2019.

Kepner, C.H. and B.B. Tregoe. *The New Rational Manager*. Princeton, NJ: Princeton Research Press, 1981.

MIL-STD-882E. *Standard Practice for System Safety*. Washington, DC: Department of Defense, 2012. Available at https://www.system-safety.org/Documents/MIL-STD-882E.pdf. Accessed June 1, 2019.

NFPA 101: *Life Safety Code*. Quincy, MA: National Fire Protection Association, 2018.

FURTHER READING

Manuele, F.A. *On The Practice Of Safety*, Fourth Edition. Hoboken, NJ: John Wiley & Sons, 2013.

Manuele, F.A. "Risk assessment and hierarchies of control." *Professional Safety*, May 2005.

CHAPTER 12

SAFETY DESIGN REVIEWS

Requirements for Design Review and Management of Change are addressed jointly in Section 8.5 of Z10.0. Although the subjects are interrelated, each has its own importance and uniqueness. Comments on Management of Change are made in Chapter 14. This chapter expands considerably on the design review requirements of Z10.0.

DESIGN AND 45001

But first, a review is provided of the references to design in 45001. There is no specifically designated section in 45001 titled design, or design requirements or design review. Under Hazard Identification—6.1.2.1—the standard says that management is to have in place processes to identify hazards arising from product and service design and from the design of the work areas. But that is all that is said about the design aspects of an occupational health and safety management system. As was said earlier in this book, 45001 is short on design aspects.

But in Annex A, those who were leaders in developing the standard show that they were aware of the importance of trying to avoid bringing hazards and risks into the workplace in the design process.

Advanced Safety Management: Focusing on Z10.0, 45001, and Serious Injury Prevention,
Third Edition. Fred A. Manuele.
© 2020 John Wiley & Sons, Inc. Published 2020 by John Wiley & Sons, Inc.

In Annex A, Provision A.6.1.2.1 is titled Hazard identification. Terminology following that caption is very close to the wording in the *Prevention through Design* standard, the designation for which is ANSI/ASSE Z590.3-2011 (R2016). This is an excerpt from A.6.1.2.1:

> The ongoing proactive identification of hazard begins at the conceptual design stage of any new workplace, facility, product or organization. It should continue as the design is detailed and then comes into operation, as well as being ongoing during its full life cycle to reflect current, changing and future activities.

Supportive wording is included in A.6.1. Item f pertains to opportunities to improve OH&S performance. One of the examples given follows.

> Integrating occupational health and safety requirements at the earliest stage in the life cycle of facilities, equipment or process planning for facilities relocation, process re-design or replacement of machinery and plant.

That is good guidance. Terminology in Annex A implies that those who adopt 45001 should apply prevention through design concepts. It would have been beneficial if such provisions were included in the standard.

DESIGN REVIEW REQUIREMENTS IN Z10.0 (8.5)

Design review and management of change provisions are under the major caption "Implementing and Operation 8.0." This is the opening sentence for 8.5.

> The organization shall establish a process to identify, and take appropriate steps to prevent or otherwise control hazards at the design and redesign stages, and for situations requiring Management of Change, to reduce potential risks to an acceptable level.

So in the design and redesign processes, designers are to prevent and otherwise control hazards. They are also to take into consideration all applicable life cycles (8.5.1) and verify that the processes are effective and ensure that risk levels achieved are acceptable (8.5.2).

In an advisory Note, the writers of Z10.0 say that: "An effective design review considers all aspects including design, procurement, construction, operation, maintenance, and decommissioning. The life cycle phases should integrate quality, health and safety, production, procurement, and consider potential impacts."

To do as Z10.0 implies, systems must be in place to avoid, eliminate, reduce, or control hazards and the risks that derive from them:

- as early as possible and as often as needed in every aspect of the design and redesign processes
- in all phases of operations.

What needs to be agreed upon in an organization is a well-understood concept, a way of thinking that is translated into a system that effectively addresses hazards and risks in the design and redesign processes, a way of thinking that is universally applied by all who are involved in the design processes.

This author holds to the premise that, over time, the level of safety achieved will relate directly to whether acceptable risk levels are achieved or not achieved in the design and redesign processes. This premise relates to the design of facilities, hardware, equipment, tooling, operations layout, the work environment, the work methods, and products. Design, as the term is used here, encompasses all processes applied in devising a system to achieve results.

Methods used to avoid bringing hazards and risks into the workplace will be broadly discussed in this chapter, with the hope that safety professionals can adopt from the materials presented to their advantage. To assist in applying the Z10.0 design provisions, this chapter includes

- A review of safety design concepts and procedures.
- A design review process.
- Comments on how some safety professionals are engaged in activities that lessen the probability of hazards and risks being brought into the workplace.
- An edited composite of procedures in place in several companies to achieve hazard avoidance and control in the design process and procedures to be followed before modified equipment is released for operation.
- A model for—A Safety Design Review and Operations Requirements Guide
- A general checklist as a reference from which a specifically tailored checklist can be developed for use in design reviews, and for equipment acceptance.

SAFETY THROUGH DESIGN/PREVENTION THROUGH DESIGN

Chapter 9, "On Hazards Analyses and Risk Assessments" and Chapter 13 "Prevention through Design" in this book are good references concerning the design review requirements in the Z10.0 standard. Also, activities undertaken by the National Safety Council and the National Institute for Occupational Safety and Health that encourage addressing hazards and risks in the design and redesign processes are directly related to Z10.0's design requirements.

NATIONAL SAFETY COUNCIL

Safety Through Design (1999) was a book created under the auspices of the Institute For Safety Through Design, an entity at the National Safety Council. The Institute's Vision was to achieve a future in which "Safety, health, and environmental considerations are integrated into the design and development of systems meant for human use." (p. xi). An Advisory committee, the members of which were drawn from industry, organized labor, academia, and others interested in the cause, agreed on the following definition.

> Safety Through Design
> The integration of hazard analysis and risk assessment methods early in the design and engineering stages and taking the actions necessary so that risks of injury or damage are at an acceptable level. (p. 3)

That definition serves well in understanding and developing the design requirements in Z10.0. The theme of *Safety Through Design* is that if decisions affecting safety, health, and the environment are integrated into the early stages of the design and redesign processes:

- productivity will be improved.
- operating costs will be reduced.
- expensive retrofitting to correct design shortcomings will be avoided.
- and significant reductions will be achieved in injuries, illnesses, and damage to the environment, and their attendant costs (p 3).

Effectiveness and economics were addressed in the following citation:

> Hazards and the risks that derive from them are most effectively and economically avoided, eliminated, or controlled if they are considered early in the design process, and where necessary as the design progresses (p 11)

All of the foregoing relates to the following premises: Risks of injury derive from hazards. If hazards are avoided, eliminated, reduced, or controlled in the design process and acceptable risk levels are achieved, the potential for harm or damage is diminished.

These premises tie in neatly with the reference in Z10 to addressing occupational health and safety management system issues, which are defined "as hazards, risks, management system deficiencies, legal requirements and opportunities for improvement" (Definitions).

In Z10.0, item 5.1.3—Responsibility and Authority—says that Top management shall, among other things, provide leadership for integrating the OHSMS within the organization's other business activities, systems, and processes.

Part II in *Safety Through Design* is titled "Integration into Business Processes." There are seven chapters in Part II that are good resources for those who attempt to integrate the Z10.0 design requirements into other business processes.

In accord with an established sundown provision, the operations of the Institute For Safety Through Design were discontinued in 2005.

NATIONAL INSTITUTE FOR OCCUPATIONAL SAFETY AND HEALTH

In 2006, several of the participants in the activities of the Institute for Safety through Design, and others, received an email from an executive at the National Institute for Occupational Safety and Health (NIOSH) encouraging our participation in a major initiative promoting Prevention through Design.

One aspect of the NIOSH initiative was to develop and approve a broad, generic voluntary consensus standard on Prevention through Design that is aligned with international design activities and practice.

This author volunteered to lead that endeavor. On September 1, 2011, the American National Standards Institute (ANSI) approved the standard *ANSI/ASSE Z590.3-2011 Prevention through Design: Guidelines for Addressing Occupational Hazards and Risks in the Design and Redesign Processes.*

Activities at NIOSH, by its charter, are limited to occupational safety and health hazards and risks, and they are the focus of Z590.3. But, by intent, the terminology in Z590.3 was kept broad enough so that the Guidelines could be additionally applicable to all hazards-based fields—product safety, environmental controls, property damage that could result in business interruption, among others. In the standard, the definition of Prevention through Design is, of necessity, applicable to work-related hazards and risks only.

> **Prevention through Design.** Addressing occupational safety and health needs in the design and redesign process to prevent or minimize the work-related hazards and risks associated with the construction, manufacture, use, maintenance, retrofitting, and disposal of facilities, processes, materials, and equipment. (p. 10)

THE SAFETY DESIGN REVIEW PROCESS

A safety design review process will not be effective until the participants on the review team have acquired knowledge of hazards and risks and have agreed upon the risk assessment methods and the risk assessment matrix to be used. Also, a safety design review process can be successfully applied only if senior management has been convinced of its value.

Bruce Main, PE, CSP (2004) is the author of *Risk Assessment: Challenges and Opportunities*. Because he is knowledgeable with respect to design reviews, he was asked for permission to excerpt the following from "Design Reviews" which is the title of Chapter 4 in his book.

Excerpts from *Risk Assessment: Challenges and Opportunities* by Bruce Main, PE, CSP. Chapter 4, "Design Reviews" with Permission

Purpose of a Design Review

Design reviews are a formal evaluation of a design to ensure that the design meets the criteria set forth for the project. Safety is typically only one element of a design review. The nature of the design and the company culture will determine the importance of safety criteria. There are some product or process designs where safety is an absolutely critical element, and others where it is a relatively minor concern. ANSI/ASSE Z590.3-2011-R2016 includes the following on design safety reviews:

> Design safety reviews are an important management process tool for integrating safety and health into the design process. This includes designs related to new facilities, processes, or operations, and for changes in existing operations. Design safety reviews are most effective when performed at an early stage when design objectives are being discussed. (p. 16)

Design engineers have the primary responsibility for making a product, machine or system work in accord with the established design criteria. This can be an extremely challenging task. Their focus tends to be creating a functional result from new and existing components, concepts or parts.

Safety practitioners have a very different focus. They tend to be concerned with how the design might "fail" by a component or user not performing as expected. In the latter case, the engineers often feel justified that the design did not "fail" at all; rather the foolish person using the design was the problem.

Each perspective is important. Without the designer's focus, a functional, working design will not result. Without the safety focus, foreseeable uses or misuses may be overlooked leading to problems after the design is in use. Design reviews offer the opportunity to make certain that an effective balance has been struck between the perspectives and resulting performance criteria.

Design Review Mechanics

Design reviews are typically conducted by a team. The members of the team will vary depending on the product or facility being designed, and the stage of the design in the development process. The design engineers and others intimately involved in creating the design should certainly be part of the review team. Specialists in safety, marketing, production, finance, quality, legal, and others should be included as appropriate. The team should be led by someone who is in a position and has the competence to take a balanced view of the process.

A design review evaluates a design against the design criteria or requirements. The review seeks to ensure that the design meets all the criteria or that trade-offs made with respect to the criteria are appropriate and necessary.

Beginning a design review without design specifications or benchmarks will almost certainly lead to frustration and lost time since issues will arise out of the review which requires further analysis and examination.

The design criteria or desired attributes are typically set by management, marketing, manufacturing, finance or other internal sources. Criteria can also come from outside sources such as standards, legal requirements, customers and others.

In conducting the design review the team will review the design and supporting analyses to reach a decision on the acceptability of the design relative to the criteria. Supporting analyses can include: structural analyses, failure mode and effects analysis (FMEA), checklists, financial, costs, compliance documents, and others. More recently, design reviews are being based on risk assessments. The design review employs a risk assessment to be certain that all hazards have been identified and that the risks associated with the hazards have been reduced to an acceptable level.

The Decision-Making Process

Design reviews integrate into a basic decision-making process. The general steps in any decision-making process are the following:

1. Identify the problem.
2. State the basic objective or goal.
3. State the constraints, assumptions, and facts.
4. Generate possible solutions.
5. Evaluate and make a decision.
6. Analyze.
7. Create a detailed solution.
8. Evaluate the solution.
9. Report the results and make recommendations.
10. Implement the decision.
11. Check the results.

In the design review process Step 8 indicates that the solution is to be evaluated. If the solution is found lacking, then the team begins to re-evaluate the constraints and generate new or differing possible solutions (returning to steps 3 and 4). If sufficient information is available at the review, the team can work through the rest of the steps to arrive at a recommendation. If additional information or analyses are needed, the team usually defers a decision and re-evaluates the design once the analyses have been completed.

In the context of safety standards, the analysis step can be straightforward when government regulations or industry standards apply. This type of evaluation is considered a compliance evaluation. The question to be answered is simply "does the design comply with the requirements of the standard?" A single design may have several standards that must be checked. In a compliance review, once the team is satisfied the design complies with the requirements of the standard and achieves acceptable risk the design progresses along the development or production process.

However, rarely are designs an exact copy of existing systems. Engineering design is a creative process that generates new and unique solutions to ever changing customer demands and desires. In many cases, industry standards do not exist to address the specific design being developed. In this case, industry standards may apply, but the new design may venture into areas not directly covered by the standard.

In situations where standards are less defined or considered a minimum that needs to be exceeded, a compliance evaluation is inappropriate. In these cases, a separate risk assessment or other safety analysis should be conducted to ensure that the design meets the required level of performance. A risk assessment will serve to identify opportunities for improvement and ensure the design under review reduces risks to an acceptable level.

EMPHASIZING CONSIDERATION OF THE WORK METHODS IN THE DESIGN PROCESS

Much has been made in this book of the need to design work methods and procedures so that they are not error-provocative or overly stressful. Considerations of hazards and risks in the design process should not be limited to the facility, equipment, and processes—i.e., to the hardware. They should also focus on the hazards and risks in the work methods prescribed, taking into consideration the

- capabilities and limitations of the workers so that the risks of injury and damage are at as low as reasonably practicable
- benefits of not creating work situations that are error-provocative.

In a paper titled "The Titanic and Risk Management," Roy Brander (2000) wrote this: "Safe design of the procedure is as important as design of the artifact" (Item 30). His point is important. It fits close to this author's observation. Too often, inadequate attention is given to avoiding hazards in the work methods, the result being that what workers are expected to do is inherently risky.

HOW SOME SAFETY PROFESSIONALS ARE ENGAGED IN THE DESIGN PROCESS

To obtain information on how safety professionals are involved in activities to avoid bringing hazards and risks into the workplace, a request for comments on the subject was made through an Internet safety server. These are some of the responses—the most unfortunate first. As they are reviewed, safety professionals may want to assess their place in the design process and look for hints on how they can improve their positions.

- From an industrial hygienist: If the engineers would only let me into the design process, I know that many of the health hazards I deal with could be better controlled, we wouldn't need to do so much testing, and our operations would be more productive because we would reduce the amount of time employees spend on testing and on the use of personal protective equipment.
- We don't have any forms or established procedures for our getting into what engineers are designing, and there is no formal method for engineering to notify us of a project. We do get copies of the engineering weekly reports, and we read those carefully. Then, we invite ourselves into the discussion.
- There was a lot of resistance by engineering to our getting into what they were designing because they had not recognized that we could be an asset. It took us a long time and quite a few money saving successes to get it ingrained into their procedures that it was to their benefit to refer capital expenditure requests to us for our input. We don't have a written procedure, but it gets done.
- Almost all of our engineering is done by outside firms. I haven't been successful in convincing the few engineers we have left that I can help them write specifications that will save them money. But I have convinced the manager of my plant that I am to have sign-off authority for new installations.

  My sign-off reviews can be embarrassing to the people who laid out the specs and I have to be diplomatic in how I do what I need to do. But the plant manager gives me support now. He has been educated.
- This is from a construction safety professional. I request to see all drawings at the 10% level while changes are still easily incorporated. I get a schedule of all construction plans and visit contractor lay down areas to inspect materials being used on the job. I attempt to educate quality assurance personnel on what to look for as safety indicators.

  I visit the engineering department and the contracting department on a daily basis when I'm in town. Yes, it is time consuming but it saves money and lives as well as equipment. It took maximum effort on my part to get where I am.
- In my company, it's in the capital expenditure procedure manual that all funding requests will receive a safety review. Managers have the same responsibility for safety as they do for productivity and quality, and when they approve the capital expenditure request, they are also signing off for safety.

  But my name appears on the capital expenditure distribution list among the management people who have to sign off as approving the request. This isn't as burdensome as it was in the beginning because I have educated a fairly stable engineering and management staff on what it takes to get my approval.
- Through our successes in ergonomics, engineering recognized our contributions not only for safety, but also for productivity. They now invite us into the design process in the idea stage. They make it plain that they look to us to see

that they don't mess up. We learned the hard way over a lot of years that it is expensive to correct hazards that are in the machines and equipment we buy. We had to do some costly equipment modifications for employee safety and health, and for environmental situations after the installations were completed and in operation. Our engineers aren't easy on us. But that's okay. We had to accept the criticism from them that our ergonomics checklists were so general that they weren't helpful.

- Our Facility Safety Manual includes extensive procedures for documented reviews for safety, health, environmental, and ergonomic before budget approval is obtained on a new project, and for equipment reviews before the equipment is released to normal production.

As you will see from what I'm mailing to you, we are deep into specifications, and the signature of a safety specialist is required in the project and budget review procedures.

Involvement of safety professionals in the design processes, specification writing, purchasing, and sign-off processes varies from nothing to being required by written procedures. In all but one of the cases cited, there is some involvement by safety professionals. And, that came about because safety professionals took the initiative and proved value.

A GOOD PLACE TO START: ERGONOMICS

Safety professionals who are not involved in the design processes should consider ergonomics as fertile ground in which to get started. It is well established that successfully applied ergonomics initiatives may result not only in risk reduction but also in improved productivity, lower costs, and waste reduction.

Furthermore, musculoskeletal injuries are a large segment of the spectrum of injuries and illnesses in all organizations. Since they are costly, reducing their frequency and severity will show notable results. Because of the prominence of musculoskeletal injuries in most all organizations, it is difficult to understand why a safety professional is not involved in ergonomics.

A SAFETY DESIGN REVIEW AND OPERATIONS REQUIREMENTS GUIDE

A composite is provided here of the procedures in place in three companies for design reviews and for a safety sign-off before new or modified equipment can be released into normal operations. In a way, these requirements represent culture statements. Managements have decided that hazards and risks are to be dealt with as equipment is designed, and before it can be placed in operation.

This composite presents a basis for thought as safety professional pursue having similar concepts and procedures adopted. It should not be adopted without changes to suit the culture and management systems in place.

REQUIREMENTS: EQUIPMENT AND PROCESS DESIGN SAFETY REVIEWS

A. Purpose

To establish procedures to ensure that hazards are analyzed and that risks are at an acceptable level when considering new, redesigned, and relocated equipment or processes.

B. Scope

These procedures apply to all equipment and processes that may present risks of injury to people or damage to property or the environment. They pertain to the design or redesign of all new, transferred, and relocated equipment and processes. For all aspects of these procedures, documentation shall be appropriate to the activity.

Safety, as the term is used here, encompasses risks of injury or damage to personnel (employees and the public), property, and the environment.

C. Responsibilities

The Location Manager. The responsibility for safety rests with the location manager in the operations for which he or she has authority.

Project Manager. The project manager is responsible for assuring that

- Corporate safety requirements are met
- Safety documentation to accompany capital expenditure requests is prepared
- Preliminary and subsequent design safety reviews are conducted
- Appropriate coordination and communication take place with outside design and engineering firms to assure that specifications are met
- Proper consideration is given to any safety problems identified by the staff during their visits to vendors and design and engineering firms prior to delivery of equipment

Design Engineers. Principle responsibility of the design engineers is to design inherently safer equipment and processes. Whether employees or contractors, design engineers shall assure that the considerations necessary for safety have been given in the design process. They will provide the Project Manager and the Safety Review Team documentation including:

- Detailed equipment design drawings
- Equipment installation, operation, preventive maintenance and test instructions

- Details of and documentation for codes and design specifications
- Requirements and information needed to establish regulatory permitting and/or registrations.

Safety Review Team. This team will conduct preliminary and subsequent design safety reviews for equipment and processes, or have design reviews made by outside consultants for particular needs. In addition to the Project Manager, members will include the project design engineers, production and maintenance personnel, the facilities engineer, selected disinterested engineers, the safety professional, and others (financial, purchasing) as needed. The Safety Review Team will also be responsible for:

- Arranging and conducting safety walk-downs of new projects
- Determining when visits are to be made at vendor design and engineering locations, and selecting the personnel with the necessary skills to make the visit

Safety Professional. The safety professional will:

- Serve as a member of the Safety Review Team and assist in identifying and evaluating hazards in the design process and provide counsel as to their avoidance, elimination, or control
- Visit design engineering firms and other vendors, when so requested by the Safety Review Team, to assure that safety problems are identified and corrected prior to shipment of equipment
- Be a signatory on Equipment Acceptance-Safety Review Form (Table 1) prior to newly installed or altered equipment or processes being released to normal production.

Department Managers and Supervisory Personnel. Department managers and supervisors will give support to the project manager for the activities that come under their jurisdiction. They will be signatories, along with engineering personnel and the safety professional, on sign-off forms before newly installed or altered equipment is released for normal operation.

D. Initial Capital Expenditure Safety Review

Capital expenditure requests at financial levels requiring divisional or corporate approval are to be accompanied by a Preliminary Safety Review Form, which is to be completed by the Safety Review Team. Comments would be included in the form, giving assurance that the hazards and risks identified can be properly dealt with.

For new or altered equipment or processes at a financial level not requiring divisional or corporate approval, formal Preliminary Safety Reviews are to be made at the discretion of the Safety Review Team.

E. Design Reviews

When designs have been completed and drawings and specifications are available, the Safety Review Team will hold one or more design review meetings to:

- Identify hazards not given appropriate attention, and recommend solutions to attain acceptable risk levels
- Assure that corporate safety requirements are being met
- Avoid the cost of risk reduction retrofitting as the project moves forward
- Assure compliance with applicable regulations, codes and standards

F. Safety Walk-Downs

When a project reaches completion at approximately a 70 % level, the Safety Review Team will arrange and conduct a safety-walk down to provide an opportunity to:

- Assure that specifications have been met
- Determine that hazards identified in the preliminary safety review and the subsequent design review have been properly addressed
- Identify hazards that may have inadvertently been built into the project, and to arrange for the necessary action to be taken

    Similarly, a final safety-walk down will be arranged as the project nears completion.

G. Equipment and Process Release Requirements

For newly purchased, redesigned, or relocated equipment or processes—before release for normal production:

- Task analyses shall be made to identify hazards and risks, and the hazards identified are to be properly dealt with
- Any revisions necessary in the written job procedures shall be made
- Retraining as required shall be given
- An Equipment Acceptance-Safety Review Form shall be completed, the signatories to which shall be the engineer-in-charge, the department manager or supervisor, and the safety professional

H. Design Reviews at Vendor and Design Engineering Locations

When considered advantageous by the Safety Review Team, arrangements will be made for personnel with the appropriate skills to visit vendors and design engineering firms to:

- Assure that vendors are building equipment to specifications
- Determine whether hazards exist that were not identified in the design review process, or by the vendor or design engineering firm that need attention
- Avoid the high cost of retrofitting for safety matters during installation, testing, and debugging

For these reviews, a specifically tailored Vendor/Design Engineering Review Form is to be created from sections of the General Design Safety Checklist that relate to the task. Reports are to be made available to the Safety Review Team and the Project Manager.

EQUIPMENT ACCEPTANCE - SAFETY REVIEW FORMS

Drafting equipment acceptance-safety review forms that are specific to every piece of equipment or process would be a mammoth undertaking. Nevertheless, industry-specific safety review checklists do exist and they should be used where applicable.

The example of an equipment acceptance review form shown in Figure 12.1 assumes the existence of a General Design Safety Checklist that can serve as a reference base for the review process. Also, the example presumes that there will be two levels of review: a preliminary review to identify items needing attention; and a final sign-off.

PRELIMINARY SAFETY DESIGN REVIEW FORMS

A Preliminary Safety Design Review goes by several names. In one company, its purpose is to meet "Fitness for Use Criteria." In another company, it is referred to as a "Hazards Screening Analysis." A Preliminary Safety Review Form is a listing of subjects pertaining to equipment, facilities, or processes that aids the reviewers in identifying hazards and risks that must be addressed. Completion of the form produces, in effect, an early design review. The review team indicates whether a subject needs further consideration.

No one list can be suitable for all needs. In drafting such a list, a safety professional should use a General Design Safety Checklist as a reference and include some, all, or more than the subjects listed in Table 12.1.

PRELIMINARY SAFETY DESIGN REVIEW FORMS

Equipment Acceptance – Safety Review Form

Dept._____ Control No. _____

Equipment Description_____

_____

This form must be completed prior to equipment being released to normal production. It is applicable to:

1. All newly installed equipment or processes
2. Changes made in the use of equipment or processes
3. Modifications of existing equipment or processes

Preliminary approval indicates that the equipment is ready for initial production trials, but needs additional work for safety as listed in a memo attachment titled "Safety items needing attention." Final approval indicates that the preliminary findings have been addressed satisfactorily and that the equipment can be released to normal production.

	Preliminary	Final
Signed:_____		
Engineer-in-charge	Date	Date
Signed:_____		
Dept. Mgr. or Supervisor.	Date	Date
Signed:_____		
Safety Manager	Date	Date

The original copy of this form shall be retained by the Department Manager or Supervisor. The Engineerin-charge and the Safety Manager will retain copies. The company's file retention policy shall apply.

FIGURE 12.1 Safety design reviews.

TABLE 12.1 Topics to Be Considered: Preliminary Design Safety Review

Ergonomics	Material handling
Machine guarding	Illumination needs
Fire protection	Means of egress
Walking and working surfaces	Confined spaces
Use of hoists, cranes, etc.	Work at heights
Electrical potentials	Lockout/tagout
Confined spaces	Temperature extremes
Hazardous or toxic materials	Noise or vibration
Personal protective equipment	Non-ionization emitters
Environmental concerns	Sanitation

GENERAL DESIGN SAFETY CHECKLIST

A detailed, specifically referenced design checklist covering all workplace safety needs would fill thousands of pages. The design safety checklist presented here is a brief composite taken from several sources. Some safety professionals will view it as excessive: others will find that it does not address all their needs. Those who use it as a reference should be aware that there are many subject-specific and industry-specific checklists to which they should also refer. For example:

- The *Guidelines for Hazard Evaluation Procedures, Second Edition* (1992), issued by the Center for Chemical Process Safety, includes a checklist-questionnaire for chemical operations that fills forty-five pages.
- The checklist in ISO 14121, *Safety of Machinery—Principles of risk assessment* (2003), a standard issued by the International Organization for Standardization, serves as a guide for those who design and manufacture equipment and machinery that goes into European workplaces.

This General Design Safety Checklist is intentionally presented as a list of questions without the boxes and lines for separation that are typically found in checklists. Also, the response that would be placed to the right of a design checklist in which users would enter check marks for "yes, no, or not applicable" have been eliminated.

GENERAL DESIGN SAFETY CHECKLIST

Preface

This checklist begins with a preface that brings attention to Haddon's unwanted energy release theory. Haddon (1970) stated in his paper "On the Escape of Tigers: An Ecologic Note" that "the concern here is the reduction of damage produced by energy transfer". But he also said that "the type of categorization here is similar to those useful for dealing systematically with other environmental problems and their ecology".

Excerpts follow from "On the Escape of Tigers: An Ecologic Note" as the article appeared in the American Journal of Public Health. (p. 2229) Note that all of the strategies have to do with facility design or work methods design, except the last two which pertain after the initiation or occurrence of an event.

The questions in section **A. Introduction: Basic Considerations** relate to Haddon's theory and are presented as general concepts to be considered when using the checklist. They emphasize that the two distinct aspects of risk are to be considered in the design process:

- Avoiding, eliminating, or reducing the *probability* of a hazard-related incident or exposure occurring
- Reducing the *severity* of harm or damage if an incident or exposure occurs.

A. Introduction: Basic Considerations

1. Prevent the marshaling of the form of energy.
2. Reduce the amount of energy marshaled.
3. Prevent the release of the energy.
4. Modify the rate or spatial distribution of release of the energy from its source.
5. Separate, in space or time, the energy being released from that which is susceptible to harm or damage.
6. Separate, by interposing a material barrier, the energy released from that which is susceptible to harm or damage.
7. Modify appropriately the contact surface, subsurface, or basic structure, as in eliminating, rounding, and softening corners, edges, and points with which people can, and therefore sooner or later do, come in contact.
8. Strengthen the structure, living or nonliving, that might otherwise be damaged by the energy transfer.
9. Move rapidly in detection and evaluation of damage that has occurred or is occurring, and counter its continuation or extension.
10. After the emergency period following the damaging energy exchange, stabilize the process.

B. Designing for Those with Disabilities

1. Do the designs take into consideration the requirements of the Americans With Disabilities Act (ADA)?
2. Are reasonable accommodations made for the disabled?

C. Confined Spaces

1. Have confined spaces been eliminated by design where practicable?
2. Are any confined spaces to be permit required? 1910.146(c)(1)
3. Have confined spaces been designed for easy of ingress, prompt egress, and where practicable, elimination of hazardous atmospheres?
4. Can confined spaces be designed with multiple, large accesses?
5. Are accesses provided with platforms that will support all required personnel and equipment?
6. Will access ports be large enough to permit entry when personnel are using personal protective equipment?
7. Will pipes or ducts limit entry to access ports?
8. Are locations of ladders and scaffolds in the space identified?
9. Are fall protection needs fulfilled (such as anchorage points)?
10. Can the necessary equipment be moved through accesses?
11. Does the design provide for isolation of the confined space from hazardous energy (i.e., electrical, chemical, etc.)?

268   SAFETY DESIGN REVIEWS

12. Does the design provide for isolation by valve blocking, spools, double blocks and bleeds, flanges, and flushing connections?
13. Can spaces be designed so that maintenance and inspection can be performed from outside or by self-cleaning systems?

D. Electrical Safety

1. Overall, will the electrical system meet OSHA/NEC standards?
2. Will the system be sufficiently flexible to allow for future expansion?
3. Will emergency power be provided for critical systems?
4. Is grounding adequate?
5. Are Ground Fault Interrupter Circuits to be installed where needed? 1910.304(f)(7)
6. Are grounding connections to piping and conduits eliminated to prevent accumulation of static electricity?
7. Is grounding provided for lightning protection on all structures?
8. Are accommodations made for special purpose or hazardous locations? 1910.307
9. Is the design adequate where there may be combustible gases or vapors?
10. Is high voltage equipment isolated by enclosures such as vaults, security fences, lockable doors and gates?
11. Are non-isolated conductors such as bus bars on switchboards or high voltage equipment connections that are located in accessible areas protected to minimize hazards for maintenance and inspection personnel?
12. Where injury to an operator may occur if motors were to restart after power failure, are provisions made to prevent automatic restarting upon restoration of power? 1910.262(c)(1)
13. Are electrical disconnect switches lockable, readily accessible, and labeled? 1910.303(f)
14. Are breakers/fuses properly sized? 1910.303(b)
15. Has the polarity of all circuits been checked? 1910.403(a)(2)
16. Do electrical cabinets and boxes have appropriate clearances? 1910.303(g) & (h)
17. Are exposed live electrical parts operating at 50 volts or more guarded against accidental contact by approved cabinets or enclosures, by location, or by limiting access to qualified persons? 1910.303(g)(2)(i)
18. Are rooms or enclosures containing live parts or conductors operating at over 600 volts, nominal, designed to be kept locked, or provisions made to be under the observation of a qualified person at all times? 1910.303(h)(2)
19. Are the electrical wiring and equipment located in hazardous (classified) locations intrinsically safe, approved for the hazardous location, or safe for the hazardous location? 1910.307(b)(1-3)

GENERAL DESIGN SAFETY CHECKLIST

E. Emergency Safety Systems—Means of Egress

1. Are means of egress adequate in number, remote from each other, properly designated, marked, lighted, and easily recognized?
2. Has emergency lighting been provided for means of egress, and elsewhere where needed?
3. Does the design contemplate emergency lighting where workers may have to remain to shut down equipment?
4. Do means of egress exit directly to the street or open space?
5. Do doors, passageways, or stairways that do not lead to an exit be marked by signs reading "not an exit" or by a sign indicating actual use?
6. Does the design provide internal refuge areas for workers who cannot escape?
7. Will reliable emergency power be provided for critical and life support systems?
8. Will emergency safety showers and eye wash stations be adequate, and properly placed?
9. Will adequate first aid stations, spill carts, and emergency stations be provided?

F. Environmental Considerations (Some Are Operational, Beyond Design)

1. Have waste products been identified and a means of disposal established?
2. Will provisions be made for responding to chemical spills (containment, cleanup, disposal)?
3. Is there an existing spill control plan for chemicals?
4. Have all waste streams been identified?
5. Are adequate pre-treatment facilities provided for process waste streams?
6. Will an adequate storage area be available for wastes held prior to treatment or disposal?
7. Will waste storage areas have adequate isolation, or containment for spills?
8. Will hazardous wastes be disposed of at approved treatment, storage, and disposal facilities?
9. Is special equipment or specially trained personnel provided for treatment operations?
10. Has the acquisition of permits been addressed for the treatment or disposal of waste streams?
11. Have state or local requirements for permitting been evaluated and factored into the project?
12. Can the facility meet regulations for reporting spills or the storage of chemicals?
13. Have adequate provisions been made for cleaning the process equipment?

14. Have provisions been made for a catastrophic release of chemicals?
15. Have provisions been made for any necessary demolition and the resulting waste?
16. Have requirements for remediation at the site prior to construction been addressed?
17. Will all practicable measures for waste minimization be implemented?
18. Have the processes that generate air pollution been evaluated for minimization potential?
19. Will adequate air pollution controls be installed (scrubbers, fume hoods, dust collectors)?
20. Have handling and cleaning of air pollution control systems been addressed?
21. Have the processes that generate wastewater been evaluated for reduction potential?
22. Will indoor spills be protected from reaching drains?
23. Will outdoor spills be protected from reaching storm water drains and sewer manholes?
24. Are adequate water disposal systems available?
25. Will pretreatment methods be necessary and provided?
26. Will the discharges of domestic and industrial wastewater be in accord with regulations?

G. Ergonomics—Work Station and Work Methods Design

1. Generally, have material handling designs considered worker capabilities and limitations, to accommodate the employee population at the 95% level?
2. Do material handling designs promote the use of mechanical material handling equipment, such as conveyors, cranes, hoists, scissor jacks, and drum carts?
3. Do design layouts reduce in so far is practicable:
   a. constant lifting
   b. twisting and turning of the back when moving an object
   c. crouching, crawling, and kneeling
   d. lifting objects from floor level
   e. static muscle loading
   f. finger pinch grips
   g. work with elbows raised above waist level
   h. twisting motions of hands, wrists, or elbows
   i. hyper-extension or hyper-flexion of wrists
   j. repetitive motion
   k. awkward postures

4. Are work stations designed to provide:
   a. adequate support for the back and legs?
   b. adjustable work surfaces that are easily manipulated?
   c. delivery bins and tables to accommodate height and reach limitations?
   d. work platforms that elevate and descend, as needed?
   e. powered assists and suspension devices to reduce the use of force?
5. Has adequate attention been given to:
   a. lighting (to Illuminating Engineering Society requirements),
   b. heat,
   c. cold,
   d. noise, and
   e. vibration?
6. Does the design accommodate the hazards inherent in servicing, maintenance, and inspection?
7. Will there be adequate clearance and ready access to equipment for servicing?
8. Will controls be efficiently located in a logical and sequential order?
9. Will indicators be easy to read, either by themselves, or in combination with others?

H. Fall Avoidance

1. Overall, has the design reduced the need for ladders and stairs in so far is practicable?
2. Where works at heights is to be done, has adequate consideration been given to providing work platforms or fixed ladders?
3. Are parapets or guardrails provided at roof edges?
4. Is equipment designed to avoid fall hazards during maintenance, inspection and cleaning?
5. Does the design provide for fall arrest measures, such as anchorage points and fall restraining systems?

I. Fire Protection

1. Overall, in the design, will national and local fire codes and insurance requirements be met?
2. Will fire pumps, water tanks/ponds, and fire hydrants be adequate?
3. Will risers and post valves be accessible, and protected from damage?
4. Will small hose standpipes be adequate?
5. Will sufficient hose racks be provided?
6. Will special fire suppression systems be provided?

7. Has containment of fire suppression water been addressed?
8. Will there be adequate external fire zones?
9. Will emergency vehicle access be adequate?
10. Will flame arresters be installed where needed on equipment vents?
11. Will fire extinguishers be of appropriate types, adequate, and mounted for easy access?
12. Will the design for location of flammables be appropriate?
13. For flammables, will storage rooms and cabinets meet national fire codes and insurance requirements?
14. For flammable liquid dispensing, will grounding, bonding and ventilation be adequate?
15. Will fire sensors, pull stations and alarms be adequate?
16. Are flooding systems designed to provide a pre-discharge alarm which can be perceived above ambient light or noise levels before the system discharges, giving workers time to exit from the discharge area?
17. Has the project been reviewed by insurance personnel?

J. Hazardous and Toxic Materials

1. In the design process, have all materials in this category been identified?
2. Have the physical properties of the individual chemicals been identified?
3. Have the most conservative exposure limits been established as the design criteria?
4. Has a determination been made to use intrinsically safe equipment?
5. Have Material Safety Data Sheets been obtained for all materials?
6. Are the reactive properties known for chemicals that will be combined or mixed?
7. Have measures been taken to eliminate, substitute for, or reduce the quantities of hazardous chemicals?
8. Does the design emphasize closed process systems?
9. Will the design properly address all occupational illness potentials, and reduce the need for monitoring, testing, and personal protective equipment in so far as is practicable?
10. Are storage facilities designed to separate hazardous from non-hazardous substances?
11. Does the design consider the chemical compatibility issues?
12. Have adequate provisions been made for chemical release, fire, explosion, or reaction?
13. Have provisions been made to contain water used in hazardous release control?
14. Are ventilation systems adequate to handle an emergency release?

15. Is the storage of hazardous chemicals below ground avoided?
16. Is storage tank location such as to reduce facility damage or damage to the public in a catastrophic event in so far as is practicable?
17. Are adequate storage tank dikes provided?
18. Will emergency ventilation be provided for accidental releases?
19. For extraordinary releases, will special ventilation, relief, deluge systems be provided?
20. Will the normal use of chemicals allow operating without personal protective equipment?
21. Will the design of bulk loading/unloading facilities contain anticipated leaks and spills?

K. Lockout/Tagout—Energy Controls

1. In the design process, has adequate attention been given to lockout/tagout requirements to prevent hazardous releases from these energy sources:
   a. electrical,
   b. mechanical,
   c. hydraulic,
   d. pneumatic,
   e. chemical,
   f. thermal,
   g. non-ionizing radiation, or
   h. ionizing radiation?
2. Are lockout/tagout devices adequate in design and number, readily accessible and operable?
3. Are lockout/tagout devices standard throughout the facility?

L. Machine Guarding

1. Overall, do the designs prevent workers hands, arms, and other body parts from making contact with dangerous moving parts? 1910.212(a)(3)
2. Have the requirements of all applicable machine guarding standards of the American National Standards Institute been identified and met?
3. Are safeguards firmly secured and not easily removed? 1910.212(a)(2)
4. Do safeguards ensure that no object will fall into moving parts? 1910(a)(1)
5. Do safeguards permit safe, comfortable and relatively easy operation of the machine? 1910.212(a)(2)
6. Can machine be lubricated without removing safeguards? 1910.212(a)(2)
7. Does the design include a system that requires shutting down machinery before safeguards are removed?
8. Are fixed machines soundly anchored?

274 SAFETY DESIGN REVIEWS

9. Are in-running nip points properly guarded? 1910.212(a)(1)
10. Will the design properly address point-of-operation exposure? 1910.212(a)(3)
11. Are all reciprocating parts properly guarded? 1910.212(a)(3)(iv)
12. Are all rotating parts properly guarded? 1910.212(a)(3)(iv)
13. Are all shear points properly guarded? 1910.212(a)(3)(iv)
14. Are exposed set screws, keyways, collars, etc. properly guarded? 1910.212(a)(3)(iv)
15. Does the design eliminate the potential for flying chips? 1910.212(a)(i)
16. Has the potential for any sparking been eliminated? 1910.212(a)(i)
17. If robots are to be used, are they designed to ANSI/RIA R15.06-1999: The American National Standard for Industrial Robots and Robot Systems—Safety Requirements?

M. Noise Control

1. In the design process, have maximum noise levels been established that are to be stipulated in specifications for new equipment?
2. Is emphasis given to controlling noise levels through engineering measures?
3. Are the size or shape of rooms and proposed layout of equipment, work stations, and break areas to be evaluated for noise levels?
4. Will workers be separated from noise by the greatest practicable distance?
5. Will barriers be installed between noise sources and workers?
6. Are enclosed control rooms to be provided for operators in areas where the noise is above trigger levels?
7. Are lower noise level processes to be selected, where feasible?
8. Have equipment and work stations been located so that the greatest sources of noise are not facing operators?

N. Pressure Vessels

1. Will all pressure vessels be designed to ASME and insurance company requirements?
2. Will pressure vessels containing flammables or combustibles meet OSHA 1910.106 and NFC standards?
3. Will pressure relief valves be:
   a. correctly sized and set,
   b. suitable for intended use, and
   c. directed to discharge safely?

O. Ventilation

1. Have all sources of emission been identified and their hazards characterized?
2. Have ways to reduce personnel interaction with the emission sources (location, work practices) been considered in the design process?
3. Have assessments been made with respect to incompatible emission streams (cyanides and acids, etc.)?
4. Has consideration been given to weather conditions and seasonal variations?
5. Will the design requirements of the ANSI Z9 series, the ACGIH Ventilation Manual, ASHRAE guidelines, and NFPA 45 and 90 met?
6. Will local ventilation effectively capture contaminates at the point of discharge?
7. Will room static pressures be progressively more negative as the operation becomes "dirtier"?
8. Will ventilation system provide a margin of safety if a system fails?
9. Will emergency power and lighting be provided on critical units?
10. Will the ventilation equipment be remote and/or "quiet"?
11. Will spray booths and degreasers meet OSHA standards?
12. Will laboratory or contaminated air be totally exhausted?
13. If contaminated air is cleaned and reused, will it meet good safety requirements?
14. Will the makeup air to hoods be clean and adequate?
15. Have flow patterns been established to prevent exposure to personnel?
16. Does the design provide for proper gauging and alarm systems with respect to a sudden pressure drop?
17. Are ventilation controls easily accessible to operators?

P. Walking and Working Surfaces, Floor and Wall Openings, Fixed Stairs and Ladders

1. Will aisles, loading docks, and through doorways have enough clearance to allow safe turns where material handling equipment is used? 1910.22(b)(1)
2. In the aisles, are people and vehicles adequately separated?
3. Are permanent aisles to be marked with lines on the floor? 1910.22(b)(2)
4. Does the design provide for floors, aisles, and passageways being free from obstruction? 1910.22(b)(1)
5. Has a logistics study been made to provide safe and efficient flow of people and materials?

6. Will the construction texture of walking surfaces be non-slip?
7. Will the floors be designed to stay dry?
8. Will water and process flows be designed to keep off the walkway?
9. Will the floors be sloped and drained?
10. Will utilities and other obstructions be routed off the walking surfaces?
11. Will the design allow future utility expansion, with added facilities not having to be above and thereby cross floors, and be obstructive?
12. Will designs for floor and wall openings meet the requirements of OSHA 1910.23?
13. Do the designs for fixed stairs, and ladders, meet the requirements of OSHA 1910.23, .24, .27?

CONCLUSION

I recommend strongly that safety professionals move toward including the design review element in Z10.0 into the safety and health management systems they influence. I stand by the premises that

- Hazards and the risks that derive from them are most effectively and economically avoided, eliminated, reduced, or controlled if they are considered early in the design process, and where necessary, as the design progresses
- Risks of injury, illness, or damage are significantly reduced if processes are in place to avoid bringing hazards and risks into the workplace.

REFERENCES

ANSI/ASSE Z590.3-2011 (R2016). *Prevention through Design: Guidelines for Addressing Occupational Hazards and Risks in Design and Redesign Processes.* Park Ridge, IL: American Society of Safety Professionals, 2011 (R2016).

ANSI/ASSP Z10.0-2019. *Occupational Health and Safety Management Systems.* Park Ridge, IL: American Society of Safety Professionals, 2019.

ANSI/ASSP/ISO 45001-2018. *Occupational Health and Safety Management Systems—Requirements with Guidance for Use.* Park Ridge, IL: American Society of Safety Professionals, 2018.

Brander, R. *The Titanic and Risk Management.* Available at http://axion.physics.ubc.ca/titanic/ (Item 30). Accessed June 1, 2019. 2000.

Center for Chemical Process Safety. *Guidelines for Hazard Evaluation Procedures,* Second Edition. New York: Center for Chemical Process Safety of the American Institute of Chemical Engineers, 1992.

Christensen, W.C. and F.A. Manuele, Editors. *Safety Through Design.* Itasca, IL: National Safety Council, 1999.

Haddon, W. "On the Escape of Tigers: An Ecologic Note." Technical Review, 1970.

ISO 14121. Safety of Machinery—Principles for Risk Assessment. Geneva, Switzerland: International Organization for Standardization, 2003.

Main, B.W. *Risk Assessment: Basics and Benchmarks.* Ann Arbor, MI: Design Safety Engineering, Inc., 2004.

OSHA. Occupational Safety and Health Administration Standard 1910. Available at http://www.osha.gov/pls/oshaweb/owastand.display_standard_group?p_toc_level=1&p_part_number=1910. Accessed June 1, 2019. 1999.

CHAPTER 13

PREVENTION THROUGH DESIGN

Several provisions in Z10.0 and 45001 relate to prevention through design (PtD) concepts. They are Risk Assessments, Design Requirements, Hierarchy of Controls, and Procurement.

Fortunately, an American National Standard exists that provides guidance on the subject. Its designation is ANSI/ASSE Z590.3-2011(R2016) and its title is *Prevention through Design: Guidelines for Addressing Occupational Hazards and Risks in Design and Redesign Processes.*

This author was privileged to be the chair of the committee that wrote the Z590.3 standard. It will be obvious that what is written here shows a bias in favor of its implementation. This chapter

- Supports application of prevention through design concepts.
- Gives a history of the safety through design/prevention through design movement.
- Comments on the activities at the National Institute for Occupational Safety and Health (NIOSH) on prevention through design—which are extensive.
- Presents highlights of Z590.3—emphasizing the applicability of the provisions of the standard to all hazards-based initiatives—environmental controls, product safety, safety of the public, avoiding property damage, and business interruption, among others.

Advanced Safety Management: Focusing on Z10.0, 45001, and Serious Injury Prevention,
Third Edition. Fred A. Manuele.
© 2020 John Wiley & Sons, Inc. Published 2020 by John Wiley & Sons, Inc.

- Discusses the culture change necessary in most organizations as prevention through design concepts are implemented within an Operational Risk Management System.
- Encourages safety professionals to become involved in prevention through design for job satisfaction and to be perceived as providing additional value in the organizations to which they give counsel.

SUPPORTING APPLICATION OF PREVENTION THROUGH DESIGN CONCEPTS

Applying prevention through design concepts—addressing hazards and risks in the design and redesign processes—is supported by the following statements.

1. Hazards and risks are most effectively and economically avoided, eliminated, or controlled in the design and redesign processes.
2. Hazard analysis is the most important safety process in that, if that fails, all other processes are likely to be ineffective (Johnson, p. 245).
3. Risk assessment should be the cornerstone of an operational risk management system.
4. Risk assessments should also be used to identify the potential for serious injuries and illnesses and fatalities occurring.
5. Achieving superior results in the prevention of serious injuries, illnesses, and fatalities requires identifying the potential for their occurrence by addressing hazards and risks in the design and redesign processes.
6. If, through the hazard identification and analysis and risk assessment processes, specifications are developed that are applied in the procurement process so as to avoid bringing hazards and their accompanying risks into a workplace, the potential for incidents that may result in harm or damage is reduced greatly.
7. In Chapter 6, "A Socio-Technical Model for An Operational Risk Management System" was presented that gives significance to applying prevention through design concepts.

If there are no hazards, there will be no potential for harm. Thus, Johnson (1980) wrote appropriately that hazard analysis is the most important safety process. Since all risks in an operational setting derive from hazards and since the intent of an operational risk management system is to achieve acceptable risk levels, it follows that risk assessment should be the cornerstone of an operational risk management system.

Figure 13.1 depicts the theoretical ideal. Prevention through design is moved upstream in the design process.

Ideally, all hazards and risks would be considered in the Conceptual and Design steps. But that requires unattainable perfection from the people involved. Hazards

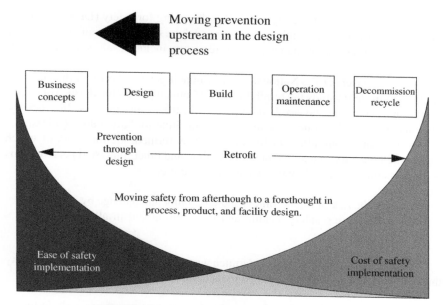

FIGURE 13.1 Prevention through design—depicted.

and risks will also be identified in the Build, Operation, and Maintenance steps for which Redesign is necessary in a retrofitting process.

HISTORY

In the early 1990s, several safety professionals recognized through their studies of investigation reports for occupational injuries and illnesses that design causal factors were not adequately addressed. For example, a study made by the author at that time indicated that although there were implications of workplace and work methods design inadequacies in over 35% of the investigation reports analyzed, the corrective actions proposed did not relate to the design implications. That study is supported by a later analysis made in Australia.

In *Guidance On The Principles Of Safe Design For Work,* issued in 2006, comments are made on the "contribution that the design of machinery and equipment has on the incidence of fatalities and injuries in Australia." They say:

> Of the 210 identified workplace fatalities, 77 (37%) definitely or probably had design-related issues involved. Design contributes to at least 30% of work-related serious non-fatal injuries. (p. 6)

To provide the needed education for designers, safety professionals are encouraged to develop their own supportive data on incidents in which design shortcomings are identified. That initiative should be followed by a major effort to have Z590.3—2011 be accepted as a design guide.

It was also noted in the early 1990s that designing for safety (DFS) was inadequately addressed in safety-related literature, and the safety and health management systems that organizations had in place infrequently included safety through design procedures.

Decision-makers at the National Safety Council gave authority to the author to form a committee to study the feasibility of the Council promoting the idea of having safety and environmental needs incorporated in the design process. In 1995, the outcome was that the Council established the Institute for Safety through Design.

An Advisory Committee for the Institute was formed, the members of which represented industry, academia, organized labor, and other interested persons. This is the Mission established by the Advisory Committee.

> To reduce the risk of injury, illness and environmental damage by integrating decisions affecting safety, health, and the environment in all stages of the design process.

A message that the Advisory Committee wanted to convey when the term Safety through Design was used is contained in its definition.

> The integration of hazard analysis and risk assessment methods early in the design and engineering stages and taking the actions necessary so that risks of injury or damage are at an acceptable level.

In the literature developed by the Institute, it was said that the following benefits would be obtained by applying safety through design concepts:

- Improved productivity
- Decreased operating costs
- Significant risk reduction
- Avoiding expensive retrofitting

Two groups were identified by the Institute to be given primary attention: Academia—professors and their students; and practicing engineers in industry and safety professionals.

Much was accomplished by the Institute. Seminars, workshops, and symposia were held. Proceedings were issued. Presentations were made at safety conferences, and a book titled *Safety Through Design* (1999) was published. Operations of the Institute were discontinued in 2005, in accord with a previously established sunset provision.

In 2006, several of the participants in the activities of the Institute for Safety through Design, and others, received an email from an executive at the National Institute for Occupational Safety and Health (NIOSH) encouraging our participation in an initiative for prevention through design. In July 2007, NIOSH held a workshop to obtain the views of a variety of stakeholders on a major initiative to "create a sustainable national strategy for Prevention through Design."

HISTORY   **283**

Some of the participants expressed the view that the long-term impact of the NIOSH initiative could be "transformative," meaning that a fundamental shift could occur in the practice of safety resulting in greater emphasis being given to the higher and more effective decision levels in the hierarchy of controls.

In 2008, NIOSH announced that one of its major initiatives was to "Develop and approve a broad, generic voluntary consensus standard on Prevention through Design that is aligned with international design activities and practice." The author volunteered to lead that endeavor. Support was obtained from the Standards Development Committee at the American Society of Safety Engineers (ASSE). It was decided to develop a Technical Report first for the learning experience that would be provided. So TR-Z790.001—Prevention through Design: an ASSE Technical Report was issued in 2009—which has been replaced, as follows.

On September 1, 2011, the American National Standards Institute (ANSI) approved the standard *ANSI/ASSE Z590.3-2011 Prevention through Design: Guidelines for Addressing Occupational Hazards and Risks in the Design and Redesign Processes*.

It must be understood that activities at NIOSH are limited to occupational safety and health. Thus, the focus of Z590.3 is work-related. But, by intent, the terminology in Z590.3 was kept broad enough so that the Guidelines could be applicable to all hazard-based fields—product safety, environmental control, property damage that could result in business interruption, among other things. In the standard, the definition of prevention through design is work-related and identical with that in the NIOSH literature.

Prevention through Design. Addressing occupational safety and health needs in the design and redesign process to prevent or minimize the work-related hazards and risks, associated with the construction, manufacture, use, maintenance, retrofitting, and disposal of facilities, processes, materials, and equipment.

NIOSH has continued its efforts to promote prevention through design. A NIOSH eNews bulletin issued in April 2019, the subject is "NIOSH Prevention through Design (PtD) Update." It came from the desk of John Howard, MD—Director, NIOSH. These are the last three paragraphs.

The continued and growing interest in prevention through design is both encouraging and full of opportunity, especially as the advantages of designing-out hazards become even more apparent with effective workplace design solutions.

There is still much to do to incorporate "prevention through design thinking" into research, practice, policy, and education. NIOSH has instituted prevention through design as one of its research programs to ensure continued focus on it. Later this year, we will introduce a "NIOSH Prevention through Design Award" to recognize successful promotion or utilization of prevention through design.

I extend my thanks to the many dedicated professionals who share responsibility for the advances made to date. We look forward to many continued collaborations and advances together.

Promoting the acquisition of knowledge of Safety through Design/Prevention through Design concepts is in concert with the ASSE Position Paper on "Designing for Safety" approved by its Board of Directors in 1994, the opening paragraph of which follows;

Designing For Safety (DFS) is a principle for design planning for new facilities, equipment, and operations (public and private) to conserve human and natural resources, and thereby protect people, property and the environment. DFS advocates systematic process to ensure state-of-the-art engineering and management principles are used and incorporated into the design of facilities and overall operations to assure safety and health of workers, as well as protection of the environment and compliance with current codes and standards.

A REVIEW OF ACTIVITIES AT NIOSH ON PREVENTION THROUGH DESIGN

Jonathan A. Bach, PE, CSP, CIH, was asked to provide comments on the activities on Prevention through Design at NIOSH. He has prepared an extensive review of those activities which follows, with his permission.

Jonathan A. Bach on Prevention through Design at NIOSH

Prevention through Design (PtD) became an Initiative by the National Institute for Occupational Safety and Health in 2007. Strategic partners included the American Industrial Hygiene Association (AIHA), the American Society of Safety Engineers (ASSE), CPWR—The Center for Construction Research and Training, Kaiser Permanente, Liberty Mutual, the National Safety Council (NSC), the Occupational Safety and Health Administration (OSHA), ORC World-wide, and the Regenstrief Center for Healthcare Engineering.

The first prevention through design workshop was held in July 2007. Attendees included representatives from industry, labor, government, and academia. Proceedings were published in 2008 in a dedicated issue of the *Journal of Safety Research*. A prevention through design Council was formed to guide the new initiative. Council members were specialists in occupational safety and health and arranged strategic goals around the themes of Research, Education, Practice, and Policy. The Plan for the National Initiative (http://www.cdc.gov/niosh/docs/2011-121/) was published in 2009.

In 2010, the prevention through design role of the designer/engineer was investigated by benchmarking PtD regulations of designers in the construction industry in the United Kingdom (UK) to further understand the potential impacts and opportunities for implementation of the PtD concept in the United States. The Education and Information Division of NIOSH coordinated a workshop, titled "Making Green Jobs Safe," which developed 48 compelling activities for including worker safety and health into green jobs and sustainable design. A summary of the workshop was published in 2011. http://www.cdc.gov/niosh/docs/2011-201/

"Prevention through Design: A New Way of Doing Business" was held in August 2011, included supportive business leaders, noted safety experts, and academic researchers. At this stage, PtD concepts were included in 12 drafts or published consensus standards. Two booklets, three textbooks, and 50 peer-reviewed papers had been published. Four case studies had been created to demonstrate the business value of PtD. Presenters shared lessons learned from a Masters-level degree program at the University of Alabama-Birmingham and a PtD course at Virginia Tech. Half a dozen universities offered courses containing PtD content. These and other success stories were highlighted at the conference.

In 2011, the American Society of Safety Engineers (ASSE) obtained approval from the American National Standards Institute (ANSI) for ANSI/ASSE Z590.3— the standard titled "Prevention through Design: Guidelines for Addressing Occupational Hazards and Risks in Design and Redesign Processes." This standard provides guidance on including Prevention through Design concepts within an occupational safety and health management system and can be applied in any occupational setting. The standard focuses specifically on the avoidance, elimination, reduction, and control of occupational safety and health hazards and risks in the design process. This standard t exists because of decisions made a NIOSH.

In 2011, NIOSH met with the U.S. Green Building Council (USGBC), agreed to collaborate on the inclusion of worker health and safety into Leadership in Energy and Environmental Design (LEED) credits for certification. In 2012, NIOSH representatives met with the Construction User's Roundtable to identify opportunities for collaboration. Additional emphasis was placed on the development of business case studies. Three papers summarizing the 2011 conference presentations in the areas of practice, policy, and research were published in the January 2013 issue of *Professional Safety*. Three more were published in the March 2013 issue, including one focused on business value, one pertaining to OSH management systems, and the third, education.

In August 2012, PtD and the NIOSH Nanotechnology Research Center (NTRC) collaborated with the State University New York at Albany, College of Nanoscale Science & Engineering to hold a Safe Nano Design workshop. The purpose was to discuss the value of applying PtD to safely synthesize engineered nanomaterials and safely commercialize nanoenabled products. Applying PtD concepts in organizations handling engineered nanomaterials assures that worker health and safety is considered at each step in the supply chain, resulting in sustainable health and safety performance.

By the spring of 2013, two additional textbooks were published. A total of 93 first-generation publications were cited by 721 second-generation publications. Fifteen consensus standards containing PtD concepts had been published. Results of a study on PtD in the United Kingdom were published in May 2013, recommending four emphases to aid the adoption of PtD in the United States: knowledge, desire (motivation), ability, and execution (https://designforconstructionsafety.files.wordpress.com/2017/07/niosh-ptd-in-the-uk-final-report-may-2013.pdf).

More than two dozen universities expressed interest in the PtD Education Modules consisting of slide decks and instructor's manuals. The first to be published was the Architectural Design and Construction module (http://www.cdc.gov/niosh/docs/2013-133). This was followed by publication of the Structural Steel Design module (https://www.cdc.gov/niosh/docs/2013-136/), the Reinforced Concrete Design module (https://www.cdc.gov/niosh/docs/2013-135/), and the Mechanical-Electrical Systems module (https://www.cdc.gov/niosh/docs/2013-134/). In December 2013, the first of a series of concise, actionable "Workplace Design Solution" documents was published, beginning with "Preventing Falls through the Design of Roof Parapets" (https://www.cdc.gov/niosh/docs/2014-108/).

In the spring of 2014, a progress report was published by NIOSH, "The State of the National Initiative on Prevention through Design" (https://www.cdc.gov/niosh/docs/2014-123/). The second Workplace Design Solution was published in May, "Preventing Falls from Heights through the Design of Embedded Safety Features" (https://www.cdc.gov/niosh/docs/wp-solutions/2014-124/).

The results of a study on the impact of building design choices on construction safety was made available as a new design tool called "SliDeRulE" at www.constructionsliderule.org. SliDeRulE (Safety in Design Risk Evaluator) helps building designers assess the construction safety risk associated with their designs. As a building is being designed, architects and engineers can use SliDeRulE to

- determine the level of safety risk associated with an entire building, a specific building system, or each of the many design elements within a building;
- compare prospective designs based on construction safety risk;
- learn about design features that increase and decrease the risk of injury; and
- create building designs that minimize the risk of construction worker injury.

The tool is free to use and intended for use by anyone involved in the design of buildings. By using SliDeRulE, hazards can be eliminated, safety risk reduced, and construction worker injuries and fatalities prevented.

In October 2014, a meeting was held in New York City with the leaders of Japan NIOSH (JNIOSH) in order to share mutual PtD efforts—concluding with a tour of the new World Trade Center being constructed with many preplanned safety and health features. In November, a PtD presentation was given at the Birck Nanotechnology Center at Purdue University, leading to ongoing collaboration with the new Purdue Process Safety and Assurance Center, P2SAC, focused on Chemical Process Safety (https://engineering.purdue.edu/P2SAC).

In 2015, PtD presentations and recommendations were provided at two of the Purdue P2SAC meetings. The year 2015 also saw the beginning of PtD collaboration with the U.S. Army Corps of Engineers (USACE) Lakes and Rivers Division

(LRD), starting with PtD training of safety and health managers from all of the LRD District offices. Collaboration with the Smithsonian Institution began with a PtD presentation at a special Smithsonian PtD event for their Facilities and Engineering teams. Nongovernmental PtD meetings and presentations in 2015 included presentations to national and international food producers, to the American Equipment Manufacturer's association, and collaboration with the Green Chemistry and Commerce Council (GC3). A PtD webinar was given for the American Association of Occupational Health Nurses (AAOHN) with over 200 in attendance.

Another Workplace Design Solution was published in August, "Supporting Prevention through Design (PtD) Using Business Value Concepts" (https://www.cdc.gov/niosh/docs/wp-solutions/2015-198/). In September, the report on the Safe Nano Design workshop was published, "Perspectives on the design of safer nanomaterials and manufacturing processes," *Journal of Nanoparticle Research* (https://doi.org/10.1007/s11051-015-3152-9). In November, a Workplace Design Solution to design-out noise was published, "Preventing Hazardous Noise and Hearing Loss during Project Design and Operation" (https://www.cdc.gov/niosh/docs/2016-101/).

After four years of collaboration with the NIOSH PtD and Construction programs, the U.S. Green Building Council (USGBC) published, in 2015, a *Leadership in Energy & Environmental Design (LEED) PtD pilot credit* for building certifications (http://www.usgbc.org/articles/new-leed-pilot-credit-prevention-through-design). The pilot credit, developed by NIOSH, prompts the use of PtD methods to design out worker hazards for both the construction phase and operations and maintenance phase of a building's life cycle. Similar to the publishing of the ANSI/ASSP Z590.3 PtD Standard, this PtD LEED credit is considered a major strategic advance for PtD, as LEED criteria are used throughout the United States, and around the world.

In 2016, NIOSH worked with a construction expert to develop model contract language that incorporates PtD roles and responsibilities into design and construction contracts. Also developed was model language for incorporating PtD into liability insurance policies for designers and builders. These will be made available in an upcoming website for PtD in Capital Projects Processes. NIOSH began development of eight industry case studies of successful PtD use, including business case analyses.

NIOSH provided PtD intervention research to OSHA's Sustainability in the Workplace group, contributing to the OSHA White Paper (http://www.osha.gov/sustainability). NIOSH collaborated on PtD integration efforts with key industry, academic, and government leaders, including OSHA, Green Chemistry and Commerce Council, American Ladder Institute, Arizona State University's Global Safety Center, Purdue Process Safety and Assurance Center, U.S. Army Corps of Engineers, and LJB Engineering. Collaboration continued with the Lakes and Rivers Division (LRD) of the Army Corps of Engineers with a PtD session presented to all chiefs of engineering for the districts in LRD. Also in 2016, NIOSH began publishing "Program Performance One-Pager" (PPOP) documents for all of their programs, including PtD (https://www.cdc.gov/niosh/docs/2016-130).

In 2017, NIOSH developed a four-hour PtD workshop to motivate, and provide workshop practice in using the methods of the ASSP/ANSI Z590.3 PtD standard to design-out hazards. This was presented for the first time at the American Industrial Hygiene Conference and Exhibition (AIHce) in Seattle, Washington.

NIOSH PtD advances were featured in two reports, the "National Occupational Research Agenda: Second Decade in Review" (https://www.cdc.gov/niosh/docs/2017-146/) and the "National Occupational Research Agenda: NIOSH Sector and Cross-Sector Program Supplement" (https://www.cdc.gov/niosh/docs/2017-147/). In 2017, NIOSH and the PtD program became heavily involved in several years of revision committee work on the influential ASSP/ANSI Z10.0 standard for Occupational Health and Safety Management Systems (OHSMS).

In addition to the first half-day PtD workshop, PtD presentations were given at three universities and a large annual conference of the Department of Defense. NIOSH contributed to the DODGE Data & Analytics SmartMarket Report, "Safety Management in the Construction Industry 2017," which included considerable focus on PtD (http://www.cpwr.com/publications/safety-management-construction-industry-2017http:/www.cpwr.com/publications/safety-management-construction-industry-2017). The PtD Program Performance One-Pager was updated in August (https://www.cdc.gov/niosh/docs/2017-180/).

In 2018, NIOSH published a suite of documents and training tools for the proposed NIOSH Occupational Exposure Banding (https://www.cdc.gov/niosh/topics/oeb/) process which quickly and accurately assigns chemicals into specific categories (bands), corresponding to a range of exposure concentrations designed to protect worker health. The occupational exposure banding process seeks to create a consistent and documented process with a decision logic to characterize chemical hazards so that timely, well-informed design and risk management decisions can be made for chemical substances that lack occupational exposure limits (OELs). NIOSH published three more Workplace Design Solutions, this time for the safe handling of nanosize materials: "Protecting Workers during Nanomaterial Reactor Operations" (https://www.cdc.gov/niosh/docs/2018-120/); "Protecting Workers during the Handling of Nanomaterials" (https://www.cdc.gov/niosh/docs/2018-121/); "Protecting Workers during Intermediate and Downstream Processing of Nanomaterials" (https://www.cdc.gov/niosh/docs/2018-122/). NIOSH collaborated with CPWR.

The Center for Construction Research and Training, to review and/or edit a number of CPWR's "Construction Solution" documents and a Toolbox Talk: "Construction Solution: Incorporating Horizontal Round Grab Bars on Fixed Ladders for Three-point Control" (http://www.cpwrconstructionsolutions.org/general_labor/solution/973/incorporating-horizontal-grab-bars-on-fixed-ladders-for-three-point-control.html); "Construction Solution: Fall Guards for Skylights" (http://cpwrconstructionsolutions.org/roofing/solution/978/fallguards-for-skylights.html); "Construction Solution: Push-Fit Heat-Free Pipe Connections" (http://www.cpwrconstructionsolutions.org/solution/979/push-fit-heat-free-connections.html); "Construction Solution: User Vibration Protection" (http://www.cpwrconstructionsolutions.org/structural_steel/solution/981/user-vibration-protection.html); and "Toolbox Talk: Roof Collapse" (https://www.cpwr.com/publications/toolbox-talks).

One of the most intensive NIOSH PtD efforts for the year involved the ongoing PtD collaboration with the Smithsonian. In addition to a PtD briefing to the architects, engineers, and managers of the Smithsonian, two half-day PtD workshops were held with dozens of Smithsonian professionals. The Smithsonian was also instrumental in helping arrange two additional half-day PtD workshops presented to professionals from 26 different federal agencies. In August, the PtD Program Performance One-Pager was updated again (https://www.cdc.gov/niosh/docs/2018-175/).

In 2019, a major advance by NIOSH researchers on an early PtD initiative was the publishing of NIOSH guidance for chemicals that do not have authoritative exposure limits (see the discussion on Health Hazard Banding below). Published in July, "The NIOSH Occupational Exposure Banding Process for Chemical Risk Management" (https://www.cdc.gov/niosh/docs/2019-132/), is accompanied by a web-based e-Tool (https://wwwn.cdc.gov/niosh-oeb/Home/Index). Two PtD workshops were held in 2019, one in New Hampshire with the OSHA Education Office, and another in Massachusetts at an annual OSHA conference, with about 95 private industry professionals participating.

A "Policy in PtD" class presentation for the University of Alabama Safety Engineering Master's program. Another highlight for 2019 is a collaboration with many construction experts (who currently use PtD methods) to present a full day of PtD education and panel discussions in New York City, hosted by the Associated General Contractors (AGCs) of New York State, "Prevention Through Design—A New Mindset" (https://www.agcnys.org/calendar/the-new-mindset-prevention-thorough-design).

NIOSH is encouraging Research in developing the business case for PtD. The focus is now on developing the process and related tools for determining business value. Areas of interest include the methods to measure the impact of PtD concepts on actually reducing injury and illness and how that increases business value.

NIOSH is currently disseminating a systematic Health Hazard Banding approach to assist safety and health professionals in providing guidance for chemicals without authoritative OELs. The NIOSH Hazard Banding process can be used with limited information and resources and can be performed quickly by in-house industrial hygienists and health and safety specialists. The outcome of the Health Hazard Banding process is an occupational exposure band (OEB).

Health Hazard Banding can be used to supplement and support OEL development by facilitating a more rapid evaluation of health risks, providing guidance for chemicals without OELs, identifying hazards to be evaluated for elimination or substitution, providing the user with recommendations when minimal data is available and a tool for the development of NIOSH Recommended Exposure Limits.

Education activities focus on providing educational materials for students, engineers, and designers. The impact of these tools must be assessed. Most graduating engineering students have little practical experience with hazard identification, risk assessment, and risk avoidance or control methods. NIOSH is working with ABET, the organization that accredits engineering and technology curricula to encourage faculty to consider the safety and health of students and staff during the discovery phase of cutting-edge research.

There are several areas where the incorporation of Prevention through Design concepts into the curriculum dovetails with ABET objectives to improve student outcomes, specifically under Criterion 3.c. Incorporating PtD concepts into the curriculum shows a commitment to continuous improvement from faculty and the learning community.

Practice activities focus on stakeholder ability to access, share, and apply successful PtD practices. In addition to our continued dialog with USGBC and CURT, NIOSH is collaborating with industry to identify Best Practices so these can be front loaded into the Capital Design Process.

When a root cause analysis uncovers a design-related hazard, the design solution should be documented for future reference during the conceptual design phase of similar facilities or equipment.

Policy activities are focused on developing a PtD culture in industry. Including PtD into consensus standards is the first step in the development of a PtD culture. The ASSP/ANSI Z590.3 PtD standard and the USGBC LEED PtD pilot credit are excellent examples.

HIGHLIGHTS OF ANSI/ASSE Z590.3

It is not the intent here to duplicate the Z590.3 standard. Highlights, only, are given. In some instances, number or letter designations appearing here may not be exactly the same as in the standard. It is expected that safety professionals who become involved in design and redesign processes will develop a familiarity with 590.3, in detail.

1. Scope, Purpose, and Application

This is the **Scope** of Z590.3: This standard provides guidance on including prevention through design concepts within an occupational safety and health management system. Through the application of these concepts, decisions pertaining to occupational hazards and risks can be incorporated into the process of design and redesign of the work premises, tools, equipment, machinery, substances, and work processes including their construction, manufacture, use, maintenance, and ultimate disposal or reuse. This standard provides guidance for a life-cycle assessment and design model that balances environmental and occupational safety and health goals over the life span of a facility, process, or product.

Although the **Purpose** statement indicates that the standard pertains principally to the avoidance, elimination, reduction, or control of occupational safety and health hazards and risks in the design and redesign processes, this important extension follows the **Purpose** statement.

Note: Incidents that have the potential to result in occupational injuries and illnesses can also result in damage to property and business interruption, and damage to the environment. Reference is made in several places in this standard to those additional loss potentials which may require evaluation and resultant action.

Largely, this standard is applicable to all hazard-based operational risks. Particularly, it relates directly to all of the four major stages of occupational risk management.

   a. *Preoperational stage*: In the initial planning, design, specification, prototyping, and construction processes, where the opportunities are greatest and the costs are lowest for hazard and risk avoidance, elimination, reduction, or control.
   b. *Operational stage*: Where hazards and risks are identified and evaluated and mitigation actions are taken through redesign initiatives or changes in work methods before incidents or exposures occur.
   c. *Postincident stage*: Where investigations are made of incidents and exposures to determine the causal factors which will lead to appropriate interventions and acceptable risk levels.
   d. *Postoperational stage*: when demolition, decommissioning, or reusing/rebuilding operations are undertaken.

The **Application** goals of utilizing PtD concepts in an occupational setting are to

   a. Achieve acceptable risk levels.
   b. Prevent or reduce occupationally related injuries, illnesses, and fatalities.
   c. Reduce the cost of retrofitting necessary to mitigate hazards and risks that were not sufficiently addressed in the design or redesign processes.

Addendum A outlines The Risk Assessment Process. Addendum B gives the Progression of Occupational Hygiene Issues Flow.

2. Referenced and Related Standards

Relative American National Standards and other standards and guidelines that pertain to the purpose of the standard are listed.

3. Definitions

The definition list grew to 27 in response to suggestions made by commenters. Since one of the standard's principle goals is to achieve acceptable risk levels throughout the design and redesign processes, the definitions relative to that goal only are listed here.

a. *Acceptable risk*: That risk for which the probability of an incident or exposure occurring and the severity of harm or damage that may result are as low as reasonably practicable (ALARP) in the setting being considered.
b. *As low as reasonably practicable (ALARP)*: That level of risk which can be further lowered only by an increase in resource expenditure that is disproportionate in relation to the resulting decrease in risk.
c. *Hazard*: The potential for harm.
    Note: Hazards include all aspects of technology and activity that produce risk. Hazards include the characteristics of things (e.g., equipment, technology, processes, dusts, fibers, gases, materials, and chemicals) and the actions or inactions of people.
d. *Hierarchy of controls*: A systematic approach to avoiding, eliminating, reducing, and controlling risks, considering steps in a ranked and sequential order, beginning with avoidance, elimination, and substitution.
e. *Probability*: An estimate of the likelihood of an incident or exposure occurring that could result in harm or damage for a selected unit of time, events, population, items, or activity being considered.
f. *Residual risk*: The risk remaining after risk reduction measures have been taken.
g. *Risk*: An estimate of the probability of a hazard-related incident or exposure occurring and the severity of harm or damage that could result.
h. *Safety*: Freedom from unacceptable risk.
i. *Severity*: An estimate of the magnitude of harm or damage that could reasonably result from a hazard-related incident or exposure.

4. Roles and Responsibilities

Top management shall provide the leadership to institute and maintain effective systems for the design and redesign processes. Key points: anticipate hazards and risks; assess risks; apply the hierarchy of controls to achieve acceptable risk levels. The following Note is significant in the Roles and Responsibilities Section.

Note: The processes of identifying and analyzing hazards and assessing risks improve if management establishes a culture where employee knowledge and experience is valued and respected and they collaborate in significant aspects of the design and redesign activities. Employees who do the work can make valuable contributions in identifying and evaluating hazards, in risk assessments, and proposing risk reduction measures.

5. Relations with Suppliers

It became apparent as suggestions were received that commenters wanted help in creating procedures to avoid bringing hazards into the workplace. This section provides that help. It outlines the discussions and arrangements organizations should have with suppliers and gives guidance on the specifics that should be expected of suppliers.

Procedures recommended relate directly to the Procurement requirements in Z10.0 and 45001. Addendum C in the standard provides procurement guidelines that are to assist in making arrangements with suppliers.

One of the provisions in this section recommends that suppliers of equipment, technologies, processes, and materials provide documentation establishing that a risk assessment has been conducted and that an acceptable risk level, as outlined by the procuring organization, has been achieved.

Addendum D in the standard is an example of a basic Risk Assessment Report that can be used as a guide.

6. Design Safety Reviews

In the design process, risk assessments would be made as often as needed on a continuum. In addition, a formal design safety review procedure should be put in place. This section is supported by Addendum E which is titled Safety Design Review Guide. Chapter 12 in this book is titled "Safety Design Reviews."

7. The Hazard Analysis and Risk Assessment Process

This is the longest section in the standard. First, an outline of the hazard analysis and risk assessment process is given. That is followed by the "how" for each element in the outline. The outline follows:

- Select a risk assessment matrix
- Establish the analysis parameters
- Identify the hazards
- Consider failure modes
- Assess the severity of consequences
- Determine occurrence probability
- Define initial risk
- Select and implement hazard avoidance, elimination, reduction, and control methods
- Assess the residual risk
- Risk acceptance decision-making
- Document the results
- Follow-up on actions taken

For many hazards, the proper level of acceptable risk can be attained without bringing together teams of people. Safety and health professionals and design engineers with the proper experience and education can reach the conclusions of what constitutes acceptable risk. For other risk situations, management should have processes in place to seek the counsel of experienced personnel who are particularly skilled in risk assessment for the category of the situation being considered.

Reaching group consensus is a highly desirable goal. Sometimes, for what an individual considers obvious, achieving consensus is still desirable so that buy-in is obtained for the actions taken. Addenda A and B in the standard serve as examples of suggested approaches to the risk assessment processes.

It is strongly urged that an appropriate risk assessment matrix be selected for the hazard analysis and risk assessment process. A risk assessment matrix provides a method to categorize combinations of probability of occurrence and severity of harm, thus establishing risk levels. A matrix helps in communicating with decision-makers on risk reduction actions to be taken. Also, risk assessment matrices assist in comparing and prioritizing risks, and in effectively allocating mitigation resources. Addendum F provides several examples of Risk Assessment Matrices and Descriptions of Terms to serve as a base for an organization to develop a matrix suitable for its operations. Addendum F is comparable in content to Chapter 9 "On Hazards Analyses and Risk Assessments."

8. Hazard Analysis and Risk Assessment Techniques

Top management shall adopt and apply the hazard analysis and risk assessment techniques suitable to the organization's needs and provide the training necessary to employees who will be involved in the process. Descriptions of eight selected techniques are presented in Addendum G. Addendum H is a Failure Mode and Effects Analysis form.

As a practical matter, having knowledge of three risk assessment concepts will be sufficient to address most, but not all, risk situations. They are Preliminary Hazard Analysis and Risk Assessment; the What-If/Checklist Analysis Methods; and Failure Mode and Effects Analysis.

9. Hierarchy of Controls

Top management shall achieve acceptable risk levels by adopting, implementing, and maintaining a process to avoid, eliminate, reduce, and control hazards and risks. The process shall be based on the hierarchy of controls outlined in this standard, which is

   a. Risk avoidance
   b. Eliminate
   c. Substitution
   d. Engineering controls
   e. Warning systems
   f. Administrative controls
   g. Personal protective equipment (PPE)

This is a "prevention" standard. Research was done to develop a variation in the hierarchy of controls to adapt to the meaning of "prevention." Elimination, which is the first action to be taken in many hierarchies of controls, did not seem to fit with meaning of prevention.

GOALS TO BE ACHIEVED   295

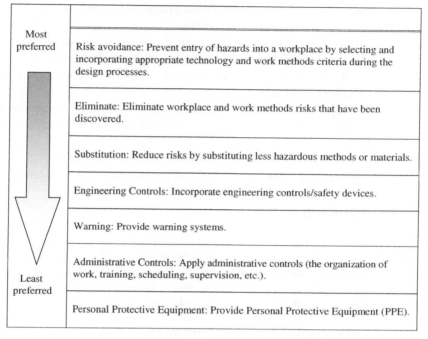

Figure 13.2   Risk reduction hierarchy of controls.

- *Elimination means*: Removal; purging; taking away; to get rid of something. To eliminate, there has to be something in place to remove.
- *Avoidance means*: To prevent something from happening; keeping away from; averting.

Designers start with a blank sheet of paper, or an empty screen in a CAD system. Designers have opportunities to avoid hazards in all design stages: conceptual, preliminary, and final. In the early design phases, there are no hazards, yet, to be eliminated, reduced, or controlled. So avoidance is a better match for a hierarchy of controls for a prevention standard. Figure 13.2 depicts the hierarchy of controls in Z90.3.

Extensive comments will be found in Addendum I for Z590.3 on each of the elements in the hierarchy. Addendum J is the Bibliography.

GOALS TO BE ACHIEVED

Z590.3 says that, in so far as is practicable, the goal shall be to assure that for the design selected:

- An acceptable risk level is achieved, as defined in this standard.
- The probability of personnel making human errors because of design inadequacies is as low as reasonably practicable.

- The ability of personnel to defeat the work system and the work methods prescribed is as low as reasonably practicable.
- The work processes prescribed take into consideration human factors (ergonomics)—the capabilities and limitations of the work population.
- Hazards and risks with respect to access and the means for maintenance are at as low as reasonably practical.
- The need for personal protective equipment is as low as reasonably practical, and aid is provided for its use where it is necessary (e.g., anchor points for fall protection).
- Applicable laws, codes, regulations, and standards have been met.
- Any recognized code of practice, internal or external has been considered.

GETTING INVOLVED

It is recommended that safety professionals use actual cases in which they are involved to support the premise that addressing hazards and risks in the design and redesign processes will result not only in achieving acceptable risk levels but also higher productivity and operational efficiency. That may be done easier in the redesign process. The author says to safety professionals:

- Try to convince designers to allow you to work with them on a project so that you can demonstrate your capability and value.
- Show the value you bring to the discussion.
- When teams are considering a given design or redesign situation, try to encourage input from all present who have knowledge of the process.
- Take a holistic, macro view, include all hazards-based subjects if you can.
- Try to involve operations personnel at all levels. Assume respectfully that those who do the work have knowledge and skill that can contribute to workplace and work methods redesign and encourage their input.
- Understand that you are proposing a culture change, that designers may presume that you are intruding into their territory and they will resist your involvement to maintain their territorial prerogative.
- Do not play down the fact that implementing a new program can be challenging and frustrating. Be patient. Utilize change management concepts. Be a good listener. Be open to comment and criticism.
- Study the incident history in the entity to which counsel is being given and that of its industry for support data showing that design shortcomings were among the contributing factors for incidents that have occurred.

A goal is to achieve buy-in by those who are involved, but particularly at a senior executive level. Training programs should be considered for the value they provide. Also, ask —will it be advantageous in a particular situation for a PtD system to be written?

PATIENCE AND UNDERSTANDING

Implementing the concepts of risk assessment and risk avoidance or reduction in the design and redesign processes is a long-term effort. On that point, Bruce Main (2012) is eloquent, as follows in the Introduction to his book *Risk Assessment: Challenges and Opportunities,* duplicated with his permission.

> This is an exciting time in risk assessment. Whether machinery uptime, process throughput, cost savings, new ideas for features, patentable innovations, or simply documenting that a company makes really good products with acceptable risks, the opportunities abound to apply the process and drive improvements. The opportunities exist because risk assessment works.
>
> The risk assessment process is a journey rather than an event. Companies that are just starting to complete risk assessments find their first efforts will require more time and will be less complete than later efforts. As personnel involved receive training and become more familiar with the risk assessment process, more hazards will be identified, more risk reduction methods deployed and the risk assessment process will improve and hasten in pace.
>
> As lessons are learned and experience is gained, risk assessment becomes more refined. However, some time and experience are required for the risk assessment process to become fully integrated into a company. How much time depends on the company and its circumstances, but it typically takes months, not weeks. Eventually the risk assessment process will become a part of normal business procedures. Until then, industry needs time to fully and formally implement these concepts (Introduction).

For safety professionals who are interested in a well-written dissertation on "Implementing and Deployment" of a risk assessment system that impacts on all design decisions, the chapter by that name, Chapter 16 in *Risk Assessment: Challenges and Opportunities* is recommended. Bruce Main has given permission to duplicate the following key points, all of which should be considered in relation to procedures in place and the culture of an organization.

KEY POINTS

1. Leadership is a key and critical factor in successfully implementing and deploying the risk assessment process.
2. Integrating risk assessment in an organization is a process that generally follows a sequence of phases. (Three frameworks are discussed).
3. Engineering design needs to change to include the risk assessment process to more effectively move safety into design. Only by changing the design process will risk assessment efforts succeed.

4. Introducing the risk assessment process will explicitly change the design process, allowing hazards to be identified and risk reduction methods to be incorporated early in the design process. As with any new process or substantial change, people may resist.
5. To be effective, the company culture must be willing to embrace the risk assessment process, and cultural acceptance stems from management leadership.
6. In industrial product or process applications, both equipment suppliers and users should perform risk assessments and be involved in the risk assessment process.
7. In consumer product and component product applications, the manufacturer is responsible for conducting the risk assessment, if applicable. Product users typically have no risk assessment responsibilities beyond using the product in conformance with the product information.
8. Practical guidance *is* shared to help companies get started and make progress in the risk assessment process. Topics addressed include when to stop risk assessment, the time to complete an assessment, leaders in me best practices, what to do in cross industry situations, when to revise an existing risk assessment, making changes to the method, results of risk assessment and others.
9. To integrate risk assessment into the design process, engineers will likely need education and training on risk assessment in some form. (p 230)

EXAMPLES

Permission has been received to duplicate a section of a construction contract for which the safety director arranged to have a section included on prevention through design, In the following, captioned **Safety and Protection**, the only changes made were to de-identify with respect to the name of the organization and the name of the person who gave permission to duplicate. This construction project is well on the way to completion as this chapter is written.

Safety and Protection

6.1 The Design-Builder shall comply with all Laws applicable to the safety of persons or property. Damage, injury, or loss to property caused by the Design-Builder, Subcontractor, or anyone directly or indirectly employed by any of them or anyone for whose acts any of them may be liable shall be remedied by the Design-Builder.

6.2 The Design-Builder shall be solely responsible for initiating, maintaining, and supervising all safety precautions and programs in connection with the Work.

6.3 Prevention through Design: In order to minimize or eliminate risk and hazards, during the design phase the Design-Builder shall assign a design safety coordinator to anticipate hazards during the construction phase(s) and during occupancy.

 A. The design safety coordinator shall develop a safety and health plan anticipating hazards applicable to the construction site, taking into account the construction activities that will take place on the site. The plan shall be provided to the construction team safety representative(s) for inclusion into their site-specific safety plan. This plan shall also include specific measures addressing hazards which may fall within one or more of these categories:

   1. Work involving engulfment hazards.
   2. Work involving falls from height hazards.
   3. Work which puts workers at risk from chemical or biological substances.
   4. Work with ionizing radiation.
   5. Work near high voltage lines.
   6. Work involving Hazardous Energy Control Procedures.
   7. Work exposing workers to the risk of drowning (work over water).
   8. Work carried out by divers.
   9. Work involving the use of explosives.
   10. Work involving the assembly or dismantling of heavy prefabricated components.
   11. Work involving an OSHA or other regulatory requirement for worker health exposure monitoring.

 B. The design safety coordinator shall also develop a safety and health plan anticipating hazards applicable to the use and maintenance of the permanent facility, ensuring the protection of employees and the public. This plan shall be provided to the Agency prior to beneficial occupancy.

6.4 The Design-Builder shall designate a qualified and experienced safety representative whose duties and responsibilities shall be the prevention of accidents and the maintenance and supervision of safety precautions and programs. This person shall be the Design-Builder's project superintendent unless otherwise designated in writing by the Design-Builder to the Agency.

6.5 The Design-Builder shall report promptly in writing to the Agency all recordable accidents and injuries occurring at the site. When the Design-Builder is required to file an accident report with a public authority, the Design-Builder shall submit a copy of the report to the Agency.

6.6 The Design-Builder shall inform the Agency of the specific requirements of the Design-Builder's safety program with which the Agency's employees and representatives must comply while at the site.

6.7 If the Agency deems any part of the Work unsafe, the Agency, without assuming responsibility for the Design-Builder's safety program, may require the Design-Builder to stop performance of the Work or take corrective measures satisfactory to the Agency, or both. If the Design-Builder does not adopt corrective measures, the Agency may perform them and deduct their cost from the Contract Sum. The Design-Builder agrees to make no claim for damages, for an increase in the Contract Sum, or for a change in the Contract Time based on the Design-Builder's compliance with the Agency's reasonable request.

6.8 The Design-Builder shall erect and maintain necessary safeguards for the safety and protection of:

A. Employees on the Work and other persons whose safety may be adversely affected by performance of the Work.

B. The Work and material to be incorporated into the Work, whether in storage on or off the site. If the Design-Builder fails to protect the Work, the Agency may, after giving notice to the Design-Builder, protect the Work and deduct the resulting cost from payment due the Design-Builder. The Agency's determination of when and to what degree such protection is necessary shall be final.

C. Other property at the site including trees, shrubs, lawn, walks, pavements, roadways, structures, and utilities not designated for removal, relocation, or replacement.

D. Adjacent property and utilities when prosecution of the Work may affect them.

6.9 The Design-Builder's duties and responsibilities for the safety and protection of the Work shall continue until the Design-Builder has completed all obligations under this Contract.

This next example is a draft of a corporate standard on prevention through design developed by David Waline. It is replicated with his permission.

PtD Corporate Standard == With Permission from Waline Consulting LTD

Document Title: Prevention through Design (PtD) Process	
Document No. CS 01	Effective Date:
Document Type: Corporate Standard—Prevention through Design	
Authority:	Revision: Draft

1.0 Purpose & Goal This global standard specifies the minimum requirements to be fulfilled by the organization to ensure that adequate consideration is given to the identification of hazards, risk assessment, and mitigation early in the design or redesign stage for all manufacturing assets with the goal of achieving Acceptable Level of Risk (ALOR). Organizational leadership will provide the necessary training, support, and resources to effectively carry out this Standard.

2.0 Organizational Benefits The derived outcomes (benefits) to be realized from fully implementing this standard is the following:

- Prevention or reduce occupationally related injuries, illnesses, fatalities, and/or damage to property and the environment
- Reduce the cost of retrofitting necessary to mitigate hazards and risks that were not sufficiently addressed in the design or redesign processes
- Minimize probability for human error
- Minimize the probability to defeat systems
- Support compliance
- Support production and maintenance standard operating procedures
- Proactively engage employees and other stakeholders in hazard identification and risk reduction effort

Prevention through Design (PtD) is a critical component of an effective
EHS Management System to achieve and sustain positive EHS performance. (Refer to ANSI/AIHA Z10-2012), Occupational Health and Safety Management Systems Standard)

3.0 Scope This standard applies to all facilities or worksites globally where the company has management control. The PtD Process is applicable to buildings, work premises, machinery, equipment, substances, process systems, tools, and work methods as it relates to their construction, manufacture, use, maintenance, and ultimate disposal or reuse. PtD is a lifecycle approach.

4.0 Implementation This PtD Standard will be deployed by the organization when any of the following events take place:

1. All capital projects (at or above $X)
2. New or redesign of facilities, equipment, products
3. Relocation of equipment or processes
4. Findings from Total Productive Maintenance (TPM) projects
5. Regulatory required
6. Injury/Illness and Loss history dictates design change
7. Management of Change (MOC)-modifications
8. Highly complex processes posing high risk

302 PREVENTION THROUGH DESIGN

9. Customer required
10. Findings from safety kaizen events
11. Carrier Required
12. Demolition, decommissioning or reuse operations undertaken

5.0 PtD Elements The following requirements will be included in all PtD related projects:

a. Safety Professional will participate in design safety reviews and risk assessments.
b. Project Manager and stakeholder roles and responsibilities to be clearly established and communicated for all projects
c. Project EHS Performance Objectives and Metrics established
d. Long-term burden costs associated with postproject management of residual risk will be captured and allocated
e. Project Execution Plan (PEP) established and tracked to project completion
f. Documented Design Safety Reviews and Risk Assessments
g. Proven design solutions library that achieve ALOR are incorporated into new designs/redesigns
h. Documented constructability review
i. Pre-approved Design/Build firms only used on projects
j. Pre-approved contractors only used on projects
k. Safety Professional review and approval of all changes impacting EHS
l. EHS specifications incorporated into procurement process
m. Safety Professional participation in vendor trials as applicable
n. Safe Occupancy requirements established and resourced
o. Documented Safety Commissioning and Start-up safety reviews
p. Performance evaluations and feedback provided to design/build firms and contractors upon completion of project
q. Lessons learned captured and shared with project team shared with other project teams for continuous improvement on future projects (document retention)

6.0 Risk Assessment Achieving Acceptable Level of Risk (ALOR) defined by the organization will be required for all projects. Any exceptions to this design rule outcome must be reviewed and approved by appropriate EHS Leadership and Operations Management. If accepted, documented effective alternative measures and controls must be put into place that protects people, property, and environment. No single individual can make the decision or has the approving authority to design outside established ALOR boundaries.

Appropriate risk assessment methodologies and tools will be selected and used by project stakeholders.

Risk Assessment tools commonly used within the organization for assessing project risk would include the following:

1. Task Based Risk Assessment (TBRA)
2. Failure Mode and Effects Analysis (FMEA)
3. What-if Analysis
4. Fault Tree Analysis (FTA)
5. Hazardous Chemical Process Hazard Analysis (HC-PHA)
6. Construction Project Hazard Analysis (C-PHA)
7. Occupational Hygiene Exposure Assessment (OHEA)

The Hierarchy of Control decision-making found in ANSI/ASSE Z590.3-2011 (R2016), Prevention through Design - Guideline for Addressing Occupational Hazards and Risks in Design and Redesign Processes" will be used by project design teams when conducting risk assessments.

The organizations PtD process will follow this order of priority in achieving ALOR (called above the line controls):

1. Risk Avoidance – prevent entry of hazards into workplace
2. Elimination – eliminate workplace hazards and work methods risks that have been discovered
3. Substitution- reduce risk by substituting less hazardous methods or materials
4. Engineering Controls – incorporate engineering controls/safety devices into new design

Warning systems, administrative controls and personal protective equipment (PPE) will not be the primary controls selected in the PtD process. They will only supplement the higher-level controls (1-4) listed above as last resort. A combination of risk mitigation controls may be required to achieve ALOR.

Residual risk must be documented as part of the PtD Process and fall within ALOR.

Special focus and emphasis will be placed on potential Serious Injury/Illness and Fatal (SIF) level risks in the design safety review process. Above the line controls will be the required control strategy for all SIF hazards, exposures, and risks.

Risk Assessment process will take into consideration all modes of operation and work activities including both normal, abnormal, nonroutine, and maintenance activities at a minimum.

PtD will begin in the earliest stages of the project planning process. PtD will be included as a minimum in these three (3) design phases:

1. Concept (risk avoidance and hazard elimination focus)
2. Preliminary (30%-50% design safety review completion stage)
3. Detailed (90% design safety review completion stage)

7.0 Stakeholders Selected project stakeholders must be familiar with ANSI/ASSE Z590.3-2011 (R2016) as this document has been incorporated by reference into this PtD standard.

Stakeholders (as applicable):

- Project Manager
- EHS Professional
- Engineer
- Designer
- Vendors
- Builder/Manufacturer
- Architect
- Construction Manager
- Exposed population – users, maintenance, operations
- Customer
- Insurance Carrier
- Suppliers
- External subject matter experts (SME's)

8.0 Supplier Agreements As part of the procurement process, equipment suppliers and manufacturers must supply the organization with a documented risk assessment.

9.0 Sustainable Control Critical-to-Safety Controls incorporated into new designs/redesigns will be incorporated into the facility's routine inspection and preventative maintenance program to maintain their effectiveness. Critical to Safety Controls would include such items (not complete listing) as:

1. Control reliable safety systems
2. Energy isolation systems
3. Presence sensing devices
4. Guarding
5. Detection systems

10.0 Document History

Section No.	Date	Description of Change	Approved By
All	X	New Standard	Draft

CONCLUSION

Involvement of safety professionals in the design processes is hugely more extensive now in relation to what it was 25 years ago. There is opportunity here for professional satisfaction through participation in the design processes and for being perceived as a valued contributor in support of operational efficiency as well as risk management.

Safety professionals who are involved in the design and redesign processes or those who want to initiate such activity and be involved would add to their skills if they became proficient, as in Chapter 3, "Safety Professionals as Culture Change Agents."

REFERENCES

ANSI/ASSE Z590.3-2011. American National Standard: *Prevention through Design: Guidelines for Addressing Occupational Hazards and Risks in Design and Redesign Processes.* Park Ridge, IL: American Society of Safety Professionals, 2011 (R2016).

ANSI/ASSP Z10.0-2019. *Occupational Health and Safety Management Systems.* Park Ridge, IL: American Society of Safety Professionals, 2019.

ANSI/ASSP/ISO 45001-2018. *Occupational Health and Safety Management Systems – Requirements with Guidance for Use.* Park Ridge, IL: American Society of Safety Professionals, 2018.

Australian Government. *Guidance On The Principles Of Safe Design For Work.* Canberra, Australia: Australian Safety and Compensation Council, an entity of the Australian Government, 2006.

Christensen, W.C. and F.A. Manuele, Editors. *Safety Through Design.* Itasca, IL: National Safety Council, 1999.

Designing for Safety. A position paper approved by the Board of Directors of the American Society of Safety Engineers. Park Ridge, IL: 1994.

Johnson, W. *MORT Safety Assurance Systems.* Itasca, IL: National Safety Council, 1980. (Also published by Marcel Dekker, NY).

Main, B.W., PE, CSP. *Risk Assessment: Challenges and Opportunities.* Ann Arbor, MI: Design Safety Engineering, Inc., 2012.

TR-Z790.001. *A Technical Report on Prevention through Design.* Des Plaines, IL: American society of Safety Engineers, 2009.

FURTHER READING

Schulte, P., R. Rinehart, A. Okun, C. Geraci, and D. Heidel . "National prevention through design (PtD) initiative." *Journal of Safety Research*, Vol. 39, pp. 115–121, 2008.

CHAPTER 14

MANAGEMENT OF CHANGE

Because of this author's belief that management of change should be a more important element in an occupational risk management system, a plea is made that readers give particular attention to the requirements for this subject in Z10.0 and 45001. They both require that management of change processes be in place for the new and the revised.

For Z10.0, design reviews and management of change are dealt with as one subject. Safety Design Reviews are the subject of Chapter 12. One of the reasons that the management of change process is addressed separately is to promote an understanding of and application of the concept on which it is based. This chapter:

- Presents the Z10.0 and 45001 management of change requirements in their entirety.
- Makes the case that having an effective Management of Change System (MOC) in place as a distinct element in a safety and health management system will reduce the potential for injuries, environmental damage, and other forms of damage at all levels of severity.
- Cites statistics in support of having effective MOC systems.
- Defines the purpose and methodology of a MOC system.
- Establishes the significance of management of change as a method to prevent serious injuries and fatalities.

Advanced Safety Management: Focusing on Z10.0, 45001, and Serious Injury Prevention,
Third Edition. Fred A. Manuele.
© 2020 John Wiley & Sons, Inc. Published 2020 by John Wiley & Sons, Inc.

- Outlines management of change procedures, keeping in mind the staffing limitations at other than large locations and their need to avoid burdensome paper work.
- Provides guidelines on how to initiate and utilize an MOC system.
- Emphasizes the significance of communication and training.
- Includes examples of six MOCs in place and provides access to four other real-world MOC systems.

In accord with the emphasis given here to management of change, the requirements in 45001 and Z10.0 are duplicated in their entirety as follows.

MANAGEMENT OF CHANGE IN 45001

8.1.3 Management of Change

The organization shall establish a process(es) for the implementation and control of planned temporary and permanent changes that impact OH&S performance, including:

a) new products, services and processes, or changes to existing products, services and processes, including
   - Workplace locations and surroundings;
   - Work organization;
   - Working conditions;
   - Equipment;
   - Work force.
b) changes to legal requirements and other requirements;
c) changes in knowledge or information about hazards and OH&S risks;
d) developments in knowledge and technology.

The organization shall review the consequences of unintended changes, taking action to mitigate any adverse effects, as necessary.

Note: Changes can result in risks and opportunities.

MANAGEMENT OF CHANGE AS IN Z10.0

8.5 Management of Change

The organization shall establish a process to identify, and take appropriate steps to prevent or otherwise control hazards for situations requiring Management of Change to reduce potential risks to an acceptable level. (See Notes 1, 2 and 3).

The process for management of change shall include:

1. Identification of tasks and related health and safety hazards;
2. Recognition of hazards associated with human factors, including ergonomics, human errors induced by the design or design deficiencies and where design makes successful job completion unnecessarily difficult;
3. Review of applicable regulations, codes, standards, internal and external recognized guidelines;
4. Application of control measures (hierarchy of controls);
5. A determination of the appropriate scope and degree of management of change; and
6. Worker participation (See Note 4)

Note 1: Management of change is the process to identify and manage changes to minimize the introduction of new hazards and risks into the work environment.

Note 2: Management of change includes changes being made in existing facilities, operations, documentation, personnel, products, or services. The management of change process should take into consideration relevant items such as:

- technology, equipment, work practices and procedures
- design specifications and raw materials
- organizational or staffing changes, standards or regulations.

Note 3: Harmonizing the management of change across all business processes and developing a means to involve affected workers is an effective method of integrating OHS into the business.

Note 4: Management of change effectiveness is enhanced when workers with knowledge and experience of equipment, processes, facilities and systems participate in the process. Such participation often includes identifying tasks, hazards, and feasibility of control measures.

8.5.1 Applicable Life Cycle Phases

During the management of change processes all applicable life cycle phases shall be taken into consideration.

8.5.2 Process Verification

The organization shall have processes in place to verify that changes in facilities, documentation, personnel and operations are evaluated and managed to ensure OHS risks arising from these changes are controlled.

Note: Business reviews can provide insight into changes that could result in creating OHS risks allowing them to be proactively controlled enhancing operational excellence, sustainability and reliability.

STUDIES THAT SUPPORT HAVING A MOC SYSTEM IN PLACE

Three significant studies establish that having a Management of Change system as an element within an Operational Risk Management System would serve well to reduce injury potential.

- Reviews made by this author of over 1950 incident investigation reports, mostly for serious injuries, support the need for and the benefit of having MOC systems. They showed that a significantly large share of incidents resulting in serious injuries occurs:
- When unusual and non-routine work is being performed
- In non-production activities
- In at-plant modification or construction operations (replacing a motor weighing 800 pounds to be installed on a platform 15 feet above the floor)
- During shutdowns for repair and maintenance, and startups
- Where sources of high energy are present (electrical, steam, pneumatic, chemical)
- Where upsets occur: situations going from normal to abnormal

Having an effective MOC system in place would have served to reduce the probability of serious injuries and fatalities occurring in the operational categories shown previously.

- A study led by Dr. Thomas Krause, then Chairman of the Board at BST in 2011 produced results in support of having MOC systems in place. (Data was provided by Dr. Krause in personal communication. BST was to publish a paper including this data. It hasn't.)

  Seven companies participated in the study. Incident investigation reports were reviewed. Shortcomings in Pre-Job Planning, another name for Management of Change, were found in 29% of incidents that had serious injury or fatality potential.

  Focusing on reducing that 29%, a noteworthy number, would be an appropriate safety and health management system goal.

- Correspondence with John Rupp, then at United Auto Workers in Detroit, confirmed the continuing history with respect to fatalities occurring in the UAW represented workplaces. A previously issued UAW bulletin (no longer available) indicated that from 1973 through 2007, 42% of fatalities occurred to skilled-trades workers, who represented about 20% of the membership.

  Additional data provided by Rupp for the years 2008 through 2011 indicated that 47% of fatalities occurred to skilled-trades workers. Skilled-trades workers are not in routine production jobs. They are often involved in unusual and nonroutine work, at-plant modifications or construction operations, shutdowns for repair and maintenance, startups, and where sources of high energy are present. That is the sort of activity for which an MOC system would be beneficial.

PURPOSE OF A MANAGEMENT OF CHANGE SYSTEM

Content of the management of change requirements as in 45001 and Z10.0 clearly establish the required provisions. This author's summation is that the MOC process is to assure that:

- On an anticipatory basis, hazards are identified and analyzed and risks are assessed.
- Appropriate avoidance, elimination, or control decisions are made so that acceptable risk levels are achieved and maintained throughout the change process.
- New hazards are not knowingly brought into the workplace by the change.
- Changes do not impact negatively on previously resolved hazards.
- Changes do not make the potential for harm of an existing hazard more severe.
- An existing occupational safety and health management system is not negatively impacted.

APPLICATION CONSIDERATIONS

In the following list of categories to which the management of change process could apply, subjects other than for the safety of employees are included to demonstrate the breadth of benefits that could be obtained from an effective MOC system. Consideration, as applicable, would be given to:

- Safety of employees making the changes.
- Safety of employees in adjacent areas.
- Safety of employees who will be engaged in operations after changes are made.
- Environmental aspects.
- Safety of the public.
- Product safety and product quality.
- Fire protection so as to avoid property damage and business interruption.

As the standards say, MOC provisions may apply when changes are made in technology, equipment, facilities, work practices and procedures, design specifications, raw materials, organizational or staffing changes impacting on skill capabilities, and standards or regulations.

OSHA MOC REQUIREMENTS

OSHA's *Rule for Process Safety Management of Highly Hazardous Chemicals, 29 CFR 1910.119,* issued in 1992, requires that an operation affected by the standard have an MOC process in place. Similar requirements do not appear in other OSHA standards.

But having an MOC system in place is a requirement to achieve OSHA's Voluntary Protection Program (VPP) designation. Item B in the VPP requirements is titled Workplace Analysis. A requirement—3. Hazard Analysis of Significant Change—is a subset of Item B. This is how requirement 3 reads:

> Explain how, prior to activity or use, you analyze significant changes to identify uncontrolled safety and health hazards and the actions needed to eliminate or control these hazards. Significant changes may include non-routine tasks and new processes, materials, equipment, and facilities.

EXPERIENCE OF OTHERS IMPLIES OPPORTUNITY

To test whether personnel in operations other than chemicals had recognized the need for and developed MOCs, assistance was sought of the staff at the American Society of Safety Professionals responsible for Practice Specialties. A request was sent to members of the Management Specialty group asking that copies of issued MOC procedures for operations other than chemicals be sent to this author. Response was overwhelmingly favorable. The number of documents received was more than could be practicably used.

Examples received demonstrate that managements in a variety of operations had recognized the need to have MOC systems in place. These MOC systems were specifically selected to show:

- The broad range of harm and damage categories covered.
- Similarities with respect to subjects covered.
- The enormity of the variations in how those subjects are addressed.

THE MANAGEMENT OF CHANGE PROCESS

As is the case with all management systems, an administrative procedure must be written to communicate what the MOC is to encompass and how it is to operate. Care should be taken to assure that the MOC system is designed to be compatible with the:

- Organization's and industry's inherent risks.
- Other and relevant management systems in place.

- Organizational structure and culture.
- Expected participation of the work force.

Although brevity is the goal, consideration should be given to the following as subjects are selected for inclusion in an MOC procedure:

1. Defining the need for and the purpose of the MOC.
2. Establishing accountability levels.
3. Specifying the criteria that are to trigger the initiation of formal change requests.
4. Making clear how personnel are to submit change requests, and specifying the change request form to be used.
5. Outlining the criteria for request reviews and responsibilities for reviews.
6. Indicating that the MOC system is to encompass:
    a) The risks to the workers who are to do the work and other employees who are affected.
    b) Possible damage to the property and business interruption.
    c) Possible environmental damage.
    d) Product safety and quality.
    e) The procedures to accomplish the change.
    f) How the results are to be evaluated.
7. Establishing that minute by minute control is to be maintained to achieve acceptable risk levels, and that risk assessments will be made as often as necessary as the work progresses.
8. Giving instruction on the action to be taken if, as the work proceeds, unanticipated risks of concern are encountered.
9. Assigning responsibility for acceptance or declination of the change request, including a management of change approval form.
10. Outlining a method to determine the actions necessary because of the effect of the changes (e.g., additional training of operators and maintenance workers; revision of standard operating procedures and drawings; updating emergency plans)
11. Indicating that a final review of the work will take place before startup of operations, and identifying the titles of the persons who are to make the review

RESPONSIBILITY LEVELS

In drafting an MOC system, responsibility levels must be defined and be in accord with an entity's organizational structure. This is a highly important step. In an entity where even minor changes in a process are considered critical as respects employee injury and illness potential, possible environmental contamination, and the quality and safety of the product, the levels of responsibility can be many.

Responsibility levels are clearly set forth in some of the examples of MOC systems included in this article, but not others.

ACTIVITIES FOR WHICH THE MOC PROCESS SHOULD BE CONSIDERED

Hazard and risk complexities of an organization in which change is to take place and the desirability of establishing an MOC system that is adequate but not overly complex should be kept in mind when selecting the activities that are to trigger activating the MOC system. Examples follow from which choices can be made.

- Nonroutine and unusual work is to be performed.
- The work exposes employees to sources of high energy.
- Types of maintenance operations for which prejob planning and safety reviews would be beneficial because of inherent hazards.
- Substantial equipment replacements.
- Introduction of new or modified technology.
- Modifications are made in equipment, facilities, or processes.
- New or revised work practices or procedures are introduced.
- Design specifications or standards are changed.
- Different raw materials are to be used.
- Modifications are made in health and safety devices and equipment.
- Significant changes occur in the site's organizational structure.
- Staffing changes in numbers or skill levels are made that may affect operational risks.
- A change is made in the use of contractors.

MANAGEMENT OF CHANGE REQUEST FORM

Content of a Management of Change Request Form should be in concert with an organization's structure and other management systems in place (e.g., capital expenditure request, work orders, purchasing procedures.) Saving the form as a computer file allows flexibility when descriptive data and comments are to be added.

Examples of such forms can be found in some of the examples that are addenda for this chapter. As would be expected, each form was developed to reflect the views and needs of a particular organization.

IMPLEMENTING THE MANAGEMENT OF CHANGE PROCESS

A review of Chapter 3—Safety Professionals as Culture Changes Agents—is suggested. It might provide some insights on achieving a culture change.

Senior management and safety professionals must appreciate the magnitude of the task they face when initiating activity to implement an MOC system. Expect push-back: egos get in the way; territorial prerogatives may be maintained; the power structure may present obstacles; and the expected and normal resistance to change can be huge since the people affected may have had little experience with the administrative systems being proposed.

Although MOC systems have been required in the chemical industries for many years, the literature indicates that difficulties have been encountered in their application. These comments are found in *Guidelines for Management of Change for Process Safety* (2008).

> Even though the concept and benefits of managing change are not new, the maturation of MOC programs within industries has been slow, and many companies still struggle with implementing effective MOC systems. This is partly due to the significant levels of resources and management commitment that are required to implement and improve such systems. MOC may represent the biggest challenge to culture change that a company faces. (p. 10).
>
> Developing an effective MOC system may require evolution in a company's culture; it also demands significant commitment from line management, departmental support organizations, and employees. (p 11)

Management commitment, evidenced by providing adequate resources and the leadership required to achieve the necessary culture change, must be emphasized. Stated or written management commitment that is not followed by providing the necessary resources is not management commitment. Because of the magnitude of the procedural revisions necessary when an MOC system is initiated, it must be recognized that methods to achieve a culture change are to be applied. The following list contains the subjects to be considered in an attempt to successfully implement an MOC/Pre-Job Planning system.

- Management commitment and leadership must be obtained and demonstrated. That means providing personal direction and involvement in initiating the procedures, providing adequate resources, and by making the appropriate decisions to achieve acceptable risk levels when there is disagreement in the change review process.
- Keep procedures as simple as practicable. An applied system that is less complicated achieves better results than an unused complex system.
- A goal is to obtain widespread acceptance and commitment.

- Communicate extensively. Inform all affected employees in advance of initiating the MOC system, solicit their input, and respect their perspectives and concerns.
- Recognize the need for and provide the necessary training.
- If practicable, go slow—step-by-step.
- Field-test a system prior to implementation. Debugging pays off in the long run.
- After the system is refined through a field test, select a job or an activity for which an MOC system would be beneficial for both productivity/efficiency and safety.
- The purposes of testing the MOC system in a selected activity are to:
  - demonstrate the value of the system,
  - achieve credibility for it, and
  - create a demand for additional application of the system.
- Emphasizing the potential efficiency benefit of an MOC system is encouraged. Doing so should result in more favorable interest in the system being proposed.
- Monitor the progress and performance of the system through periodic audits and through informal inquiry of employees on their perspectives.

RISK ASSESSMENTS

Some of the MOC examples require that risk assessments be made at several stages of the change activity. The intent is to achieve and maintain acceptable risk levels throughout the work process. Thus, risk assessments are to be made as often as needed as changes occur and particularly when unexpected situations arise. In this regard, safety professionals who become skilled in making risk assessments can provide a consultancy that demonstrates significant value.

ANSI/ASSE Z590.3-2011(R2016) is titled *Prevention through Design: Guidelines for Addressing Occupational Hazards and Risks in Design and Redesign Processes*. Risk assessment is its core. Its content is applicable to management of change whether the contemplated change involves new designs or redesign of existing operations.

THE SIGNIFICANCE OF TRAINING

To emphasize the significance of training in achieving a successful MOC system, reference is made again to the *Guidelines for Management of Change for Process Safety* (2008).

Training for all personnel is critical. Many systems failed or encountered severe problems because personnel did not understand why the system was necessary, how it worked, and what their role was in the implementation. (p. 58)

Achieving a culture change to put a successful MOC system in place will not take place without a training program that helps supervisors and workers understand the concepts to be applied. Where the MOC system relates to many risk categories (occupational, the public, environmental, fire protection and business interruption, product quality and safety), the training provided must be more extensive.

DOCUMENTATION

The importance of maintaining a history of operational changes needs emphasis. It is vital that all modifications be recorded in drawings, prints, and appropriate files. They become the historical records that would be reviewed when changes are to be made at a later date. Comments on *changes made that were not recorded* in drawings, prints, and records are found too often in reports on incidents resulting in serious consequences. Examples found in incident investigation reports of unrecorded changes are as follows:

- The system was rewired.
- A blank was put in the line.
- Control instruments were disconnected.
- Relief valves for higher pressures had been installed.
- Sewer line sensors to detect hazardous waste were removed.

ON THE MOC EXAMPLES

To display the substance and variety of the MOC systems organizations have in place, very little change was made in the examples this author received. They vary greatly in content and purpose. Some are contained in one page. Others require several pages to cover the complexity of exposures and procedures. Some MOC procedures have introductory statements on policy and procedure. Some don't. Nevertheless, it is apparent that an MOC system need not necessarily have to meet a theoretical ideal to provide value.

Examples are to serve as references. It is highly recommended that none of the examples be adopted as presented. An MOC system should be drafted in accord with an organization's particular needs and its culture.

As each of the MOC examples received was reviewed, it became apparent that some of the terms used in them were not readily understandable. But more than likely the terms are understood in the organization that developed the MOC system. It was decided to leave those terms in the examples to emphasize that: whatever terminology is included in an MOC procedure, the terms must be in concert with the language commonly used within an organization; and the terms must be understood by all who become involved in an MOC initiative.

Six out of ten MOC examples are included in this chapter as addenda. Because of self-imposed space limitations, the other four are not. A few are lengthy. But all are available on the Internet and are downloadable. Access can be obtained at https://www.assp.org/docs/default-source/psj-articles/manuele_moc_0712.pdf.

This is an American Society of Safety Professionals site for which open access is provided. An earlier and briefer version of this chapter was published in the July 2012 issue of *Professional Safety*, journal of the American Society of Safety Professionals.

Example 1 – For an Operation Producing Mechanical Components

The Pre-Job Planning and Safety Analysis system shown in Example 1, a one-page outline, was developed because of adverse occupational injury experience in work that was often unusual or one-of-a-kind or required extensive and complicated maintenance activity. (Example 1 is an addendum for this chapter.)

Its relative simplicity in relation to other examples will be obvious. But it was successfully applied for its purposes. It would be well to note that safety professionals:

- Prepared the data necessary to convince management and shop floor personnel to try the pre-job planning system they proposed.
- Said that the training sessions held were highly significant in achieving success.
- Emphasized that work situations discussed in the training sessions were real to that organization.
- Addressed the benefits for both productivity/efficiency and risk control in their proposal and in the training session.

For whoever initiates a MOC system, the procedure described in the following will be of interest.

At a location where the serious injury experience was considered excessive for non-routine work, safety professionals decided that something had to be done about it. As they prepared a course of action and talked it up at all personnel levels, from top management down to the worker level, they encountered the usual negatives and push-back: e.g., it would be time consuming, the workers would never buy into the program, and the supervisors would resist the change. The safety professionals considered the negatives as normal expressions of resistance to change.

Their program consisted, in effect, of indoctrinating management and the work force in the benefits to be obtained by doing pre-reviews of jobs so that the work could be done effectively and efficiently while at the same time controlling the risks.

Eventually, management and the line workers agreed that classroom training sessions could be held. Later, the safety professionals said that the classroom training sessions and follow-up training were vital to their success.

At the beginning of each of those sessions, a management representative introduced the subject of Pre-Job Planning and Safety Analysis and discussed the reasons why the new procedure was being adopted. Statistics on accident experience prepared by safety professionals were a part of that introduction.

Then, safety professionals led a discussion of the outline shown in MOC Example 1. It set forth the fundamentals of the pre-job review system being proposed. After discussion of those procedures, attendees were divided into groups to plan real-world scheduled maintenance jobs that were described in scenarios that had been previously prepared.

At this location, supervisors took to the pre-job planning and safety analysis system when they recognized that the system made their jobs easier, improved productivity/efficiency, and reduced the risks. And they took ownership of the system.

As one of the safety professionals said: "Our supervisors and workers have become real believers in the system." And a culture change had been achieved.

Note the requirements under the caption "Upon Job Completion" in the Pre-Job Planning and Safety Analysis Form. The detail of the requirements reflects particular incidents with adverse results that occurred over several years.

It is recommended that every MOC system include similar procedures to be followed before the work can be considered completed.

Example 2 – A Specialty Construction Contractor

This Field Work Review and Hazard Analysis system is encompassed within two pages. It was provided by a safety professional employed by a specialty construction contractor that has several crews active in various places at the same time. Note that the names of employees on a job are to be recorded as having been briefed on the work to be done. The checklist included in the form pertains to occupational, public, and environmental risks.

When asked what gave impetus to the development of the change procedure, the safety professional responded—we learned from costly experience. It was said that the procedures required by the change system are now imbedded in the company's operations and that it is believed that the procedure results in greater efficiency. It is known that fewer costly incidents have occurred.

This example has a direct relation to the purposes of ANSI/ASSE 10.1-2011, a standard for Construction and Demolition Operations titled *Pre-Project & Pre-Task Safety and Health Planning*. This standard is an excellent resource for contractors and for organizations that establish requirements for contractors on their premises. Note the distinctions: Pre-Project Planning and Pre-Task Planning.

Example 3 – A System That Emphasizes Permits and Requires Employee Signatures

This Pre-Task Analysis form makes much of obtaining required Permits and of assuring that supervisors brief employees on the order of activities and of the risks to be encountered. Employees are required to place their signatures on the form, indicating that they have been so briefed. This is the only Example requiring employees to acknowledge, by their signatures, that they have been informed about hazards and risks prior to commencement of work. (Example 3 is an addendum for this chapter).

Example 4 – A Paper Setting Forth a Management Policy and Procedure

This is a basic guidance paper on MOC concepts and procedures, condensed to three pages. It is a composite of several MOC policies and procedures in place issued by organizations in which operations were not highly complex. Reference is made in this example to a MOC "Champion". The point made is that someone has to be assigned the responsibility for the change to be accomplished and manage it through to an appropriate conclusion. (Example 4 is an addendum for this chapter).

Example 5 – A MOC System With a Specifically Defined Prescreening Questionnaire

This MOC system, in three pages, commences with an interesting prescreening questionnaire. If the answer is "No" to all of the questions asked, the formal Management of Change Checklist and Approval form need not be completed for the work being proposed. The question often arises in discussions of Management of Change systems—To what work does the system apply? This organization developed a way to answer that question for its own operations.

Example 6 – A High Risk Multiproduct Manufacturing Operation

This Management of Change Policy and Procedure reflects the high hazard levels in the organization. Captions in this four-page system are: safety; ergonomics; occupational health; radiation control; security/property loss prevention; clean air regulations; spill prevention and community planning; clean water regulations; solid and hazard waste regulations; environmental, safety, and health management systems; and an action item tracking instrument. This appears in the procedure paper announcing the system:

> If a significant change occurs with respect to key safety and health or environmental personnel, the matter will be reviewed by the S&H Manager and the Environmental Manager and a joint report including a risk assessment and their recommendations will be submitted to location management.

Example 7 – A Food Company

This four page Management of Change policy includes product safety and product quality among the subjects to be considered. Detail on the subjects listed is highly technical and extensive. The safety director at this location informed this author that discipline in the application of this MOC system is rigid and that it reflects management's determination to avoid damaging incidents affecting personnel and the environment and variations in product quality and taste. Provisions for pre-start-up and postmodification are extensive. A risk assessment is to be made after changes are made and prior to start-up.

Example 8 – A Conglomerate

Iota Corporation has a five page management of change procedure outlined in four sections. It is worthy of review.

Section I requires completion of a Change Request and Tracking System Requirements form in which the change is described, a tracking number is assigned, and approval levels are established. Approval levels are numerable, including headquarters in some instances.

Section II outlines a Change Review and Approval Procedure which is extensive with respect occupational safety and health and environmental concerns.

Section III is a Pre-Implementation Action Summary Form which lists subjects for which actions are necessary before work on the change can be commenced and the persons responsible for those actions.

Section IV lists eleven points in a Post Completion Form.

Example 9 – An Extensive MOC System

Reading this Management of Change standard is recommended for educational purposes because of its structure and its content. It is somewhat different in relation to the other examples. In its seven pages:

- Requirements for technical changes and organizational changes are dealt with separately and extensively.
- Much is made about organizational changes for which risk assessments are required.
- Technical changes to which the standard applies are outlined in detail.
- Risk assessment is a separately listed item—pertaining to all operations.
- After a lengthy discussion of general considerations, requirements for a six point Management of Change Standard are outlined. They are: Management Process; Capability; Change Identification; Risk Management; The Change Plan; and Documentation. Each of these subjects is thoroughly discussed.

This is a concept and procedural paper. It does not include the forms used to implement the procedures.

Example 10 – An International Multi-Operational Entity

Application of this MOC system, titled Management of Change Policy for Safety and Environmental Risks, extends the activities of safety and environmental professionals beyond that of any other example. It is an informative read. The example shown here covers ten pages. (Example 10 is an addendum to this chapter.) The bulletin, issued by the Safety and Industrial Hygiene entity within the organization, is much longer than is shown here. All exhibits could not be made available because of proprietary reasons. Brief comments follow on the unique aspects of this MOC system.

- Due Diligence is included in a list of Definitions. Safety and environmental professionals are to assess acquisitions et al.
- Global Franchise Management Board Members are listed under Responsibilities. They are to ensure compliance with the standard.
- A Preliminary Environmental, Safety & Health Assessment Questionnaire shall be initiated during the project planning stage.
- Under a Section titled Evaluating Change (Risk Assessment Guidelines) these subjects are included, which may not be included in other examples, at least not as extensively.
  - New Process Product and Development
  - Capital Non-Capital Project
  - External Manufacturing
  - Business acquisitions
  - Significant downsizing/hiring
- Conducting Risk Analyses is a major section.

This is a noteworthy MOC system because of its breadth. It is interesting that the system was issued by the Safety & Industrial Hygiene unit. That implies management support for personnel in that unit and superior operational risk management.

CONCLUSION

It is the intent of this chapter to provide a primer that can serve as a base from which an MOC system suitable for a particular entity can be crafted. This author recommends that safety professionals consider whether the organizations to which they give counsel could benefit from having MOC systems within their overall Operational Risk Management Systems. Having such a system in place for changes that should be prestudied because of their inherent hazards and risks and which may impact on safety, productivity, and environmental controls is good risk management.

REFERENCES

ANSI/ASSE A10.1-2011. *Construction and Demolition Operations: Pre-Project & Pre-Task Safety and Health Planning*. Park Ridge, IL: American Society of Safety Engineers, 2011.

ANSI/ASSE Z590.3-2011 (R2016). *Prevention through Design: Guidelines for Addressing Occupational Hazards and Risks in Design and Redesign Processes*. Park Ridge IL: American Society of Safety Engineers, 2011 (R2016).

ANSI/ASSP Z10.0-2019. *Occupational Health and Safety Management Systems*. Park Ridge, IL: American Society of Safety Professionals, 2019.

ANSI/ASSP/ISO 45001-2018. *Occupational Health and Safety Management Systems— Requirements with Guidance for Use*. Park Ridge, IL: American Society of Safety Professionals, 2018.

The Center for Chemical Process Safety. *Guidelines for Management of Change for Process Safety*. Hoboken, NJ: A joint publication of the Center for Chemical Process Safety of the American Institute of Chemical Engineers (New York, NY) and John Wiley& Sons (Hoboken, NJ), 2008.

MOC Examples. Access to all ten of the MOC systems can be obtained at https://www.assp.org/docs/default-source/psj-articles/manuele_moc_0712.pdf Accessed June 1, 2019.

OSHA's Rule for Process Safety Management of Highly Hazardous Chemicals, 29 CFR 1910.119, issued in 1992. Available at https://www.osha.gov/Publications/osha3132.html#moc. Accessed June 1, 2019. 1992.

OSHA's Voluntary Protection Program (VPP) requirements. Available at https://www.osha.gov/dcsp/vpp/application_sitebased.html. Accessed June 1, 2019. 1982b.

ADDENDUM A

MOC EXAMPLE 1

Alpha
Corporation Form 2668
Issue Date June 1, 2018
Effective June 8, 2018

ALPHA CORPORATION

Pre-Job Planning and Safety Analysis Outline

1. Review the work to be done. Consider both productivity and safety:
   a. Break the job down into manageable tasks.
   b. How is each task to be done?
   c. In what order are tasks to be done?
   d. What equipment or materials are needed?
   e. Are any particular skills required?
2. Clearly assign responsibilities.
3. Who is to perform the pre-use of equipment tests?
4. Will the work require: a hot work permit; a confined entry permit, lockout/tagout (of what equipment or machinery), other?
5. Will it be necessary to barricade for clear work zones?

Advanced Safety Management: Focusing on Z10.0, 45001, and Serious Injury Prevention,
Third Edition. Fred A. Manuele.
© 2020 John Wiley & Sons, Inc. Published 2020 by John Wiley & Sons, Inc.

ALPHA CORPORATION **325**

6. Will aerial lifts be required?
7. What personal protective equipment will be needed?
8. Will fall protection be required?
9. What are the hazards in each task? Consider –
   - Access
   - Work at heights
   - Work at depths
   - Fall Hazards
   - Worker position
   - Worker posture
   - Twisting, bending
   - Weight of objects
   - Elevated load
   - Fire/Explosion
   - Electricity
   - Chemicals
   - Dusts – Noise
   - Weather
   - Sharp objects
   - Welding
   - Vibration
   - Stored energy
   - Dropping tools
   - Pressure
   - Hot objects
   - Forklift trucks
   - Conveyors
   - Moving equipment
   - Machine guarding
10. Of the hazards identified, do any present severe risk of injury?
11. Develop hazard control measures, applying the Safety Decision Hierarchy.
    - Eliminate hazards and risks through system and work methods design and redesign
    - Reduce risks by substituting less hazardous methods or materials
    - Incorporate safety devices (fixed guards, interlocks)
    - Provide warning systems
    - Apply administrative controls (work methods, training, etc.)
    - Provide personal protective equipment
12. Is any special contingency planning necessary (people, procedures)?
13. What communication devices will be needed (two-way, hand signals)?
14. Review and test the communication system to notify the emergency team (phone number, responsibilities).
15. What are the workers to do if the work doesn't go as planned?
16. Considering all of the foregoing, are the risks acceptable? If not, what action should be taken?

Upon Job Completion

17. Account for all personnel
18. Replace guards
19. Remove safety locks
20. Restore energy as appropriate
21. Remove barriers/devices to secure area
22. Account for tools
23. Turn in permits
24. Clean the area
25. Communicate to others affected that
26. Document all modifications to the job is done prints and appropriate files

ADDENDUM B

MOC EXAMPLE 2

Beta Company Form 1812
Issue Date June 1, 2008
Effective June 8, 2008

BETA COMPANY

Field Work Review and Hazard Analysis

All work done by our field crews involves change, whether for new system installations or modifications of existing systems. To assure that assignments are carried out efficiently and that our crew members are not in situations where the risks are unacceptable, the crew supervisor shall complete Form 1812 prior to the start of an assignment to assure adequate planning.

As the supervisor considers the work to be done, a discussion/briefing will be held with crew members to obtain their input and to assure that all crew members understand the assignment and their responsibilities. Crew members will be reminded that if hazardous situations arise or work conditions change substantially, they are encouraged to call a halt to the work and look out for other crew members. If that occurs, a new Form 1812 must be prepared.

Nearly all of our field assignments require several days for completion. Form 1812 will be reviewed at the start of each day for that day's briefing on the work to be done.

Each crew member will have communication equipment and the phone numbers to be called in the event of an emergency.

Lindsey Wolfe
Division Manager

Advanced Safety Management: Focusing on Z10.0, 45001, and Serious Injury Prevention, Third Edition. Fred A. Manuele.
© 2020 John Wiley & Sons, Inc. Published 2020 by John Wiley & Sons, Inc.

BETA COMPANY **327**

Form 1812: Field Work Review and Hazard Analysis

Daily, crew supervisors are to review the job plan with crew members, seeking their input, and whenever the scope of the job changes.

Beta Company Assigned Number _____ Job Description _____

_____

_____

Job Location _____

_____

Briefing Date _____

Start Time _____

Completion Date _____

_____

Supervisor _____

Cell Phone Number_____

Names of Crew Members for This Job   Emergency Phone Numbers

_____

_____

_____

_____

_____

_____

List Each Job Step, the Potential Hazards, and the Actions to be Taken to Achieve Acceptable Risk Levels.

1. _____

_____

2. _____

_____

3. _____

_____

MOC EXAMPLE 2

Check Each Applicable Subject	To Be Considered
_____ Housekeeping, trash removal	_____ Clean up procedure and responsibility assigned
_____ Public exposure	_____ Isolating work area, prevent public assess
_____ Environment discharge potential	_____ Assessment, permits, supervision
_____ Capabilities of crew for work to be done	_____ Call for special subject skill, additional training
_____ Working at heights or while elevated	_____ Fall protection, Ladder safety
_____ Cranes, hoisting equipment	_____ Certified personnel, conditions, pre-use inspection
_____ Entering a confined space	_____ Confined space permit, testing, trained crew
_____ Welding, soldering, cutting	_____ Hot work permit, supervisory controls
_____ Machining, grinding, drilling, cutting	_____ Guarding, health evaluation by industrial hygienist
_____ Chemicals	_____ MSDS, industrial hygienist counseling
_____ Heavy or awkward lifting	_____ Use of mechanical equipment, teamwork
_____ Hazardous energy: electrical, hydraulic pneumatic, thermal, chemical	_____ Lockout/Tagout, personnel skills, knowledge of hazards, supervision

_____ Crew has been reminded that if hazardous situations arise or work conditions change so that additional risks are encountered, the work must stop.

The work steps, sequentially, have been reviewed on the job site and discussed with the crew.

Supervisor's Signature _____

Date _____ Phone _____

Copy to Group Manager: Name _____

Date _____ Phone _____

ADDENDUM C

MOC EXAMPLE 3

GAMMA CORPORATION

Pre-Task Analysis

Completed form must be submitted and work must be authorized before activity is commenced

Submitted by _____ Date _____ Location _____

Description of work _____
_____
_____
_____

Commencement date _____ Expected completion date _____

Permits, special skills and licenses: Work must not be commenced if required permits have not been received or if arrangements have not been made for the special skills and licenses required.

Advanced Safety Management: Focusing on Z10.0, 45001, and Serious Injury Prevention,
Third Edition. Fred A. Manuele.
© 2020 John Wiley & Sons, Inc. Published 2020 by John Wiley & Sons, Inc.

MOC EXAMPLE 3

Permits	**Required**		**Received**	
Confined space	___ Yes	___ No	___ Yes	___ No
Electrical	___ Yes	___ No	___ Yes	___ No
Combustion equipment	___ Yes	___ No	___ Yes	___ No
Excavation	___ Yes	___ No	___ Yes	___ No
Hot work	___ Yes	___ No	___ Yes	___ No
Lifting-rigging	___ Yes	___ No	___ Yes	___ No
Fire systems	___ Yes	___ No	___ Yes	___ No

			Obtained	
Will special skills be required	___ Yes	___ No	___ Yes	___ No
Will special licenses be needed	___ Yes	___ No	___ Yes	___ No

Check off each of the following that apply. Each checked subject is to be addressed in the pre-task planning.

___ Lockout/Tagout	___ Pinch points	___ Inadequate access	___ Hot/cold/burns
___ Chemical exposure/spill	___ Electrical shock	___ Sharp objects	___ Excavations
___ Equipment loading/unloading	___ Asbestos	___ Falling objects	___ Particles in eye
___ Elevated work	___ Fall from height	___ Manual lifting	___ Lighting
___ Lifting or rigging	___ Mobile equipment	___ Isolated area	___ Radiation
___ Fire/explosion	___ Ladder usage	___ Confined space	___ Heat stress
___ High noise levels	___ Scaffolding	___ Inhalation hazard	___ Other
___ Other	___ Other	___ Other	___ Other

Specify personal protective equipment needed _____
_____
_____

Specify special equipment needed if any _____
_____
_____
_____

If unexpected and unacceptable risks are encountered, it must be understood that work is to be stopped until the situation is resolved. This provision must be made clear in pre-job discussions and in employee briefings.

Employee briefings. Supervisors are to assure that all employees involved in the work are briefed on the order of and the procedures to be followed and of risks considered in the pre-job discussions.

Name	Signature	Date	Name	Signature	Date

Additional space is provided for the briefings of personnel added after the work is commenced.

Name	Signature	Date	Name	Signature	Date

Notifications: Names of persons to be notified before the work is commenced and after it is completed.

Pre-work commencement

Names	Department
_____	_____
_____	_____
_____	_____

Upon completion

Names	Department
_____	_____
_____	_____
_____	_____

For emergencies encountered, list names and phone numbers of the people who should be notified.

Names	Department	Phone Number
_____	_____	_____
_____	_____	_____
_____	_____	_____

Approval signatures

Safety and environmental personnel, who are to assure that foreseeable hazards have been identified and risk control methods are appropriate

_____ Date _____
_____   _____
_____   _____

Manager with authority to approve the project

_____ Date _____

Upon completion

All personnel, tools and equipment accounted for ____ Yes ____ No
Operation successfully restored   ____ Yes ____ No

Signature: Project initiator _____ Date _____

ADDENDUM D

MOC EXAMPLE 4

A COMPOSITE MANAGEMENT SYSTEM GUIDE FOR MOC SYSTEMS

Using several resources on management of change and reflecting on this author's experience, the following general statement on a management of change policy and procedures is offered as a guide for reflection by those who initiate MOC systems. Note that it encompasses both occupational safety and environmental considerations. It is a brief management guide only and is presented as such. It does not contain any forms. This composite guide is worthy of thought. It will serve well as a resource in initiatives to adopt the management of change provisions in Z10.0. Forms to which reference is made are not included.

Management of Change Policy and Procedures: Occupational Safety and Health and Environmental Considerations

1. Overview

This policy defines the requirements for a Management of Change process with respect to occupational safety and health and environmental considerations.

2. Purpose
This policy establishes a process for evaluating occupational safety and health and environmental exposures when operational changes are made so as to control the internal risks during the change process and to avoid bringing new hazards and risks into the workplace.

3. Scope
This policy applies to all operations at this location: there are no exceptions

4. Responsibility
  a. The Management Executive Committee, with counsel from the senior safety professional, is responsible for establishing processes to determine when these management of change procedures shall apply, how they are to be implemented and by whom, and to follow the processes through to an effective conclusion.
  b. Facility management is responsible for ensuring that these processes and procedures to address the safety, health and environmental implications of operational changes are implemented within their areas of responsibility.
  c. Employees at all levels – division managers, supervisors, line operators and ancillary personnel – after receiving training on this policy and process, are responsible within their domains of influence for initiating communications on operational changes that may impact on safety, health and environmental considerations and for the implementation of this policy and process.
  d. All safety professionals have responsibility to identify operational changes that require study as to hazard and risk potential and to bring their observations to management's attention through their organizational structures.

5. Application
  a. This policy applies to all operational changes that may potentially impact on the safety and health of employees and on our environmental controls.
  b. This policy will be implemented as an adjunct to all issued Safety, Health an Environmental policies and procedures, but particularly to our Design and Procurement of Equipment and Facilities Procedure.
  c. The senior safety professional, and other safety professionals with particular skills, shall participate routinely in management staff meetings when operational changes are discussed.
  d. A safety professional will sign-off on the change plans considering the provisions of our Design and Procurement of Equipment and Facilities Procedure and on the New Product Development process.
  e. Examples of operational changes to which this policy and process may apply include:
    - Unusual, non-routine, non-production work, work where high energy exposures are contemplated, and maintenance projects for which the

scope of the work requires a determination that pre-job planning and safety analysis would be beneficial

- Revisions in operating methods and procedures
- Revised production goals
- Plans to lower operating costs
- Revisions in staffing levels, upward or downward
- Organizational restructuring
- Revisions in the environmental management system
- New product development
- Adoption of new information technology that has an impact on operations
- Changes in safety, health or environmental regulations
- Acquisitions, mergers, expansions, relocations or divestitures

6. Management of Change System

When an operational change is identified that requires study as respects its impact on occupational safety, health or environmental controls, a person at an appropriate management level shall be appointed the Management of Change Champion to chaperon the review process to a conclusion.

    a. That person, having obtained counsel from safety professionals, will determine how extensive the review procedure will be and decide on whether:

- Completion of a Management of Change Request Form is necessary (Figure 1 in this chapter) within which the accountability and sign-off levels are set forth
- Multidisciplinary group discussions are to be held to encompass the content of the Pre-Job Planning and Safety Analysis Outline (Addendum B in this chapter)
- The hazards and risks that may result as the operational change moves forward are of greater significance and require appointment of an ad hoc Management of Change Committee to oversee the project. Safety professionals having the necessary skills are to be members of such committees.
- Our Capital Expenditure Request Procedure is to be initiated

7. The Management of Change Champion shall:

    a. Assure that input on the operational change has been obtained from all who might be impacted

    b. Arrange for resources, staffing and scheduling to accomplish the change

c. Schedule the necessary risk assessment
   d. Obtain comments from line operating personnel on their views on how the hazards and risks can be ameliorated, and their concerns
   e. Get a sign-off from safety personnel
   f. Follow the review process to a logical conclusion
   g. Arrange a final review of the changes made to assure that hazards and risks have been properly addressed
   h. Determine that residual risks, after the risk reduction and control measures have been taken, are acceptable
   i. See that documentation is appropriate

   - The Management of Change Champion is to give emphasis to documenting all changes made that should be recorded in prints and appropriate files so that persons who make further changes at a later date will know precisely what was done.

   - In making decisions on what documentation is to be made, this principle is to apply: be super cautious and consider later needs.

   - Risk assessments are to be retained in accord with our document retention policy

   j. Have Standard Operating Procedures modified as necessary?
   k. Determine whether additional training is necessary

8. Training

   a. Personnel responsible for safety, health and environmental training at management, supervisory, line worker, and ancillary personnel levels shall incorporate this policy and process into the training curricula.
   b. Employees made shall receive training on the revised Standard Operating Procedures before changes are finalized.
   c. In the training process, employees will be properly informed with respect to their assuming responsibility for the aspects of safety, health and environmental matters over which they have control.

ADDENDUM E

MOC EXAMPLE 5

Advanced Safety Management: Focusing on Z10.0, 45001, and Serious Injury Prevention,
Third Edition. Fred A. Manuele.
© 2020 John Wiley & Sons, Inc. Published 2020 by John Wiley & Sons, Inc.

Epsilon Corporation
Management of change Pre-Screening Questionnaire (FormS8861)

Date: _____ Change Originator _____
Project Title: _____ Project Location _____
Description of Work _____
_____
_____

Instructions

Answer the following questions to determine if the full MOC process needs to be followed. If the answer to all of these questions is "No" then the MOC Checklist does not have to be completed. If the answer to any question is "Yes", the MOC Checklist Form (S8862) and a Job Safety/Risk Assessment Form (S8864) must be completed.

Yes	No	
___	___	Will process pipe work, plumbing or electrical equipment be installed, relocated or removed?
___	___	Will work be done on equipment or ductwork that contains asbestos or lead-based paint?
___	___	Will ductwork be installed, relocated or removed?
___	___	Will there be any changes to environmental controls?
___	___	Will equipment be installed that requires more than plugging into a 110-volt outlet?
___	___	Will there be any changes to HVAC systems?
___	___	Will there be additions/removals/changes to pollution control devices?
___	___	Will the changes have proposed impact on existing emergency exit routes?
___	___	Will temporary installations of plumbing, electricity or other process lines be made?
___	___	Will new chemicals, chemical waste or waste water streams be introduced to the site?
___	___	Will the change impact on life safety equipment and fire protection?
___	___	Will this change require new permits or revisions of existing permits?
___	___	Will additional air emissions, water discharges or hazardous streams be generated?

It is understood that additional pages may be necessary to describe changes of substance. For emphasis, it is stated again that a risk assessment determination shall be completed for all Yes answers. A signed MOC checklist, when necessary, is to be provided by the Change Originator to the MOC Coordinator and the ESH Manager.

Epsilon Corporation
Management of Change Checklist Form S8862

Change Originator _____ **Date Initiated** _____ **Date Closed** _____

Change Type

_____ Permanent _____ Temporary _____ Days Required

_____ Emergency: To Prevent

___ Injury ___ Release ___ Equipment Damage ___ Process Shutdown

To Be Completed By Change Originator

Description of and Reason for the Change (Include possible process and ESH impacts of the work.)

_____
_____
_____
_____
_____

Are drawings, sketches, revised procedures attached? ___ Yes ___ No ___ Number of attachments

Complete the Following Checklist. Discuss with ESH, MOC Coordinator, Maintenance, Operations, and Engineering as needed, based on the level of risk. Attach additional information if the situation requires.
Check Yes, No, or Unsure for each issue.

Y	N	U		Y	N	U	
__	__	__	New or modified equipment	__	__	__	Uses new chemical or creates chemical hazard
__	__	__	New or modified procedure	__	__	__	Change in PPE requirements
__	__	__	New or modified facilities infrastructure	__	__	__	Bypass of safety systems, alarms or safeguards
__	__	__	New or modified utilities	__	__	__	Change of fire alarm settings required
__	__	__	Potential for process reverse flow	__	__	__	Requires new JSA or risk assessment
__	__	__	Change in emergency exit routes	__	__	__	Involves facility demolition, renovation or repair
__	__	__	Requires new or revised permits	__	__	__	Results in modification or relocation of work area

MOC EXAMPLE 5

___ ___ ___ Impairs Fire Protection Systems and Requires insurance carrier notification

___ ___ ___ Other

___ ___ ___ Other

___ ___ ___ May create tripping or ergonomics hazards

___ ___ ___ Requires insurance carrier notification and review

___ ___ ___ Other

___ ___ ___ Other

Submit comments on all Yes answers and any other safety concerns. A risk assessment shall be completed for all Yes answers. Provide a copy of the signed off Checklist to the Epsilon Corporation MOC Coordinator and to ESH.

Type of Training Required for Epsilon Personnel and Contractors Performing the Work. A Record is to be Made of All Training Provided.

_____ Verbal _____ Hands on/demonstrated _____ Written and documented

Required Signatures/Approvals

Change Originator	Date	MOC Coordinator	Date
ESH Manager	Date	Facility Manager	Date
Other:	Date	Other	Date

Items Required Before MOC Checklist is Considered Completed

With Respect to the Project
_____ Work as described has been completed
_____ Pre-startup ESH and project/task manager completion review and documentation

Required training has been completed
_____ Operations
_____ Maintenance
_____ ESH

Process Information Has Been Updated
- _____ MSDS's
- _____ Site engineering drawings, block diagrams, plans etc. have been updated
- _____ Piping/plumbing/process diagrams have been revised
- _____ Above and below ground utility changes have been documented
- _____ Insurance companies, municipalities and other authorities have been informed

Procedures Have Been Updated
- _____ Operating
- _____ Maintenance
- _____ ESH
- _____ Other

ADDENDUM F

MOC EXAMPLE 6

Zeta Procedure K3334

Issue Date: May 16, 2017

Effective Immediately

Zeta Corporation

Management of Change Policy and Procedure

This procedure is to provide a systematic approach for evaluating and addressing environmental, safety and health risks when changes are made at our facilities. It is applicable in all areas when changes are made with respect to, but not limited to, the following:

- Changes that affect the process or equipment
- Significant changes to operating or maintenance procedures
- Development and introduction of new products
- Addition of new raw materials or feedstock
- Significant changes to existing raw materials or feedstock.
- New business ventures
- Discontinuation of operations or processes
- Changes that result in new "stand alone" equipment or process modifications
- **Changes affecting staffing** – additions needed and as respects departing personnel
- Changes that affect the community, and environment situations as a result of facility expansion, modifications and/or new process sites
- Shutdown activities (e.g. annual maintenance shutdown, special shutdown to install new equipment, etc.)

Management of Change Checklist Form J8865 has been developed to assist personnel in fulfilling their obligations.

If a significant change occurs with respect to key safety and health or environmental personnel, the matter will be reviewed by the S&H Manager and the Environmental Manager and a joint report including a risk assessment and their recommendations will be submitted to location management.

Bruce Wayne
President

Management of Change Checklist: Form J8865

Section 1.

Procedure Number _____ Issue Date _____
Revision Date _____ Expiration Date _____
Names of: Originator _____
Department Manager _____
Location Manager _____
Group Director (if required) _____
Description of Change _____
_____
_____

Basis for the Change: Essential to Operation _____
Safety Improvement _____ Environmental Control _____ Quality _____
Fire Protection _____ Productivity _____ Other _____

A Check Mark Shall Be Entered for Each of the Following Items. For Items Marked "Action Needed", Comments on Their Resolution Are to Be Made in Section 12. The Originator will have risk assessments made as necessary as the change proceeds.

	Action Needed	No Action Needed		Action Needed	No Action

Section 2. Safety

2.1 Machine Guarding	__	__	2.11 Exits/Access and Evacuation Plans	__	__
2.2 Lockout/Tagout	__	__	2.12 Emergency Response Planning/Training	__	__
2.3 Need for Agency Approvals	__	__	2.13 Signage	__	__
2.4 Fall Protection and Tie Off Points	__	__	2.14 Electrical Devices/ Conductors/ Disconnects	__	__
2.5 Confined Space Entry	__	__	2.15 Arc Flash Analysis	__	__
2.6 Contractor Safety	__	__	2.16 Single Line Drawing	__	__
2.7 Walking/Working Surfaces/Ladders	__	__	2.17 Safety Management System Changes	__	__
2.8 PPE Assessment/ Certification	__	__	2.18 Job Safety Analysis	__	__
2.9 Powered Industrial Trucks	__	__	2.19 Mobile Equipment	__	__
2.10 Material Handling Equipment	__	__	2.20 Crane or Hoist Usage	__	__

	Action Needed	No Action Needed		Action Needed	No Action

Section 3. Ergonomics

	Action Needed	No Action Needed		Action Needed	No Action Needed
3.1 Repetitive Motion	—	—	3.4 Vibration/Employee Exposure	—	—
3.2 Handling Objects/over 50 lbs	—	—	3.5 Grip Strength over 20 lbs	—	—
3.3 Awkward Body Motions	—	—			

Section 4. Occupational Health

	Action Needed	No Action Needed		Action Needed	No Action Needed
4.1 Industrial Hygiene Assessment	—	—	4.6 Personal Protective Equipment	—	—
4.2 Exposure to New Materials	—	—	4.7 Medical Surveillance	—	—
4.3 Emergency Shower/Eye Wash	—	—	4.8 MSDS for New Materials	—	—
4.4 Radiation	—	—	4.9 Exposure Monitoring	—	—
4.5 Noise Control	—	—	4.10 Warning Systems	—	—

Section 5. Radiation Control

	Action Needed	No Action Needed		Action Needed	No Action Needed
5.1 Permit or License	—	—	5.3 State or federal Reporting	—	—
5.2 Inventory Status	—	—	5.4 Posting/Labeling Requirements	—	—

Section 6. Security/Property Loss Prevention

	Action Needed	No Action
6.1 Emergency Power Back-up Systems	—	—
6.2 Plant Security Provisions	—	—
6.3 Traffic, Roadways, Emergency Egress	—	—
6.4 Flammable and Hazardous Materials Storage and Handling	—	—
6.5 No Smoking signs in Appropriate Areas	—	—
6.6 Fire Resistant Structures, Firewalls, Construction or Processes Requiring Special Protection	—	—
6.7 Temporary Perimeter Security During Construction	—	—
6.8 Additional Permanent Perimeter Security	—	—
6.9 Explosion or Implosion Hazard Protection	—	—
6.10 Building Permits	—	—
6.11 Water Supply or Fire Protection Systems Extension	—	—
6.12 Special Startup/Shutdown Requirements	—	—
6.13 Seismic Evaluation	—	—
6.14 Property Insurance Company Standards and Necessary Review	—	—

Section 7. Clean Air Regulations

	Action Needed	No Action
7.1 Will the changes being made impact on existing applications for and permits received?	—	—
7.2 Construction and Operating Permits Needed?	—	—
7.3 Will Emissions Controls Be Affected?	—	—
7.4 Revisions in Operations and Maintenance Plan?	—	—
7.5 Will EPA Industrial Boiler Maximum Control Technology (MACT) Be Affected?	—	—
7.6 Additional Monitoring Requirements?	—	—
7.7 EPA Risk Management Planning (RMP) Impacted?	—	—
7.8 New Source Review (NSR) Applicability or Permitting?	—	—
7.9 Will EPA Prevention of Serious Deterioration (PSD) Rules Be Impacted	—	—

Section 8. Spill Prevention and Community Planning

8.1 Spill prevention control and counter-measures	—	—
8.2 Emergency response and coordination with community agencies	—	—
8.3 Toxic release inventory requirements	—	—
8.4 Hazardous chemical inventory /regulatory requirements	—	—
8.5 New chemicals or raw materials requiring regulatory notification	—	—

Section 9. Clean Water Regulations/ Potable Water Waste Regulations

	Action Needed	No Action Needed
9.1 Process Water Permitting	—	—
9.2 Storm Water Permitting	—	—
9.3 Strom Water Pollution Prevention	—	—
9.4 Wastewater Treatment System	—	—
9.5 Water System distribution	—	—
9.6 Back Flow Prevention/PotableWater	—	—
9.7 Land Disturbance Permit Required	—	—
9.8 Publicly Owned Treatment Works	—	—

Section 10. Solid and Hazardous

		Action Needed	No Action
10.1	Generator Status Impacted	—	—
10.2	Accumulation and Storage Areas	—	—
10.3	New Waste Stream Created	—	—
10.4	Disposal Options	—	—
10.5	Waste Minimization Programs	—	—
10.6	Asbestos Exposure and Abatement	—	—
10.7	Empty Container Handling	—	—
10.8	Recycling and Reuse Requirements/Limitations	—	—
10.9	Substitutions for toxic materials	—	—
10.10	Rag /Wipe Storage and Cleaning	—	—
10.11	Demolition Contamination & Disposal & Recovery Act Regulations	—	—
10.12	EPA Resource Conservation	—	—

Section 11. Environmental, Safety, and Health Management System

		Action Needed	No Action
11.1	Will environmental, safety, or health work procedures be impacted?	—	—
11.2	Are training materials adequate for this project?	—	—
11.3	Will training be necessary for our employees and contractors for this project?	—	—
11.4	After-project-completion, will additional training be necessary?	—	—

Section 12. Action Item Tracking Instructions. Enter the section number where "Action Needed" was recorded, a brief description of the action to be taken, the name of the person to whom the action is assigned, expected completion date, and the signature of the person verifying completion of the action, and the date.

12.1 **Planning /** Action items that must be completed during the planning and design phase of the project.
12.2 **Pre-Construction.** Action items that must be completed prior to the construction phase of the project.
12.3 **Pre-Start-Up.** Action items that must be completed prior to the start-up phase of the project.
12.4 **Normal Operation.** Action items that must be completed when normal operations are commenced.

For each of the sub-parts 12.1 through 12.4 of Section 12, the following format applies.

348   MOC EXAMPLE 6

Sec. No.	Give a Brief But Adequate Description of the Action Needed	Responsible Person	Due Date	Signature, Completion	Completion Date

Section 13. Approval Signatures and Dates

Title	Signature	Date
Initiator		
Department Manager		
Location Manager		
Group Director		
Environmental Manager		
Safety and Health Manager		

CHAPTER 15

THE PROCUREMENT PROCESS

Both the Z10.0 and 45001 standards require that what is purchased conforms to the organization's occupational health and safety management system. In reality, the purpose of the procurement process is to avoid bringing hazards and their accompanying risks into the workplace. To assist safety professionals as they give advice on implementing those provisions, this chapter:

- Duplicates the procurement requirements of the standards.
- Comments on prevalent purchasing practices.
- Establishes the significance of the procurement processes.
- Discusses the pre-work necessary to include safety specifications in the procurement process.
- Provides some resources.
- Comments on available occupational health and safety purchasing specifications.
- Gives examples of design specifications that become purchasing specifications.

Advanced Safety Management: Focusing on Z10.0, 45001, and Serious Injury Prevention, Third Edition. Fred A. Manuele.
© 2020 John Wiley & Sons, Inc. Published 2020 by John Wiley & Sons, Inc.

PROCUREMENT REQUIREMENTS IN 45001 AND Z10.0

45001—8.1.4 Procurement

In the 45001 standard, the procurement requirements are stated in this one sentence: The organization shall establish, implement and maintain process(es) to control the procurement of products and services in order to ensure their conformity to its OH&S management system.

But the guidance given in Annex A on procurement is more precise. Organizations are advised to have systems in place "to determine, assess and eliminate hazards and reduce occupational health and safety risks" for a large variety of subjects. Those processes are also to ensure that purchased goods or services are safe to use and are "delivered according to specifications and (are) tested to ensure that (they) work as intended."

Z10.0—8.6 Procurement

In Z10.0, the organization shall establish a process to:

A. Identify and evaluate the potential OHS risks associated with purchased products, raw materials, and other goods and related services before introduction into the workplace;
B. Establish requirements for supplies, equipment, raw materials, and other goods and related services purchased by the organization to control potential OHS risks, and
C. Ensure purchased products, raw materials, and other goods and services conform to the organization's OHS requirements.
D. Address outsourced arrangements where there is an impact on the performance of the OHSMS.
E. Ensure that contractors are capable of meeting the requirements of its OHS management system.

WHY EMPHASIZE HAVING SAFETY-RELATED SPECIFICATIONS IN PROCUREMENT DOCUMENTS?

In Item B, above in the Z10.0 requirements, the organization is to "Establish requirements." In 45001, it is assumed that there are specifications for procurement. Because of my experience, I make much of safety professionals participating in establishing safety-related specifications for procurement documents. An example of that experience follows.

> When our safety through design committee discussed the possible benefits of adopting the design concepts, we were promoting in the 1990s, our chair person, who was the head of industrial engineering in a major company, listened

until we thought we were finished and then moved to the board on which we had written the benefits.

Tactfully, he let us know that we had omitted one of the most important benefits which was to avoid the cost of retrofitting. He made it clear that in his experience, retrofitting can be costly. That was a new thought to me. I called eight colleagues who were in corporate safety positions and asked whether they had experienced a situation in which equipment was brought into a location that required considerable safety-related retrofitting.

Every person I called had a story to tell. And sometimes fitting safety-related needs into equipment that had already been built was notably expensive. They acknowledged that having safety-related specification in all purchasing agreements where they were needed would be beneficial.

For equipment in place, retrofitting to accommodate safety needs starts with evaluating the deficiencies in the equipment in operation—that is, identifying what was overlooked in the design or purchase processes. Unfortunately, the resulting level of risk when safety requirements are addressed through retrofitting may be higher than would be the case if appropriate safety specifications were included in the bid or purchasing papers.

As retrofitting proceeds, it is easy for the decision-makers to rationalize acceptance of higher risk levels. Nevertheless, the goal should be to achieve acceptable risk levels.

NEED FOR A CULTURE CHANGE

Getting managements and purchasing personnel to adopt the procurement provisions in Z10.0 or 45001 will not be easy. A safety professional who proposes adding procurement provisions as an element in safety and health management systems and to have safety specifications included in a company's purchasing practices should expect the typical resistance to change. In most places, a culture change will be necessary for success.

An interpretation of the procurement requirements in 45001 and Z10.0 could be—safety and health professionals, you are assigned the responsibility to convince managements and purchasing agents that, in the long term, it can be very expensive to buy cheap.

SIGNIFICANCE OF THE PROCUREMENT PROVISIONS

Is it obvious that this author places great emphasis on having the procurement provisions in Z10.0 and 45001 becoming an element in safety and health management systems because doing so prevents introducing hazards and risks into the workplace. This is the thinking that supports that position:

352   THE PROCUREMENT PROCESS

Risks of injury derive from hazards which are defined as the potential for harm. If hazards are properly addressed and avoided, eliminate, reduced or brought under control in the design process so that the risks deriving from them are at an acceptable level, the potential for harm or damage and operational waste is decreased.

A logical extension of addressing hazards and risks in the design process is to have the design specifications the organization decides upon included in purchase orders and contracts so that suppliers and vendors know what safety specifications are to be met. That reduces the probability of bringing hazards into the workplace.

Although having extensive safety specifications included in purchase orders or contracts is not a broadly applied practice, safety professionals are encouraged to consider the benefits to be achieved if they are. If the ideal is attained in the purchasing process and hazards and risks brought into the workplace are as low as reasonably practicable, the result is significant risk reduction, which means fewer injuries.

Unfortunately, the only safety-related terminology in purchase orders and contracts may be to—meet all Occupational Safety and Health Administration (OSHA) and other governmental requirements.

PRE-WORK NECESSARY FOR PROCUREMENT APPLICATIONS

As is said in Chapter 12, "Safety Design Reviews," there is a close relationship between establishing safety design specifications and including safety specifications in procurement documents. The latter cannot be successfully achieved until the former has been accomplished. Once safety design specifications are established, the next step is to have them applied internally for facilities and equipment in place. Then, an appropriate extension is to have them incorporated into purchase orders and in contracts.

It is common practice for vendors and suppliers of equipment to use their own, and possibly inadequate, safety specifications if a purchaser has not established its own requirements. That could end up being costly for the purchaser, especially if production schedules are delayed and the retrofitting expense to get systems operating as designed and for safety purposes is substantial.

AVAILABLE HEALTH AND SAFETY PURCHASING SPECIFICATIONS

Several safety-related texts were reviewed to determine whether they give guidance on including safety specifications in the procurement process. They do not. But the Z10.0 committee is preparing a handbook for the application of the standard, and one of the chapters is on procurement. Quite probably, the British Standards Institute will issue a comparable publication related to 45001.

Several examples of purchasing documents that include safety specifications are posted on the Internet. It seems that a large share of them was issued by educational institutions. In the following list, the first is a paper issued by the European Agency for Safety and Health at Work. It is a 76-page, undated guideline on safety and health and the supply chain. It is on the Internet, downloadable, and, they say, it can be reproduced provided the source is acknowledged. This is what is said in the Executive Summary:

> In today's global and national economies, businesses increasingly rely on the outsourcing of parts of their activities and processes. Companies thus function and compete more and more on a supply-chain level, in specific networks with their suppliers and service providers. This outsourcing trend and growing importance of supply chains has its implications for the working conditions and health and safety of workers of supplier and contracting companies.
>
> This report therefore sheds light on occupational safety and health (OSH) within these complex supply chain networks. Based on a literature, policy and case study review it attempts to give an overview of how OSH can be managed and promoted through the supply chain, and which drivers, incentives and instruments exist for companies to encourage good OSH practices among their suppliers and contractors.

- European Agency for Safety and Health at Work
  Promoting occupational safety and health through the supply chain (Undated)
- University of Wollongong, New South Wales, Australia.
  WHS Purchasing Guidelines (WHS – Workplace Health and Safety), 2016.
- SafeProd
  Safety Requirements Specification Guideline – 19 pages, 2015
- Environmental services Association
  Designing Health and Safety into Procurement and Contract Management, undated
- Delphi
  Design-In Environmental Health and Safety Specification, 2013
- OSHWIKI
  Occupational safety and health in the supply chains, 2013

Examples of safety-related design specifications that become purchasing specifications are not easily acquired. Most companies consider their specifications proprietary and do not make them available to others freely. For the moderate-sized company with a limited engineering staff, writing design and purchasing specifications will not be easy to do.

It seems appropriate to suggest that organizations prevail upon the business associations of which they are members to undertake writing generic design specifications and purchasing specifications that relate to the hazards and risks inherent in their operations. Safety professionals should consider this inadequacy as an opportunity to be of service and make their presence felt.

OPPORTUNITIES IN ERGONOMICS

Safety professionals who are not involved in the design or purchasing processes could consider ergonomics as a productive field in which to get started. Some comments made in Chapter 12, "Safety Design Reviews," are repeated here because they also apply to procurement provisions in Z10.0 and 45001.

Musculoskeletal injuries, ergonomically related, are a large segment of the spectrum of injuries and illnesses in all industries and businesses. Since they are costly, reducing their frequency and severity will show notable results. Furthermore, it is well established that successful ergonomics applications result not only in risk reduction but also in improved productivity, lower costs, and waste reduction.

GENERAL DESIGN AND PURCHASING GUIDELINES

Addendum A in this chapter is a combination of design and purchasing guidelines currently in use. Note again that design specifications were developed that became purchasing specifications. This document is presented here as a reference from which engineering personnel and safety professionals can make selections and add, subtract, or alter items to suit location needs.

It would be inappropriate to implement these Guidelines without study and adjustment to reflect the hazards and risks inherent in a particular entity.

Adoption of a modification of the Guidelines will usually require persuasive discussion with the purchasing staff. Safety professionals must understand the enormity of what is being undertaken when they try to have safety specifications included in purchasing documents when that is not the current practice.

Nevertheless, the productivity, risk reduction, and waste saving benefits of a process that avoids bringing hazards and risks into the workplace cannot be refuted.

The Guidelines commence with Sections on General Safety Requirements, Machine Guarding, Industrial Hygiene, Ergonomics, Machine and Process Controls, and Environmental Impact/Hazard Evaluation. A Section 8 sets forth a Procedure for which the instruction is "Use this document as a guide whenever purchasing new (or modifying existing) equipment."

Major parts of that Section are captioned: Codes and Standards; Equipment/Fixture Design; Mechanical—Design and Construction; Electrical—Design and Construction; Pneumatics – Design and Construction; Software; and Machine Guarding.

These Guidelines are quite broad. They relate to occupational safety and health, environmental concerns, productivity, and avoiding events that result in business interruption.

CONCLUSION

Too much stress cannot be placed on the significance of the Procurement provisions in Z10.0 and 45001 and the benefits that will derive from their implementation. It stands to reason—if the purchasing process limits bringing hazards and risks into the workplace, the probability of incidents resulting in injury or illness will be diminished. That is what Z10.0 and 45001 are all about: To reduce the risk of occupational injuries, illnesses, and fatalities.

Although Addendum A to this chapter is lengthy, it is recommended that safety professionals read it to gain an appreciation of how extensive design and purchasing specifications can be and how they can serve well to avoid bringing hazards and risks into the workplace.

Addendum B is a find. It is brief but inclusive for the entity that issued it. Purchasing guidelines developed at the University of Wollongong in Australia are an excellent resource for safety professionals and highly recommended.

REFERENCES

CIPS Knowledge Works. *Risk Management in Purchasing and Supply Management*, Undated. Available at https://www.cips.org/Documents/Resources/Knowledge%20Summary/Risk%20Management.pdf. Accessed June 1, 2019. n.d.

Delphi. *Design-In Environmental Health and Safety Specification, 2013*. Available at https://www.delphisuppliers.com/vendor_documents/delphi-a/ems_reqs_for_ehsim/Design%20in%20Environmental%20Health%20%20Safety%20Specification%20Version%202%20%20%202012--2013.pdf. Accessed June 1, 2019. n.d.

Environmental Services Association. *Designing Health and Safety into Procurement and Contract Management*, Undated. Available at http://www.hse.gov.uk/waste/services/assets/esa-contractor.pdf. Accessed June 1, 2019. n.d.

European Agency for Safety and Health at Work. *Promoting Occupational Safety and Health Through the Supply Chain, 2012*. Available at https://osha.europa.eu/en/tools-and-publications/publications/literature_reviews/promoting-occupational-safety-and-health-through-the-supply-chain. Accessed June 1, 2019. n.d.

OSHWIKI. *Occupational Safety and Health in the Supply Chains, 2013*. Available at https://oshwiki.eu/wiki/Occupational_safety_and_health_in_the_supply_chains. Accessed June 1, 2019. n.d.

SafeProd. *Safety Requirements Specification Guideline* – 19 pages, 2015. Available at https://sp.se/sv/index/services/functionalsafety/Documents/Safety%20requirements%20specification%20guideline.pdf. Accessed June 1, 2019. n.d.

University of Wollongong, New South Wales, Australia. WHS Purchasing Guidelines,2016. Available at https://staff.uow.edu.au/content/groups/public/@web/@ohs/documents/doc/uow017048.pdf. Accessed June 1, 2019. n.d.

FURTHER READING

ANSI/ASSE Z590.3-2011(R2016). *Prevention Through Design: Guidelines for Addressing Occupational Hazards and Risks in Design and Redesign Processes*. Des Plaines, IL: American Society of Safety Professionals, 2011 (R2016).

ANSI/ASSP Z10.0-2019. *Occupational Health and Safety Management Systems*. Park Ridge, IL: American Society of Safety Professionals, 2019.

ANSI/ASSP/ISO 45001-2018. *Occupational Health and Safety Management Systems – Requirements with Guidance for Use*. Park Ridge, IL: American Society of Safety Professionals, 2018.

ADDENDUM A

GENERAL DESIGN AND PURCHASING GUIDELINES

1. **Purpose**

    1.1 This document provides general technical requirements and guidelines for the design, build and purchase of equipment and fixtures.

2. **Scope**

    2.1 This document applies to the Arlington manufacturing facility in Campbell, IL.

3. **References**

    3.1 AR-15 – General Safety Standards

    3.2 AR-19 – Procedure for Processing Purchase Requisitions and Purchase Orders

4. **Definitions**

    4.1 None.

5. **Material and Equipment**

    5.1 Any relevant material or equipment as needed.

Advanced Safety Management: Focusing on Z10.0, 45001, and Serious Injury Prevention, Third Edition. Fred A. Manuele.
© 2020 John Wiley & Sons, Inc. Published 2020 by John Wiley & Sons, Inc.

GENERAL DESIGN AND PURCHASING GUIDELINES   357

6. **Responsibilities**

   6.1 All Arlington associates in the Campbell manufacturing facility must adhere to the guidelines provided in this document when purchasing new equipment or modifying existing equipment.

7. **Requirements**

   7.1 Purchasing - Follow the guidelines for purchasing as found in AR-19.

   7.2 Safety

       7.2.1 Machine Guarding - Refer to AR-57 - Machine Guarding Procedure for Safety.

       7.2.2 Guarding for Construction – Follow AR-62

           7.2.2.1 Give consideration to the best appropriate finish of all parts including functionality and aesthetic purposes.

           7.2.2.2 Use stainless steel or ceramics for parts that contact the product. Approval of any parts/material which contact product must be obtained on a case by case basis.

           7.2.2.3 Anodize aluminum parts.

           7.2.2.4 Use industrial paint or primer when the painting of steel is necessary.

           7.2.2.5 Finish all surfaces to prevent corrosion.

           7.2.2.6 Consider material compatibility for long component life and the prevention of corrosion.

           7.2.2.7 Allow no direct contact of the product with aluminum, bronze, lead, or oil.

           7.2.2.8 Use steel or aluminum framing for large transparent doors in high stress or vibration areas.

           7.2.2.9 Use only polycarbonate materials such as RAND or a similar material for transparent doors and guards since Plexiglas materials often become brittle and shatter upon impact.

           7.2.2.10 Make all exposed edges (i.e. where contact with a body part may occur during operation of equipment) with a 3/16 inch radius (minimum).

           7.2.2.11 Method of fastening panels to framework:

               1. Use only "TORX" screws.

               2. Use through bolt, flat washers, and anti-vibratory nut, if possible - Tapping into transparent panels is NOT ACCEPTABLE.

               3. Place bolts spaced not more than 6 inches center to center.

4. Use hinged panels with captive style fastener for the latch.

7.2.3 Signage - Clearly identify all controls and devices with appropriate warning signs, labels and tags.

7.3 Industrial Hygiene

7.3.1 Noise

7.3.1.1 Make the maximum noise level of the equipment 80dBA, 8 Hour Time Weighted Average when measured on the "A" scale of a standard sound level meter or noise dosimeter within 3 feet of the equipment.

7.3.1.2 Make the maximum peak noise level 115 dBA when measured on the "C" scale of a standard sound level meter, within 3 feet of the equipment.

7.3.2 Ventilation - Use a systematic approach when designing or modifying exhaust ventilation systems.

7.3.2.1 Refer to "Industrial Ventilation - A Manual of Recommended Practices", published by the American Conference of Governmental Industrial Hygienists (ACGIH), Section 6.0, which presents a systematic approach for designing or modifying exhaust ventilation systems.

7.3.2.2 As a baseline, design to a minimum of 90 fpm and a maximum of 150 fpm face velocity at point source of exhaust.

7.3.2.3 Equip each exhaust flow vent system with a continuous air flow monitoring device to detect any loss in air flow: Tie the system into the site emergency power system.

7.3.3 OSHA Hazard Communication

7.3.3.1 Submit a Material Safety Data Sheet to site safety representative for approval prior to design and build of any new or upgraded equipment or process that requires the use of solvents, chemicals, or other fluids.

7.3.3.2 An investigation will be conducted to determine if the equipment or process will pose any physical or health hazard.

7.3.3.3 Measures will then be recommended to minimize the risk. This includes descriptions of features and safeguards protecting the operator from direct or fugitive exposure to chemicals, solvents or generated waste streams.

7.4 Ergonomics

7.4.1 General Workstation Design - Consider the following ergonomic guidelines for general workstation design as optimal dimensions and are not intended to restrict or limit your ability to design effective workstations. Above all, general workstation design should factor in the amount of risk employees will be subjected to when using workstations.

7.4.1.1 Typical risk factors associated with industrial designs include but are not limited to:
1. excessive forces
2. poor body postures
3. high repetition
4. vibration

7.4.1.2 Design workstations, when possible, to allow operators to work in both sitting and/or standing positions.

7.4.1.3 Make work surface height for a sit/stand station 38" to 40".

7.4.1.4 Design "seated only" work surface height shall be 28-30" (28" is preferred) when "sit/stand" workstation is not feasible.

7.4.1.5 Minimize work surface thickness, including underside support members (1" preferred, 2" maximum).

7.4.1.6 Design work fixtures to allow hands to be positioned no higher than 4" above the work surface. The total dimension from the underside of the work station to the point where the work is performed should not exceed 6". If visual requirements necessitate higher work positioning, arm rests should be provided.

7.4.1.7 Provide unobstructed legroom for the operator to sit comfortably, (24" wide, 26" from the floor to the underside of work surface, 18" deep from the front edge of work surface).

7.4.1.8 Allow a 4" x 4" toe cutout at the bottom of the station when Standing Stations are used.

7.4.1.9 Design equipment/fixtures to be adaptable for convenient use by right or left handed operators, wherever possible.

7.4.1.10 Round all leading edges which the operator may come into direct contact with whenever possible. Try to recess the external hardware like hinges, door pulls, knobs, etc. as much as possible to avoid contacting the operator.

7.4.1.11 Minimize equipment/fixture size to limit forward bending or reaching, as much as possible to allow for tote pans and/or parts trays to be positioned in front of the operator. All operator/equipment interaction issues must be considered.

7.4.1.12 When fixed (nonadjustable) features are included into the design, the following principles are important to remember:
1. Design clearance dimensions for a tall operator (74" in height).
2. Design reaching dimensions for a short operator (60" in height).
3. Design fixed height dimensions designed for an average operator (66" in height).

7.5 Machine and Process Control

7.5.1 Master Control Relay - Provide Master Control Relay for emergency shutdown.

7.5.1.1 Hardwire emergency Stop pushbuttons and all Safety Switches in series to the Master Control Relay.

7.5.1.2 If any of these devices opens, the Master Control Relay should de-energize and remove power from the control circuit.

7.5.2 Emergency Stop - Provide Emergency Stop pushbuttons in an obvious location and within easy reach of either hand of the operator.

7.5.2.1 Design the E-stop such that to restart after Emergency shutdown, it is necessary to pull out the E-STOP button and then press the appropriate buttons to initiate normal operation.

7.5.2.2 Design the E-STOP to interrupt power from the outputs, drives, and other powered devices.

7.5.3 Interrupt DC Power Supply - Use DC power supply, interrupted on the DC side for faster response.

7.5.4 PLC Power - Wire power to the PLC outputs through a set of master control relay contacts.

7.5.5 Interlocks - Design machine to not be capable of running in continuous RUN mode with interlocked guards out of position or removed.

7.5.6 Mechanical Design:
1. No sharp edges or corners.
2. No shear points or pinch points.
3. Design turntable machines to not catch arms, hands, fingers or clothes and with filled access areas.
4. Consider torque or force limiting devices for any part-moving device such as turntables, carriages and slide assemblies, etc. (For instance, magnetic couplings used for emergency breakaway on linear shuttle assemblies.)

5. Use four-way, spring-centered, pneumatic valves for equipment working off of two hand controls.

7.6 Environmental Impact/Hazard Evaluation

7.6.1 Have materials used for all new or upgraded equipment and processes that are to be located in the facility evaluated to assure that they comply with Arlington's Policy and Governmental Regulations concerning such materials. This applies particularly to chemicals and generated waste streams.

7.6.1.1 Environmental Impact -Have any equipment or process which may release any chemicals to the environment evaluated for environmental impact. All information necessary for this impact study must be submitted to the site environmental coordinator early enough in the development stage to avoid costly rework or delays.

7.6.1.2 Material Safety Data Sheets - Have any chemical proposed for a new process approved by the site safety representative. A Material Safety Data Sheet (MSDS) and any other hazard data information must be submitted for evaluation prior to release of any purchase order for new equipment.

8. Procedure

8.1 Use this document as a guide whenever purchasing new (or modifying existing) equipment.

8.2 Codes and Standards

8.2.1 At a minimum, all equipment shall comply with the latest revisions of the applicable specifications, codes, and standards. When deemed applicable by Engineering, documentation/labeling of said compliance shall be provided. Note: In case of conflicting specifications, the more stringent shall apply.

8.3 Equipment/Fixture Design

8.3.1 Make all surfaces that may come into contact with the operator's body are free of sharp edges and corners (minimum 3/16" radius).

8.3.2 Tilt or orient fixtures to allow the operator to perform all work with a neutral body posture.

8.3.2.1 A general guideline is to tilt the fixture 15^0 toward the operator to enhance access and visibility, and to minimize awkward postures.

8.3.2.2 Specific recommendations will be made upon review of the job function in question.

8.3.3 Locate frequently accessed controls in front of and close to the operator to minimize reach distances.

- 8.3.4 Minimize repetitive reaches in front of the body to never exceed 16".
- 8.3.5 Repetitive reaches above chest height, below work surface height or behind the body are not acceptable.
- 8.3.6 Design repetitively used control buttons (e.g., cycle start buttons) to require nominal activation of one pound or less. Where possible, it is also desirable for the push button to be 2-3" in diameter.
- 8.3.7 Provide control knobs and handles with a nominal diameter of 1-1/4 inch. Clearance must be provided in equipment to avoid bumping the fingers or hands when parts are being positioned.

8.4 Mechanical - Design and Construction

- 8.4.1 Design equipment in a manner to prevent operator mistakes. Note: This includes preventing incorrect installation of tooling for setup, incorrect loading of parts, installing wrong parts or incomplete parts. The goal is to make the machine capable of producing zero defects without relying on correct operational procedure.
- 8.4.2 Materials/Finishes
    - 8.4.2.1 Materials "IN" Contact with the Product Approval of any parts/material which contact product must be obtained on a case by case basis. Listed below are some guidelines to aid in choosing a material.
        1. Use nonabrasive and non-marring materials.
        2. Use Stainless Steel - 300 series is preferred, 400 series in specific applications. Passivation and/or electropolish finish.
        3. Use Aluminum - hard anodize only.
        4. Use Plastics - fluorocarbons, polycarbonates, acrylics, ABS, polypropylenes, polyethylenes and nylons, with approval of site project coordinator.
    - 8.4.2.2 Materials "NOT" In Contact With the Product
        1. Use Carbon Steels - properly prepared and painted, flash chromed or electroless nickel plated.
        2. Use Stainless Steels - anodize.
        3. Use Aluminum - anodize.
    - 8.4.2.3 Miscellaneous Finishes
        1. Cover any table top surface or equipment that comes in contact with product with either stainless steel or laminate (see specifications for requirements). Approval of any parts/material which contact product must be obtained on a case by case basis.
        2. Do not use Wood Products in the manufacturing area.

8.4.3 Equipment Size and Weight Restrictions
  8.4.3.1 Free Standing Equipment - No restriction
8.4.4 Tooling Requirements
  8.4.4.1 Locating Datum - Dimension all tooling from and locate the part from the primary datum of the component part. Features - Design tooling with the following features.
   1. Interchangeability - Use dowels, bushings, and standard dowel patterns for locating to the equipment when practical all like tooling.
   2. Quickchange - Use quickchange features on tooling to accomplish the changeover within the time requirement and in a repeatable manner. Quickchange techniques or devices such as single fastener size, quarter turn fasteners, locking knobs, slotted holes, setup blocks, special tools, etc. may be used.
   3. Adjustment/Installation - Design tooling to be fixed and not adjustable to the equipment. Adjustment features should be incorporated into the equipment (locking slide) rather than the tooling (slotted holes) if possible.
   4. Design tooling to assure the correct installation of the tooling onto the equipment, using asymmetrical dowel patterns or similar feature.
8.4.5 Hardware Items
  8.4.5.1 Include one (1) resettable stroke counter and one (1) nonresettable stroke counter with each piece of automatic equipment.
8.4.6 Working Environment
  8.4.6.1 Design all equipment to operate in a Class 100,000 working environment unless otherwise specified.
8.4.7 Preferred Mechanical Components
  8.4.7.1 The following Preferred Components are listed in two groups, "A" and "B". "A" components are preferred as "first choice" and "B" components as "second choice".
  8.4.7.2 Use of "B" components shall require the approval of the Equipment Engineer.
  8.4.7.3 Use Preferred Components except when their use is not practical or jeopardizes project delivery.
  8.4.7.4 Obtain approval from Equipment Engineer prior to substitution for Preferred Components.
  8.4.7.5 Use standard commercially available components wherever possible.

8.4.7.6 General Equipment and Machinery
[In a resource for this Addendum, a chart appears listing equipment items and brand names categorized in the aforementioned groups "A" and "B". It is not duplicated here.]

8.5 Electrical - Design and Construction

8.5.1 Enclosures

8.5.1.1 Use enclosures properly rated for the environment for which they will be exposed.

8.5.1.2 Equip the main control enclosure with a fusible and lockable disconnect. Alternately, if appropriate, a fusible lockable disconnect may be provided directly upstream of the main control enclosure. It shall be readily accessible as stated in the NEC.

8.5.1.3 **If any live circuit should exist by design in the system after opening the main disconnect, the main disconnect and the enclosure(s) containing that live circuit shall be obviously labeled stating this condition and the source and voltage of the live circuit.**

8.5.1.4 Provide spare space in the electrical enclosure for future additions.

1. Allow for panels less than 500 square inches: a minimum of 40% spare capacity shall be provided.
2. Allow for panels greater than 500 square inches, a minimum of 15% spare capacity shall be provided.

8.5.1.5 Do not mount control equipment on the door or sides of the enclosure except devices such as push buttons, pilot lights, selector switches, meters, and instruments.

8.5.1.6 Mount terminal strips on inside enclosure sides and door, when appropriate. All terminal strips must be permanently, mechanically affixed - not glued or attached with a sticky strip.

8.5.1.7 Appropriately protect all wiring against excessive flex and pinching as a result of enclosure door operation.

8.5.1.8 During operation, the control enclosure interior temperature shall not experience a rise in temperature greater than 40° Fahrenheit over ambient temperature. Should the installed equipment cause this condition to occur in a static environment, filtered forced air shall be provided to maintain this temperature requirement. Appropriate alarms on this air flow must be provided.

8.5.2 Transformers

8.5.2.1 Use control transformers, where applicable, sized for 10% to 25% spare capacity (but not less than 100 VA). Transformers 2 KVA and larger shall be of the dry type and shall be mounted externally.

8.5.2.2 Supply Source for the main control transformer shall be taken from the load side of the main disconnecting device. The isolated secondary of the main control transformer shall provide, nominally 120 VAC, single phase and shall provide a bonded ground.

8.5.2.3 Fusing - Provide on primary and secondary.

8.5.3 Component Mounting

8.5.3.1 Use meters and operating controls placed so as to conform with Arlington ergonomic and AR-19 methods. Discuss with the electrical engineering representative.

8.5.3.2 Make all panel mounted components removable without having to remove the control panel.

8.5.3.3 Identify control equipment mounted on the machine by use of engraved Lamicoid type labels. Labels shall be permanently secured to the machine in order to facilitate their identification and location on wiring diagrams.

8.5.3.4 Identify control equipment mounted internal to the control cabinet by Brady labels or its equivalent. Dymo type labels shall not be used.

8.5.3.5 Permanently label all components (relays, transformers, fuses, terminal blocks, etc.) mounted on the equipment and in the control enclosures according to a scheme agreed upon by an electrical engineering representative. Labels must be affixed so that replacing parts does not remove or change the intended identification.

8.5.4 Push Buttons

8.5.4.1 Use "Start" of the fully guarded momentary contact type on equipment or operations sensitive or very dangerous in the event of inadvertent start-up. Otherwise, flush momentary type buttons shall be used.

8.5.4.2 Use Normal "Stop" buttons of the extended momentary contact type.

8.5.4.3 Locate the "Stop" button near the start push button.

8.5.4.4 Use Emergency stop push buttons of the illuminated, maintained type when depressed.

1. Arrange the circuit so that depressing an emergency stop push button will inhibit all machine motions and drop out the hard wired start/stop circuit. The E-stop

light will be wired independent of other E-stop lights so that any depressed E-stop can be identified regardless of the position of any other E-stop push button.

8.5.4.5 Pushbutton Colors

COLOR	TYPICAL FUNCTION	EXAMPLE
Red	Stop	Stop motors
	Emergency Stop	Master stop
Green	On/Start	Start of a cycle or motor
	Autocycle	
Black	Inch, Jog	Inching
	Horn Silence	Jogging
Yellow	Return, Reset	Return of machine to safe or normal condition
White	By approval only	
Gray	By approval only	
Blue	By approval only	
Orange	By approval only	

8.5.5 Position Sensors – Digital: Solid-state optical or proximity sensors are preferred to electromechanical limit switches. Sensors with built-in status indicating lights are preferred to those without.

8.5.6 Sensors/Signal Conditioners – Analog: Use solid-state sensors or signal conditioners with a 4-20mA DC output.

8.5.7 Power Distribution

8.5.7.1 Power Supply Protection - Provide suitable power supply protection/conditioning equipment if the equipment is susceptible to power brownouts, voltage surges/spikes, or line conducted electromagnetic interference.

8.5.7.2 Surge Suppression - Include an MOV or diode in parallel located as close to the load as practical on all inductive loads - motor starters, solenoids, relays, transformers, etc., require

8.5.7.3 Overloads - Use overloads of the manual reset type. Overload reset buttons shall be installed in the panel enclosures to allow manual reset, unless specifically provided for in separate motor starter control boxes.

8.5.8 Wiring Practices

8.5.8.1 Remote Interlocks - See Section 8.5.13.1, for labeling and Section 8.5.11 for color coding, for installations and equipment containing different sources of power.

8.5.8.2 Failsafe Operation - Utilize sensors, controls, and logic in such a manner that their failure or loss of power would produce the least undesirable consequences. In case of power failure, circuits must be manually re-energized to restart. The overall start and stop function must be hard-wired which eliminates the dependence on an electronic device or controller.

8.5.8.3 Push-To-Test Devices - Use pilot lights of the push-to-test type. However, if they are originating from a programmable logic controller (PLC) output, then a dedicated push button input to the PLC shall be utilized to test all of such PLC output pilot lights. All other devices such as displays and horns must be tested by a push-to-test function, should be wired to the dedicated push-to-test button where possible.

8.5.8.4 Interconnections/Terminals

1. Provide all components such as sensors, temperature sensors, heaters, and small motors which require frequent replacement with an appropriately designed quick disconnecting means. Terminals are one acceptable means.

2. Package and terminate all panel-mounted solid state devices and/or components (resistors, diodes, etc.) to be easily removable with the use of hand tools. They must be installed to allow easy access for troubleshooting and testing. They shall not be mounted near large heat-producing components such as transformers. They shall not be soldered and/or "butt-spliced".

3. Calibrate critical process control variables on a regular basis. Where applicable and practicable, provide banana jacks/switches or accessible quick disconnects to allow convenient access for calibration. Discuss calibration requirements with the Electrical Engineer or an Instrument Mechanic. A calibration label shall be affixed to the device (or in close proximity), indicating the equipment I.D., calibration date, calibrator, and calibration due date.

4. Terminate devices external to any control enclosure at a terminal block in the control enclosure. Wiring from external devices shall not be connected directly to a device in the control enclosure. However, 440 volt motors may be wired directly to the motor starter. Thermocouples and similar very low voltage signal carrying conductors may be terminated directly on the device to which it interfaces. If terminals are used, they must be of a special type appropriate to signals being conducted. One terminated and identified wire shall be returned for test purposes from a connection between limit switches, push buttons or other devices connected in series. This test point termination may be in field junction boxes where easily accessible if appropriate.
5. Wire control circuit voltage reference points to a terminal in each terminal enclosure external to the control panel. All terminals shall be marked to correspond with terminal markings as specified on wiring diagrams.
6. Protect sensor devices with cord leads such that the leads cannot be damaged. Open wiring is not permitted.
7. Do not make electrical connection to control devices with soldered connections or wire nuts. Sensors shall be terminated on terminals in a junction box. A field junction box should be mounted to provide terminals if necessary.
8. Use terminal blocks with terminal clamp screws with no more than two wires under one screw.
9. Wire terminal blocks and mount so that internal and external wiring does not cross over the terminals.
10. Provide spare terminals in each terminal enclosure including enclosures external to the control panel. The number shall be at least ten percent of the total in use or a minimum of six, whichever is greater. If fewer than 10 terminals are used and space prohibits, then less than 6 spares are adequate. There shall be spares appropriately spaced on each terminal block. A general rule would be to include 2 to 3 spares for each 10 terminals.
11. While maintaining functional appearance, placement of conduit and sealtite runs shall not restrict access for repair or replacement of machine parts.

All conduit and sealtite shall be secured to permanent fixtures, walls, or bracing to avoid loose and damageable runs.

8.5.9 Circuit Installation

8.5.9.1 Separate AC wiring from DC wiring on both inside control panels and in wire runs, and signal wiring shall be separated from power wiring. Signal wiring shall be properly shielded. Thermocouple wiring shall be run separately unless shielded in which case it may be run with signal wiring. Where AC and DC wiring must be crossed, the wires or wire bundles shall cross at 90 degrees to each other.

8.5.9.2 Sufficiently loop wiring to components mounted on doors to allow easy opening of the door and protect from excessive flex or pinching.

8.5.9.3 Use stranded copper of type MTW or THHN on all control wiring.

8.5.9.4 Wireways, sealtight, and conduit must meet NEC requirements and provide for adequate spares without over filling. Conduit or sealtight is required for machine wiring.

8.5.9.5 Use appropriately sized cable/wire anchors that are permanently affixed to the surface. Self-adhesive anchors are not acceptable unless mounted by screws.

8.5.9.6 All cable/wire straps shall be appropriately sized, spaced, and tensioned to avoid cable/wire insulation deformation/damage and provide adequate support.

8.5.9.10 Replace nicked or cracked wire insulation.

8.5.9.11 Use a plug-type removable terminal block at all locations to be separated for shipment where large equipment with interconnecting control panels are required, in order to minimize installation wiring.

8.5.9.12 Protect all wiring running alongside or over sharp edges.

8.5.9.13 Shield and ground all low voltage DC (signal) wires and cable to prevent noise interference.

8.5.9.14 Utilize a ground fault circuit Interrupter (GFCI) whenever the process is a wet process. (See the NEC)

8.5.10 Maintenance Considerations.

8.5.10.1 **Give consideration to accessing electrical equipment for maintenance troubleshooting.** This may involve maintenance bypass switches for certain interlocks. However, the machine should not be allowed to run production in the maintenance bypass mode. The

370 GENERAL DESIGN AND PURCHASING GUIDELINES

equipment should be so designed to provide access to sensors, switches, and motors etc. for preventive maintenance procedures performed on a regular basis.

8.5.10.2 Provide all wiring with a service loop sufficient to remake the connection at least three times.

8.5.10.3 Label all wire ends as indicated on the wiring diagrams. Where no wiring label is indicated, a to/from designation shall be used. Jacketed power wiring which is color coded can have the labeling on the jacket.

8.5.11 Wire Colors

8.5.11.1 Three-Phase Power and Motor Wiring

VOLT	A=L1	B=L2	C=L3	NEUTRAL	G=GROUND
480	BROWN	ORANGE	YELLOW	GRAY	GREEN
240	BLACK	RED	BLUE	WHITE	GREEN
208	BLACK	BLACK	BLACK	WHITE	GREEN

8.5.11.2 One-Phase Power and Motor Wiring

VOLT	PHASE	N=NEUTRAL	G=GROUND
480(277)	APP PHASE	GRAY	GREEN
240(120)	APP PHASE	WHITE	GREEN
208(120)	APP PHASE	WHITE	GREEN
120	BLACK	WHITE	GREEN

8.5.11.3 AC Control Wiring

VOLT	PHASE	N=NEUTRAL	G=GROUND
5 to 60	PINK	WHITE/PINK	GREEN
61 TO 120	RED	WHITE/RED	GREEN

8.5.11.4 DC Control Wiring and DC Motor Leads

VOLT	(+)	(−)
0 TO 11	PURPLE	WHITE/BLUE
5 to 60	BLUE	WHITE/BLUE
61 TO 120	BLUE/BLACK	WHITE/BLUE

8.5.11.5 Use green or green w/yellow tracer for grounding circuits ONLY.

8.5.11.6 Cable sheathing and noise suppression conductors do not suffice for fault carrying conductors. A separate ground shall be used.

8.5.11.7 For external power or control wiring which is not de-energized when the equipment main disconnect is opened, color code is as follows: use a wire with a yellow tracer - the base color being that which corresponds to the circuit voltage in use.

8.5.11.8 Make any temporary wiring or jumpers yellow or orange and installed in a manner indicating that this is obviously the purpose.

8.5.11.9 Select appropriate conductor and jacket insulation for the environment and the service intended. All low voltage conductor insulation shall be 300 volts minimum while high-voltage conductor insulation shall be 5X the nominal conductor voltage or 600 volts minimum. The insulation shall be appropriately selected to withstand the minimum bend radii expected and the effects of the environment (UV, ozone, oil, etc.).

8.5.11.10 "HI-POT" test all power wiring should be at the appropriate voltage/duration for the nominal conductor voltage carried.

8.5.11.11 Allowed Exceptions

1. Intrinsically safe wiring may require all conductors to be blue. Under certain conditions, sheathing may be the acceptable fault current conductor in intrinsically safe wiring. (See the NEC.)

2. If AC and AC neutrals or AC and DC neutrals are tied together, then the color of the corresponding AC neutral of the highest voltage will be used on all those common neutrals.

3. Wiring on devices purchased completely wired.

4. Where insulation is used that is not available in the colors specified.

5. Equipment for use outside of the United States when the above color coding is not in agreement with the established local electrical codes.

8.5.12 Methods of Grounding

8.5.12.1 Use a separate grounding conductor for grounding of equipment. A stranded or braided copper conductor shall be used for grounding where subject to vibration.

8.5.12.2 Grounding by attaching the device enclosure to the machine with bolts or other approved means shall be considered satisfactory if all paint and dirt are removed from joint surfaces. Moving machine parts having metal-to-metal bearing surfaces shall not be considered as a grounding conductor.

8.5.12.3 Do not use the grounded neutral conductor of a circuit for the grounding of equipment. No neutral shall be grounded, unless electrically isolated from the plant power distribution system in or immediately adjacent the equipment in question. This is to prevent current imbalances in the power distribution system which would affect ground fault detection.

8.5.12.4 Do not mix signal and power grounds.

8.5.12.5 Terminate signal grounds at one central location whenever possible to eliminate ground loops and noise.

8.5.13 Remote Interlocks

8.5.13.1 Attach caution labels, for installations and equipment containing different sources of power, adjacent to the main disconnects stating:

CAUTION

THIS PANEL CONTAINS MORE THAN ONE SOURCE OF POWER

DISCONNECT THE FOLLOWING SOURCES BEFORE SERVICING:

(Identity and location of all disconnects shall be shown on the label.)

8.5.13.2 Use labels of red Lamicoid with white lettering or of equivalent quality, secured in place by screws or rivets. Where practical, the remote sources of power shall be interlocked with the disconnect means. If not, a manual means shall be provided to quickly disconnect these sources.

8.5.13.3 Make all ungrounded wiring which contain remote sources of power yellow throughout the control panel.

8.5.14 Operator Interface - Apply Human Engineering criteria (Ergonomics) to the design of the operator interface. Except on the smallest systems, message board type status and alarm indicators are preferred to an array of indicator lights and associated nameplates.

8.5.15 Variable Speed Motor drives - Above 1/4HP, solid-state variable frequency AC motor drives are preferred to solid-state variable voltage DC motor drives. Below 1/4HP stepper motor drives are

GENERAL DESIGN AND PURCHASING GUIDELINES   **373**

preferred to variable voltage DC drives. The selection should be discussed with xxxxx electrical engineering representative to provide standardization when possible.

8.5.16 Programmable Controllers

8.5.16.1 Discuss the type of PLC, CPU version, and I/O selection with xxxxx electrical engineering representative.

8.5.16.2 Insure that the supplier follows all design criteria established by the PLC manufacturer for installation of PLCs.

8.5.16.3 Ground I/O cards according to function (i.e. DC inputs, AC input, AC outputs, etc.), and spare slots should be left between these function groupings.

8.5.16.4 At least one hard-wired Emergency Stop function shall be generated to create an emergency shutdown independent of the PLC and shall function even if a component of the PLC fails. The Emergency Stop function shall be interlocked into the PLC software. The E-Stop hardwiring shall open appropriate power circuits to PLC outputs.

8.5.16.5 Shield low voltage DC wiring with the following: Below 15 volts which interfaces to I/O modules, sink or source currents less than 8 ma, or input signal delay times less than 10 ma.

8.5.16.6 Maintain the shield continuity throughout the system when shielded cable is used.

8.5.16.7 Provide a minimum of 20% spare slots for future expansion.

8.5.16.9 PLC Digital Inputs - Use input modules of 110VAC or 24V (first preference is for AC, with DC being the second preference). Discuss with the xxxxx representative.

8.5.16.10 PLC Digital Outputs - Use 24V for indicator lights and alarms. AC is first preference, DC is second preference. For pneumatic or hydraulic solenoids and motor starters, 115V AC modules can be used. Solenoids should be rated for continuous duty.

8.5.16.11 PLC Analog Inputs/Outputs - The preference is 4-20 ma DC. Should 1 - 5 VDC input/output be required, precision 250 OHM resistors (1%) shall be used and installed at the terminating location.

8.5.16.12 Communication - In order to achieve the goal of integrating manufacturing machine raw data, components and process parameters should be shareable to the higher level of the computer system, a communication system shall be provided. It includes an interface module, communication software, and communication ports. Coaxial cable and twisted pair wire are common for the communication media; however, the optic fiber cable is preferred for the working environment with high electrical noise.

8.5.17 Instrumentation - All instruments and measuring devices used must be approved by an electrical engineering and/or instrumentation technician prior to use. Special considerations must be given to standardization, precision and accuracy, calibration requirements and maintenance. All original manuals and specifications shall be provided.

8.5.18 Machine Installation Drawings

    8.5.18.1 General - Include an installation drawing showing physical dimensions for mechanical, electrical, and service requirements with respect to the space needed for proper installation for each machine that is to be installed.

    8.5.18.2 Service Requirements - Clearly indicate the appropriate electrical service (i.e. 480V, 3 phase) for operation on the drawing.

8.5.19 Machine Documentation

    8.5.19.1 ANSI/USAS Y32.10-1967 drawing symbols should be used. The electrical designation and descriptions on related pneumatic/hydraulic piping diagrams shall match those on the electrical diagrams.

    8.5.19.2 Provide a P & ID diagram if appropriate.

    8.5.19.3 Provide electrical control drawings in AUTOCAD in all cases. Formats, numbering systems, symbology, details, annotation to be discussed with xxxxx electrical engineering representative. Examples will be provided. Standard symbols generally follow standard ABCD.

    8.5.19.4 All programs for drives, PLCs, etc. shall be sufficiently annotated and correspond accurately to electrical drawings.

    8.5.19.5 Include all original manuals for components with machinery such as the following: motor drive manuals, operator interfaces, special programming devices, message displays, etc.

    8.5.19.6 Provide a Bill of Materials with detailed parts description, manufacturer, and part number. The format should be discussed with xxxxx electrical engineering representative. Examples can be provided.

8.5.20 Preferred Electrical Components

    8.5.20.1 The following Preferred Components are listed in two groups, "A" and "B". "A" components are preferred as "first choice" and "B" components as "second choice" if no "A" components are applicable.

    8.5.20.2 Use of "B" components shall require the approval of the equipment engineer.

    8.5.20.3 Obtain approval for substitution of Preferred Components from the equipment engineer.

    8.5.20.4 Electrical Components
[This section contains a list of electrical components and brand names for "A" and "B" choices. It is not duplicated here.]

8.6 Pneumatics - Design and Construction

    8.6.1 Air Supply - Operate equipment from a single incoming air supply drop.
        Maximum Supply Pressure - 110 psig
        Design Pressure - 60 psig recommended where possible.

    8.6.2 Hardware/Circuit

        8.6.2.1 Non-Lubricated System - Use pneumatic devices and circuits which do not require lubrication.

        8.6.2.2 Disconnect - Design equipment/fixture to operate from a single quick disconnect (with exhausting, locking valve).

        8.6.2.3 Cylinders
            1. Use ports with lockable flow control valves.
            2. Cylinders used in the vertical orientation and which could pose a hazard to the operator during an air dump shall have a pilot operated check valve in the exhausting port to keep the cylinder from lowering during an air dump.
            3. Use cylinders of the permanently lubricated type.
            4. Adjustable cushioning at both ends is preferred.
            5. Use rods with self-aligning rod end coupling.

        8.6.2.4 Exhaust - to be reclassified, filtered, and directed away from the operator and conform to Class 100,000 working environment.

        8.6.2.5 Directional Valves - Modular packages or manifold style are preferred.

        8.6.2.6 Do not use Air Over Oil or Air Pressure Intensifiers.

        8.6.2.7 Fittings - Use plastic, stainless or brass.

8.6.2.8 Ports - Make all ports NPT wherever possible. If NPT is not available, then BSP shall be used. Devices with ports other than NPT shall be labeled to indicate the port type.

8.6.3 Labels - Clearly identify all devices. Labels shall be located on the structure of the equipment next to the device so that the label remains when the device is changed. Bilingual labeling considerations are to be addressed at the discretion of Site Project Coordinator.

8.6.4 Preferred Pneumatic Components

8.6.4.1 The following Preferred Components are listed in two groups, "A" and "B". "A" components are preferred as "first choice" and "B" components as "second choice" if no "A" components are applicable.

8.6.4.2 Use of "B" components shall require the approval of the equipment engineer.

8.6.4.3 Use Preferred Components except when their use is not practical or jeopardizes project delivery.

8.6.4.4 Obtain approval for substitution for preferred components of the equipment engineer.

8.6.4.5 Use standard commercially available components (ISO Standard Preferred) wherever possible.

8.6.4.6 Pneumatic Components
[This section lists pneumatic components and the names of "A" and "B" supplier companies. It is not duplicated here.]

8.7 Software

8.7.1 PLC Software Control Philosophies

8.7.1.1 Logic Location/Recovery - Locate all start/stop logic and equipment control logic in the PLC. This logic shall be retained in PLC memory while the system is down due to normal or emergency stop, or due to power failure. The goal of the PLC logic will be zero recovery; i.e., after a failure has been corrected, restarting the system shall be accomplished by pushing the "start" button.

8.7.1.2 Event Driven - Write software to sequence on events (switch closures) rather than time. Counters, Timers, and One-Shots shall be used only if necessary. If they are used, use as many conditions as practical.

8.7.1.3 Structure - Write PLC software in modules, using tables wherever possible, and arranged in the following order:

1. System start-up
2. Nonmotion logic (lights, horns, alarms)

GENERAL DESIGN AND PURCHASING GUIDELINES 377

        3. Motion logic (solenoids, motors)

        4. CIM/HMI interface logic

    8.7.1.4 All alarms shall be latched. Alarms shall be cleared by use of an acknowledge button.

  8.7.2 Testing Objectives: Test software to ensure proper operation in the following areas.

    8.7.2.1 The system performs in compliance with the statement of requirements.

    8.7.2.2 The software is error free and executes correctly as defined by the process specifications.

    8.7.2.3 That operating faults, alarms, interlocks and error conditions are detected and recover as specified.

    8.7.2.4 That automatic and manual abort and recovery functions perform as specified.

    8.7.2.5 That operator interfaces are correct as specified.

8.8 Machine Guarding

  8.8.1 Refer to MG2468 - 18 for guarding requirements and suggested types.

9. Appendices

  9.1 Appendix I - Operation and Maintenance Manual

  9.2 Appendix II - Codes and Standards

Appendix I

Operation and Maintenance Manual [To be provided by the equipment supplier.]

    An Operation Maintenance Manual is required for all Automatic Equipment to the extent appropriate to communicate proper operation and maintenance activities. The scope of the manual may range from one page of instructions for simple equipment to a full comprehensive manual for complex equipment.

Appendix II

This appendix lists certain Codes and Standards to which the supplier is to adhere.

ADDENDUM B

UNIVERSITY OF WOLLONGONG

WHS Purchasing Guidelines (WHS = Workplace Health and Safety)
A footnote on the first page says "Hardcopies of this document are considered uncontrolled." This is the latest version, having a 2016 November date.

Contents
1. Introduction
2. Scope
3. Responsibilities
4. Determining WHS Requirements

    4.1 Purchasing plant and equipment

    4.2 Purchasing hazardous chemicals
5. Process for identifying WHS requirements

    5.1 Risk Control Measures

    5.2 Supplier's Capacity to Comply

    5.3 Verification of WHS Requirements

    5.4 Repeat Purchases

    5.5 Standing Orders

Advanced Safety Management: Focusing on Z10.0, 45001, and Serious Injury Prevention,
Third Edition. Fred A. Manuele.
© 2020 John Wiley & Sons, Inc. Published 2020 by John Wiley & Sons, Inc.

5.6 Credit Card/Petty Cash Purchases

   5.7 Consultation

6. Related Documentation
7. Program Evaluation
8. Version Control Table
9. Appendix 1 Examples of WHS Specifications and Control Measures

1 INTRODUCTION

The University of Wollongong (UOW) is committed to the provision of a safe and healthy workplace for all workers, students and visitors. Purchasing items can introduce health and safety risks to the workplace and by incorporating risk management activities into purchasing processes this can help to ensure new hazards are not introduced to the workplace or suitably controlled.

In essence, there are three questions which need to be asked during the procurement process to ensure potential risk of goods and services are identified and controlled before being introduced into the workplace:

- Are there any risks associated with goods or services being purchased?
- Can these risks be eliminated from the purchase altogether?
- If we cannot eliminate the risks how can we manage the risks for the intended item to ensure health & safety? Or in other words what is the supplier or University required to implement in order to minimize the risks associated with the proposed purchase?

2 SCOPE

The purpose of this guideline is to ensure that suitable consideration is given when purchasing items which may have an adverse impact on health and safety. This guideline is to be applied in conjunction with the University's Purchasing and Procurement Policy and related procedures and applies to any item purchased by the University via orders, tenders, contracts, petty cash and credit card transactions. The guidelines apply to any purchase of plant and equipment, materials or substances including those which are hired, leased, or donated to the University.

The purchase of services and/or labor hire should refer to the Contractor WHS Guidelines. The purchase of non-hazardous items such as stationery, office supplies, books, journals, travel, conference and other membership fees, accommodation, computer software, minor hardware items and non-hazardous chemicals are not subject to the requirements of the guideline.

3 RESPONSIBILITIES

Executive Deans, Directors, Heads and Managers of Units are to ensure that these guidelines are implemented within their area of responsibility.

Any person in the University who purchases, leases or hires goods is responsible for identifying WHS requirements for any item which may pose a reasonably foreseeable injury in the workplace and upon receipt, verify that WHS specifications or control measures are in place to eliminate or reduce the risk.

Persons responsible for purchasing activities and payments are to complete the internal training on the University's purchasing system, Basware. Persons undertaking the purchase of goods should have the capability to ensure that item being received is fit for purpose by identifying any WHS requirements prior to purchase and then verifying those requirements upon receipt.

Knowledge of this document shall be the base skill, experience or qualifications used for the safe purchase of products. Other skills, experience or qualifications for the identification of WHS requirements of specific goods are outlined in Appendix 1.

4 DETERMINING WHS REQUIREMENTS

The underlying principle of determining WHS requirements before the purchase of items is to ensure that hazards are not introduced into the workplace without consideration and to ensure that any controls needed for the use of the item are in place prior to the item being used in the workplace.

Any Hazardous items to be purchased should be evaluated to determine what WHS requirements and controls are needed to ensure that the hazards are controlled prior to the items being used in the workplace. It is much more efficient and less costly to identify safety controls before an item is purchased rather than to wait till it arrives and discover that additional costly controls measures are required to use it safely.

If the item to be purchased is covered by existing risk control measures and meets industry standards, codes of practice and legislative requirements then a pre-purchase checklist and/or a more detailed risk assessment is not required prior to purchase. Where a pre-purchase checklist and/or risk assessment is not required the person purchasing the goods must still ensure the goods do not pose a risk to health and safety and check the goods meet any required WHS specifications before the item is used in the workplace.

If the item to be purchased is not adequately covered by existing risk controls or does not meet industry standards, codes of practice and legislative requirements then a pre-purchase checklist and/or more detailed risk assessment must be completed prior to purchase (any new hazardous items purchased from overseas must have a pre-purchase checklist and/or risk assessment completed).

4.1 Purchasing plant and equipment

Sufficient information must be sought by the person requesting the goods (not the person raising the purchase order) before new plant or equipment is introduced into the workplace so that the health and safety implications can be assessed in advance. It is necessary to consider impacts such as:

- where the equipment will be located;
- whether there is enough space for the item;
- whether the item requires modification in order for it to meet industry standards, codes of practice or legislative requirements;
- whether workers will need extra training;
- whether Safe Work Procedures will need to be updated;
- whether the equipment produces noise, fumes, extra heat etc.;
- if additional PPE might be required

This will enable any risk control measures required for its safe use to be in place prior to arrival. Such information gathering will also identify whether the relevant legislation, standards or codes of practice applicable to the equipment can be complied with e.g. notification, licensing, certification etc.

4.2 Purchasing hazardous chemicals

Prior to the purchase of any Hazardous chemicals the purchaser is to source a Safety Data Sheet (SDS) from the supplier (if the SDS is not available on Chem Alert). The purchaser must review the content of the SDS and verify if the listed controls are suitable for the intended storage and usage processes for the particular chemical. The pre-purchase checklist can also be used to assist with determining any other WHS requirements associated with purchasing a new chemical.

For the WHS requirements associated with purchasing scheduled drugs or poisons refer to the Scheduled Drugs and Poisons Guidelines. Red chemicals require a detailed risk assessment to be completed prior to the use of the chemical.

5 PROCESS

The diagram below illustrates the procedure for the identification and verification of WHS requirements in the procurement process.

5.1 WHS Requirements and Risk Control Measures

Requirements and/or risk control measures for WHS are to be detailed on the internal purchase order form, or purchasing system Basware ('Item description' or

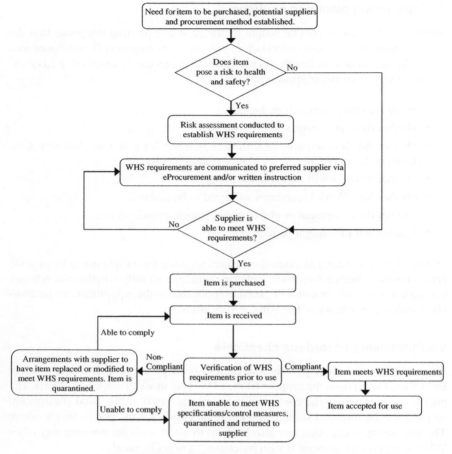

FIGURE 15.1 Procurement process.

'Additional info' fields) and provided to the supplier. See Appendix 1 for examples of WHS requirements for common items.

When items are purchased outside the purchasing system the supplier is to be provided with details of any WHS requirements or risk control measures in writing. This may include WHS requirements listed on either of the following:

- purchase order
- via a completed risk assessment, or
- a letter outlining specifications.

A list of applicable legislation and WHS specifications is provided in Appendix 1 Examples of WHS Specifications and Control Measures.

5.2 Supplier's Capacity to Comply

It is necessary to ensure that the supplier can meet the requirements as stated in the WHS specifications or control measures. This can be derived through discussions with the supplier on the required WHS specifications in the prepurchase stage of procurement to ensure the item is fit-for purpose.

The process of measuring the capacity of a supplier to meet the WHS specifications related to an item shall be documented on the purchase requisition form or other supporting documentation. The process of measuring a supplier's capacity to meet WHS specification includes checking to ensure that the supplier can meet or exceed the WHS specifications or control measures through the risk assessment process or specifications as outlined in Appendix 1.

Should the supplier not have the capacity to comply with identified WHS requirements the purchase is to be made through another supplier capable of meeting the requirements. Alternatively, the selection of another item that is capable of meeting the WHS requirements could be explored.

5.3 Verification of WHS Requirements

The verification of WHS requirements is required upon arrival of goods to ensure that the WHS requirements or control measures have been met as detailed on the purchase order. Verification should be conducted by the person who ordered the item and/or who determined any the WHS requirements. Verification of WHS specifications are to be documented by the person receiving the goods.

Examples of verification may include:

- checking to ensure that containers are clearly labeled
- checking that an item is labelled to indicate that it has been made to comply with the relevant Australian Standard or legislative requirement
- checking the compatibility of the item to be stored in compliance with the dangerous goods requirements
- ensuring that an item is fitted with physical control measures such as guarding of moving parts.

The post purchase declaration on the pre-purchase checklist should be signed to confirm that specifications have been met or the purchase order can be signed upon receipt of the goods stating that the goods have met the requirements specified.

When WHS requirements or control measures cannot be verified, the item must be quarantined and/or tagged out until the verification is complete. Items that are unable to be verified must be returned to the supplier.

If a hazard is not identified prior to purchase but becomes apparent once the item has been received or used a hazard report shall be lodged using SafetyNet. The hazard report shall detail the corrective actions required to eliminate or minimize the risk of injury to an acceptable level.

5.4 Repeat Purchases

A risk assessment can be re-used for repeated purchases of the same item or where the supplier has previously demonstrated compliance to WHS requirements. However, if the use or quantity of the item differs and has a greater impact on health and safety, the risk assessment should be reviewed and modified accordingly.

5.5 Standing Orders

Where a standing order has been raised with a supplier, the supplier must indicate in writing via a Memorandum of Understanding that all products being supplied to the University will conform to applicable legislation, codes of practice or Australian Standards not limited to those outlined in Appendix 1 of these Guidelines.

Where WHS requirements are identified outside of the Memorandum of Understanding, these shall be outlined in writing to the supplier prior to purchase.

Products received by the University from a supplier with a standing order are required to be verified for compliance to WHS requirements prior to use.

5.6 Credit Card/Petty Cash Purchases

To ensure compliance with these guidelines the preferred purchase method for items with WHS considerations is the online purchasing system rather than credit card or petty cash transactions.

When materials or substances are required to be purchased using a credit card or petty cash, the person purchasing the item shall consider the potential for the equipment, material, facility or substance to pose a risk to health and safety. Where it is identified a risk assessment is required it shall be conducted in accordance with the WHS Risk Management Guidelines.

5.7 Consultation

Any proposed changes to the working environment through the purchase of goods and services that could place a risk to health and safety must be consulted with workers who are likely to be affected. This can be achieved by being raised with the Supervisor or local area Workplace Advisory Committee.

6 RELATED DOCUMENTATION

Use the links below for the following related documentation:

- Purchasing and Procurement Policy
- WHS Risk Management Guidelines
- SafetyNet– Risk assessment form
- Pre-Purchase Checklist

7 PROGRAM EVALUATION

In order to ensure that these guidelines continue to be effective and applicable to the University, these guidelines will be reviewed regularly by the WHS Unit in consultation with the WHS Committee.

Conditions which might warrant a review of the guidelines on a more frequent basis would include:

- reported hazards or injuries
- non-conforming systems
- WHS Committee concern.

Following the completion of any review, the program will be revised/updated in order to correct any deficiencies. These changes will be communicated via the WHS Committee.

8 VERSION CONTROL TABLE

Editorial Note from this author: This section pertains to the controls in place concerning the purchasing guidelines. Only a few entries are shown as examples.

Version Control	Date Released	Approved By	Amendment
1	December 1997	WHS Manager	Document created
2	February 2003	WHS Manager	Modification to reflect current process
7	August 2010	Manager WHS	Document updated to incorporate the Personnel name change to Human Resources
11	May 2015	Manager WHS	Aligned to risk management methodology outlined in the Business Assurance Risk Management Policy. Removed references to extreme risks
14	November 2016	Manager WHS	Updated Section 5: Process

9 APPENDIX I WHS REQUIREMENTS

Requirements are for applicable Australian standards. They fill 4 ½ pages and are not duplicated here. What is important is that whatever is procured must meet the standards. Standards listed are as follows.

- Personal protective equipment – for which there are 7 subsets
- Hazardous chemicals and dangerous goods – for which there are 4 subsets
- Plant and equipment – for which there are 20 subsets

CHAPTER 16

EVALUATION AND CORRECTIVE ACTION

In Z10.0, Clause 9.0 contains specific provisions to achieve both a performance evaluation and to establish corrective action mechanisms as is said in its first two bulleted items.

This section defines requirements for processes to

- Evaluate the performance of the occupational health and safety management systems (OHSMS) through Monitoring, Measurement and Assessment, Incident Investigation, and Audits and
- Take corrective action when nonconformances, system deficiencies, hazards, and incidents that are not being controlled to an acceptable form of risk are found in the OHSMS.

In 45001, the purpose principally of Clause 9 is for performance evaluation. But as will be seen later in this chapter, the outputs of a management review are to include opportunities for continual improvement, any needed changes to the OH&S management system, and actions if needed etc.

Having processes for taking corrective action are a requirement of Clause 10 in 45001, the title of which is Improvement. This is its opening General statement for Clause 10 in 45001:

The organization shall determine opportunities for improvement (see Clause 9) and implement actions necessary to achieve the intended outcomes.

Advanced Safety Management: Focusing on Z10.0, 45001, and Serious Injury Prevention,
Third Edition. Fred A. Manuele.
© 2020 John Wiley & Sons, Inc. Published 2020 by John Wiley & Sons, Inc.

Rather extensive comments on these subjects were made in Chapter 1—an Overview—and some of those comments are duplicated here.

MONITORING, MEASUREMENT, AND ASSESSMENT IN Z10.0 AND MONITORING, MEASUREMENT, ANALYSIS, AND PERFORMANCE EVALUATION IN 45001

This section of 45001 opens with statements indicating that an organization shall establish, implement, and maintain a process(es) for monitoring, measuring, analysis, and performance evaluation—and determine what needs to be measured in relation to legal provisions, its health and safety activities, progress made, and effectiveness of its controls. It is also required that an organization determines the methods to be used in the required processes and when the data is to be analyzed.

In Z10.0, they say that measures of performance may include various indicators for system effectiveness. Examples of system elements may include the following:

- management leadership
- worker participation
- measurement of progress toward achieving objectives and targets
- support provided
- quality of hazard identification and control
- quality and completeness of risk assessment and risk reduction activities
- indicators from occupational health risk assessment and risk indicators
- occupational injury and illness rates
- workplace inspections and testing.

A caution is given on the use of occupational injury and illness rates as a precise indication of system effectiveness. Yes, they are a qualified measure of effectiveness, they are broadly used and should be used with particular care. As it was said in Chapter 7—Innovations in Serious Injury, Illness, and Fatality Prevention—it has been determined that having good injury and illness rates does not assure that an organization has good preventive measures in place with respect to avoiding low probability/serious consequence events.

For both standards: Using the data developed, the organization is to evaluate its health and safety performance and communicate broadly on the results and retain documentation.

In 45001, a subsection titled Evaluation of compliance follows the subsection on Monitoring, measurement, analysis, and performance evaluation. Its opening statement indicates that an organization shall have processes in place to assure that legal and other requirements are met.

INCIDENT INVESTIGATION

Although incident investigation is the next subject addressed in Z10.0, comments are not made in 45001 on this subject until requirements for Clause 10—Improvement—are discussed.

Thus, all the following comments on incident investigation pertain to the requirements set forth in Clause 9 of Z10.0.

Requirements in Z10.0 are that an organization establishes a process to report, investigate, analyze, document, and communicate incidents to learn from the event, improve the management system, address OHSMS nonconformances, and other factors that may be causing or contributing to the occurrence of incidents. Investigations shall be performed by personnel demonstrating competence in investigation procedures, conducted in a timely manner, and include worker participation.

Incident investigation processes should define what needs to be investigated and when, who should participate, and how recommendations to prevent recurrence should be generated and communicated. For incidents to be investigated, they must be reported. Organizations should have all barriers to incident reporting removed.

They say that findings from incident investigations should be used for developing and implementing corrective action plans. Lessons learned from these investigations can then be fed into the planning or corrective action processes.

It should be understood that incidents may be symptoms of deficiencies in the occupational health and safety management system. If the ideal could be reached, hazards and their underlying system deficiencies would be identified before any injury or illness occurs.

They say in Annex A for Z10.0 (E9.2) "There are many techniques for incident investigation. Only a few of the tools used in business management systems are mentioned here":

- The Five Why Problem Solving Technique.
- Fishbone Analysis.
- Critical Incident Technique or CIT.
- 8D Analysis.
- FMEA—Failure Mode and Effects Analysis.
- Bow-tie Analysis.
- CAST—Causal Analysis using Systems Theory.

Wherever appropriate, the investigation process used to identify causal factors for incidents should be the same as the organization uses for incidents that occur in other disciplines.

Understanding the gap between work as imagined and work-as-done helps in an incident investigation. A just culture is needed to insure that workers feel secure when they disclose how work is actually performed.

390 EVALUATION AND CORRECTIVE ACTION

There is value in feeding lessons learned from investigations into the planning and corrective action processes. That fits well with the emphasis given in this book to serious injury, illness, and fatality prevention.

This author's research indicates that incident investigations are too often not done well. For those safety professionals who are deep into leading and lagging indicators, he now asks, if improving occupational risk management systems is their goal, should incident investigation reports be considered a good source for leading indicators?

This subject is explored in depth in Chapter 19, Incident Investigation, in this book.

AUDITS IN Z10.0: INTERNAL AUDITS IN 45001

A summation of what is said in both standards about audits is given here. An organization shall:

- define the criteria for audits
- achieve an understanding that audits are management system audits rather than compliance audits
- include management shortcomings with respect to relative legal requirements
- have audits made periodically with respect to application of the provisions in the OHSMS
- ensure that audits are made by competent persons not attached to the location being audited (but those persons can be in the same organization)
- have auditors communicate immediately on potentials for serious injuries, illnesses, or fatalities so that swift corrective action can be taken
- document and communicate the results
- have management action taken on deficiencies mentioned in audit reports.

Audits are to measure the organization's effectiveness in implementing the elements of the OHSMS. Thus, audits are to determine whether the management systems in place do or do not effectively identify hazards and control risks.

Audits can be highly beneficial. They can be next to valueless. An expanded treatise on audits is presented in this book in Chapter 17, Audit Requirements.

CORRECTIVE ACTION IN Z10.0

Following Audits in Z10.0, still under Clause 9.0—Evaluation and Corrective Action—there is a section titled Corrective Action. Users are to achieve an understanding that utilizing common business processes to get corrective actions taken will be more effective. An example given is to use an existing maintenance work order system.

They also say in the standard that involvement of other parts of the organization, such as engineering, procurement, maintenance, and R&D to correct occupational health and safety issues will foster relationships that will support the integration of Occupational Safety and Health (OSH) into the business.

An additional purpose is to make it clear that when corrective actions for an identified hazard require a significant period of time to implement, immediate interim corrective action should be taken. They say that expedited actions should include efforts to discontinue the exposure by removing the person at risk, discontinuing the operation, or reducing the risk. After the implementation of corrective actions, a review of the effectiveness is to be conducted.

Because of the particular importance of taking corrective action, the specific requirements in Z10 are cited here.

Specific Corrective Actions in Z10.0

The organization shall establish a corrective action process(es) to (see Note 1).

A. Address nonconformances, system deficiencies, hazards, and incidents that are not being controlled to an acceptable level of risk (see Note 2).
B. Identify and address new and residual risk associated with corrective actions that are not being controlled to an acceptable level of risk.
C. Expedite action on high-risk hazards that could result in fatality or serious injury/illness (FSII) that are not being controlled to an acceptable level of risk.
D. Account for worker participation.
E. Review and ensure effectiveness of corrective actions taken.

Note 1: An organization needs to address all identified system deficiencies and inadequately controlled hazards through the corrective action process, regardless of how those deficiencies and hazards were identified.
Note 2: Risk cannot typically be eliminated entirely, though it can be substantially reduced through application of the hierarchy of controls. Residual risk is the remaining risk after controls have been implemented. It is the organization's responsibility to determine whether the residual risk is acceptable. Wherever the residual risk is not acceptable, additional actions to reduce the risk are needed.

To emphasize: If, after preventive measures have been taken and the risk is not acceptable, additional actions must be taken so that the risk is as low as reasonably practicable and acceptable.

MANAGEMENT REVIEW IN 45001—SECTION 9.3

In 45001, management review is a subset of Performance Evaluation. For this section, this is the opening sentence:

EVALUATION AND CORRECTIVE ACTION

Top management shall review the organization's OS&H management system, at planned intervals, to ensure its continuing suitability, adequacy and effectiveness.

Purposes of the management review requirements in 45001 are comparable to those of Management Review—Cause 10 in Z10.0.

In 45001, reviews shall include consideration of:

- the status of open items from previous reviews
- audit results
- input from workers.
- resources needed.
- risks and opportunities that should be discussed
- changes in external and internal issues that are relevant to the occupational health and management system
- extent to which the OH&S policy and objectives have been met
- trending of performance
- notable incidents that have occurred since the previous review
- opportunities for continual improvement
- opportunities to improve integration of the OH&S management system with other operating processes
- suitability, adequacy, and effectiveness of the OH&S management system and need for changes.

Just how much preventive action is to be taken because of the management reviews made in accord with the requirements of 45001 is debatable. But it seems that the required content of the management review reports indicates that actions necessary are at least to be recorded. Those requirements follow, verbatim from the 45001 standard.

The outputs of the management review shall include decisions related to:

- the continuing suitability, adequacy, and effectiveness of the OH&S management system in achieving its intended outcomes
- continual improvement opportunities
- any need for changes to the OH&S management system
- resources needed
- actions, if needed
- opportunities to improve integration of the OH&S management system with other business processes
- any implications for the strategic direction of the organization.

Top management shall communicate the relevant outputs of management reviews to workers, and where they exist, workers' representatives.

The organization shall retain documented information as evidence of the results of management reviews.

FEEDBACK AND ORGANIZATIONAL LEARNING IN Z10.0

Throughout Z10.0, there is an emphasis on communicating continuously about hazards, risks, deficiencies in management systems, legal issues, and opportunities for improvement. Although the requirement, as in the following, is stated in one sentence, consider the extent of what is proposed in the following Note.

> The organization shall establish processes to create, retain, and transfer knowledge within the organization in order to support the OHSMS.
>
> *Note*: The findings and lessons arising from all elements of the OHSMS activities become part of the information that feeds back to context of the organization:
> (Section 4) strategic considerations.
> (Section 5) management leadership and worker participation.
> (Section 6) planning.
> (Section 7) support.
> (Section 8) implementation and operation.
> (Section 9) evaluation and corrective action.
> (Section 10) management review.

This feedback loop is used to help determine the underlying causes and other factors contributing to system deficiencies, system improvements, or risk control failures and is an essential component of the continual improvement of the OHSMS.

CONCLUSION

Plan-Do-Check-Act cycle is a caption within the Introduction for 45001 (p. vii). They say in the opening sentence that "the OH&S management system approach applied in this document is founded on the concept of Plan-Do Check-Act (PDCA)." Then they say that "this document incorporates the PDCA concept into a new framework." Their depiction of the revision retains the letters P–D–C–A and relates the titles of the major clauses in 45001 to the letters.

Similarly, in the Introduction to Z10.0 (p. iii) they say "the design of ANSI Z10.0 encourages integration with other management systems to facilitate organizational effectiveness using the elements of Plan-Do-Check-Act (PDCA) model as the basis for continual improvement." And there is a depiction of the P-D-C-A concept in the Introduction. It is followed by a complex model that makes no mention of P-D-C-A. (p. iv)

Nevertheless, when applying the Plan-Do-Check-Act continual improvement process, an important element is to determine whether the management systems put in place achieve what is intended. That is the purpose of the evaluation of performance clauses in 45001 and Z10.0. They provide evaluation mechanisms so that deficiencies in the systems instituted can be identified and, subsequently, acted upon. This is an important continual improvement function.

REFERENCES

ANSI/ASSP Z10.0-2019. *Occupational Health and Safety Management Systems*. Park Ridge, IL: American Society of Safety Professionals, 2019.

ANSI/ASSP/ISO 45001-2018. *Occupational Health and Safety Management Systems—Requirements with Guidance for Use*. Park Ridge, IL: American Society of Safety Professionals, 2018.

FURTHER READING

Manuele, F.A. *On the Practice of Safety*. Fourth Edition. Hoboken, NJ: John Wiley & Sons, 2013.

CHAPTER 17

MANAGEMENT REVIEW/ IMPROVEMENT

In Z10.0, the title for Clause 10.0 is Management Review. Comparable provisions for management reviews in 45001 are in its Clause 9, which is titled Performance evaluation. In 45001, the title for Clause 10 is Improvement. Provisions for improvement appear throughout Z10.0, but specifically in its clause 9.0, which is titled Evaluation and Corrective Action.

Since the Clauses—although similarly numbered—pertain to different subjects, an attempt will not be made here to present composites of their requirements.

First, the entirety of the requirements of Clause 10 as in 45001 is duplicated. There is a peculiarity in Clause 10 in which incidents are in the same category as nonconformities. It could be argued that incidents are nonconformities. Nevertheless, the requirements when incidents or nonconformities occur are very much the same as for the investigation of worker incidents. Allthe Clause 10 requirements in 45001 are as follows.

CLAUSE 10—IMPROVEMENT—AS IN 45001

10.1 General

The organization shall determine opportunities for improvement (see Clause 9) and implement necessary actions to achieve the intended outcomes of the OH&S management system.

Advanced Safety Management: Focusing on Z10.0, 45001, and Serious Injury Prevention,
Third Edition. Fred A. Manuele.
© 2020 John Wiley & Sons, Inc. Published 2020 by John Wiley & Sons, Inc.

10.2 Incident, Nonconformity, and Corrective Action

The organization shall establish, implement, and maintain a procedure(es) including reporting, investigation, and taking action to determine and manage incidents and nonconformities.

When an incident or nonconformity occurs, the organization shall:

a) react in a timely manner to the incident or nonconformity, and as applicable:
   1. take action to control and correct it and
   2. deal with the consequences.
b) evaluate, with the participation of workers and the involvement of other relevant interested parties, the need for corrective action to eliminate the root cause(s) of the incident or nonconformity, in order that it does not recur or occur elsewhere by:
   1. investigating the accident or reviewing the nonconformity;
   2. determining the cause(s) of the incident or nonconformity; and
   3. determining if similar incidents have occurred, if nonconformities exist, or if they could potentially occur.
c) review existing assessment of OH&S risks and other risks, as appropriate.
d) determine and implement any action needed, including corrective action, in accordance with the hierarchy of controls (section 8.1.2) and management of change (section 8.1.3).
e) assess OH&S risks that relate to the new or changed hazards, prior to taking action.
f) review the effectiveness of any action taken, including corrective action.
g) make changes to the OH&S management system, if necessary.

Corrective actions shall be appropriate to the effects or potential effects of the incidents or nonconformities encountered.

The organization shall retain documented information as evidence of:

- the nature of the incidents or nonconformities and any subsequent actions taken;
- the results of any action and corrective action, including their effectiveness.

The organization shall communicate this documented information to relevant workers, and, where they exist, workers' representatives, and other relevant interested parties.

Note: The reporting and investigation of incidents without undue delay can enable hazards to be eliminated and associated OH&S risks to be minimized as soon as possible.

10.3 Continual Improvement

The organization shall continually improve the suitability, adequacy, and effectiveness of the OH&S management system, by:

a) enhancing OH&S performance;
b) promoting a culture that supports an OH&S management system;
c) promoting the participation of workers in implementing actions for the continual improvement of the OH&S management system;
d) communicating the relevant results of continual improvement to workers, and, where they exist, workers' representatives;
e) maintaining and retaining documented information as evidence of continual improvement.

One requirement for continual improvement is of particular interest to this author because of his strong beliefs that safety is culture driven. Note that item b, aforementioned, indicates that an organization shall promote a culture that supports an OH&S management system.

This author has written previously that investigations should be performed by personnel demonstrating competence in investigation. He also wrote that people making investigations often were not competent, and causal factors were not identified.

Incident investigations should generate the recommendations necessary to prevent recurrence, and those recommendations should be communicated to all who should be informed. To emphasize : It should be understood by all who do incident investigations that incidents may be symptoms of deficiencies in the safety management systems. If the ideal could be reached, hazards and their underlying system deficiencies would be identified before any injury or illness occurs.

Where appropriate, the investigation process used to identify causal factors for incidents should be the same as the organization uses for incidents that occur in other disciplines. There is value in feeding lessons learned from investigations into the planning and corrective action processes.

For those safety professionals who are into leading and lagging indicators, this author now asks, if improving occupational risk management systems is their goal, should incident investigation reports be considered a good source for leading indicators?

With respect to the Management Review requirements in Clause 10 in Z10.0, an excerpt from Chapter 2, Organizational Culture. Management Leadership and Worker Participation is pertinent. It follows.

Section 5 in Z10.0 is captioned "Management Leadership and Worker Participation". In 45001, the title of Section 5 is "Leadership and worker participation". These sections pertain to the most important element in a safety and health management system.

Safety professionals will surely agree that top management leadership is crucial and that success cannot be achieved without it. As management provides the leadership and makes decisions directing the organization, the outcomes of those decisions establish its safety culture. Also, continual improvement processes cannot be successful without sincere top management direction.

For emphasis, this statement is repeated: Top management leadership and involvement is the most important element in a safety and health management system. And top management cannot provide the leadership necessary to create and maintain a positive safety culture without reviews being made on a scheduled basis of the status of the elements in the organization's operational risk management system.

It is the purpose of Management Review requirements in Z10.0 to provide the necessary status reports. These are the requirement of Clause 10 in Z10.0.

CLAUSE 10 – MANAGEMENT REVIEW IN Z10.0

This section defines the requirements for periodic management reviews of the OHSMS.

10.1 Management Review Process

The organization shall establish a process for top management to review the OHSMS at least annually and to recommend improvements to ensure its continued suitability, adequacy, and effectiveness. (See Notes)

Inputs to the management review process shall include, among other information:

A. Progress in the reduction of risk;
B. Effectiveness of processes to identify, assess, and prioritize risk and system deficiencies;
C. Effectiveness in addressing underlying causes of risks and system deficiencies;
D. Learning from system feedback loops;
E. Input from workers, workers' representatives and interested parties;
F. Status and effectiveness of corrective actions and changing circumstances;
G. Follow-up actions from OHSMS audits and previous management reviews;
H. The progress made towards meeting organizational objectives and targets; and
I. The performance of the OHSMS relative to expectations, taking into consideration changing circumstances, resource needs, alignment with the business plan and consistency with the OHS policy.

J. Information from top management relative to changes in the organization and its activities with a potential impact on OHS

Note 1: Management reviews are a critical part of the continual improvement of the OHSMS. This review is not just a presentation or a noncritical review of the system but should focus on results and opportunities for continual improvement. It is up to the organization to determine appropriate measures of OHSMS effectiveness.

Note 2: Results of the management review should be made available to affected individuals.

10.2 Management Review Outcomes and Follow-Up

At the conclusion of the review, top management shall determine the:

A. Future direction of the OHSMS and opportunities for integration with business processes; and
B. Need for changes to the organization's policy, priorities, objectives, resources, or other OHSMS elements.

Action items shall be developed from the findings of the management review. Results and action items from the management reviews shall be documented, communicated to affected individuals, and tracked to completion.

Note: Affected individuals include those impacted by or responsible for addressing findings of the management review, so appropriate action may be taken. Examples of affected individuals include workers, contractors, workers' representatives, and any existing OHS committee(s), as applicable.

In the Introductions to both standards, it is said that they incorporate the concepts on which the Plan-Do-Check-Act (PDCA) model is built. As shown in the following, the management review process commences with the "Check" step in the PDCA model and provides input to senior management so that, in the "Act" step:

- processes previously put in place can be accepted as satisfactory or revised and
- action can be initiated to establish processes that are missing.

PLAN-DO-CHECK-ACT

Plan: Identify the problem(s) (hazards, risks, management system deficiencies, legal requirements and opportunities for improvement—as in OHSMS Issues, Definitions for Z10.0 (p. 4)
Plan: Analyze the problem(s) and Develop solutions
Do: Implement solutions
Check: Evaluate the results to determine that:

1. problems were resolved, only partially resolved, or not resolved.
2. actions taken did or did not create new hazards.
3. acceptable risk levels were or were not achieved.

Act: Accept the results, or take additional corrective action, as needed.

Reviews by top management are required because personnel at that level have the authority to make the decisions necessary about actions to be taken and the application of resources. To be effective, the review process should ensure that the necessary information is available for top management to evaluate the continuing suitability, adequacy, and effectiveness of the safety management system.

This Management Review section gives safety and health professionals an opportunity to assist in providing objective summary reports on the status of the safety management systems and to present managements with proposals to overcome shortcomings.

In accord with the PDCA concept, the overriding theme of the Management Review is to achieve continual improvement. Thus, having action items for improvement in the review process and follow through are vital.

In many companies, a major Management Review process is conducted annually, and a summary progress report carrying the signature of the chief executive officer is published. Such reports may be made available broadly.

REFERENCES

ANSI/ASSP Z10.0-2019.*Occupational Health and Safety Management Systems*. Park Ridge, IL: American Society of Safety Professionals, 2019.

ANSI/ASSP/ISO 45001-2018. *Occupational Health and Safety Management Systems – Requirements with Guidance for Use*. Park Ridge, IL: American Society of Safety Professionals, 2018.

FURTHER READING

Manuele, F.A. *On The Practice of Safety,* Fourth Edition. Hoboken, NJ: John Wiley & Sons, 2013.

CHAPTER 18

AUDIT REQUIREMENTS

As is the case with every aspect of an organization's endeavors, making periodic reviews of progress with respect to stated goals is good business practice. Stated goals, in this instance, would be to have processes in place that meet the requirements of 45001 or Z10.0.

Internal audit requirements in 45001 are at 9.2 within Clause 9—Performance evaluation. In Z10.0, they are at 9.3 within Clause 9.0—Evaluation and Corrective Action. A composite of their requirements is as follows.

> An organization shall periodically have audits made of its health and safety management system to determine whether the requirements of this standard are implemented and maintained.
>
> - Findings shall take into consideration the content of previous audits.
> - Processes for identifying and controlling hazards, risks, system deficiencies, and opportunities for improvement shall be evaluated.
> - Audits are to be performed by competent persons who are to be objective and impartial and who are independent of the location being audited.
> - Audits are to focus on the processes necessary to have and maintain an effective health and safety management system.
> - Audits are to take into consideration legal or other requirements but they are to be systems audits and not compliance audits.

Advanced Safety Management: Focusing on Z10.0, 45001, and Serious Injury Prevention,
Third Edition. Fred A. Manuele.
© 2020 John Wiley & Sons, Inc. Published 2020 by John Wiley & Sons, Inc.

- If situations are noted that could be expected to be the causal factors for serious injury, illness, or fatality, communication to decision-makers on those situations shall be immediate so that corrective action can be taken.
- Results of audits shall be documented, and the documents shall be distributed to all who should be informed of their content: Documents shall be retained in accord with the organization's document retention policy.
- Corrective action is to be taken by management to address nonconformities and to continually improve the OH&S management system.

Audits of safety and health management systems perform a valuable function in that they determine the effectiveness or ineffectiveness of systems in place—or their absence. This author believes that audits are an exceptionally important management device.

To assist safety professionals in crafting or recrafting safety and health audit systems to meet the requirements of Z10.0 or 45001, this chapter:

- Establishes the purpose of an audit
- Discusses the implications of observed hazardous situations
- Explores management's expectations with respect to audits
- Establishes that safety auditors are also being audited during the audit process
- Comments on auditor qualifications
- Discusses the need to have safety and health management system audit guides related to the hazards and risks in the operations at the location being audited
- Provides information and resources for the development of suitable audit guides

THE PRINCIPLE PURPOSE OF A SAFETY AUDIT: TO IMPROVE THE SAFETY CULTURE

Throughout this book, emphasis has been given to the premise that safety is culture driven. Results achieved with respect to safety are a direct reflection of an organization's culture. Kase and Wiese (1990) drew an appropriate conclusion in their book *Safety Auditing: A Management Tool* when they said early in a chapter titled "Successful Auditing" that:

> Success of a safety auditing program can only be measured in terms of the change it effects on the overall culture of the operation and enterprise that it audits. (p 36)

The Kase and Wiese observation can be supported logically. For a safety and health management system audit, the paramount goal is to have a beneficial effect on an organization's decision-making. Thus, a safety audit report is to serve as a base for improvement: in how things are done; on an organization's *system of expected performance*; and, thus, on an organization's culture.

A safety audit report provides an assessment of the outcomes of the safety-related decisions made by management over the long term. Those outcomes are determined by evaluating the adequacy of what really takes place with respect to the application of existing safety policies, standards, procedures, and operating processes.

SIGNIFICANCE OF OBSERVED HAZARDOUS SITUATIONS

Hazardous situations observed in operations during a safety audit should be viewed principally as indicators of inadequacies in the safety management processes that allowed them to exist. Assume that management takes corrective action to eliminate every hazardous situation noted in an audit report. Still, little will be gained if no change is made in the overall decision-making to improve the management systems relative to the existence of hazardous situations.

REASONABLE MANAGEMENT EXPECTATIONS: THE EXIT INTERVIEW

Although safety audits are exceptionally valuable processes, they can be time consuming and expensive. Safety professionals should not be surprised if informed managements expect noteworthy results from the audit process that benefit their operations. Safety professionals who make audits should prepare well for the exit interview. That means:

- having been objective in their evaluations of management systems.
- having good justification for their findings.
- being able to support the prioritizing in management system improvements they propose.

In an exit interview with informed management personnel, the auditor or the audit team should anticipate and prepare beforehand to respond to questions comparable to the following:

- what are the most significant risks?
- what improvements in our management systems do we need to make?
- in what priority order should I approach what you propose?
- are there alternative risk reduction solutions that we can consider?
- will you work with me to determine that the actions we take and the money we spend attain sufficient risk reduction?

Audit systems fail if they do recognize management needs and if they are not looked upon as assisting management in attaining their operational goals. Safety auditors will not be perceived favorably if their work is not considered an asset to managements who seek to improve their safety and health management systems and their safety culture.

Unfortunately, safety auditors cannot absolutely assure managements that every hazard and risk has been identified. Some hazard/risk situations remain obscure, and humans have not yet developed the perfection necessary to identify all of them.

As examples, the negative impact of less-than-adequate decisions effecting design and engineering, purchasing, and maintenance may not be easily observable because their effect may not be felt for several years.

It should be made clear to management that applied safety auditing is based on a sampling technique and that it is patently impossible to identify 100% of the hazard/risk situations and shortcomings in safety management systems.

EVALUATIONS OF AUDITORS BY THOSE AUDITED

Safety professionals should also recognize that the time spent by auditors, the impressions they create, and the time expenditures required of the personnel at the location being audited are also being evaluated. Speculate on the possible comments made upward by location management personnel to executive management for the following situation. It happened.

> Four safety and health auditors spent a week making an audit of a 370-employee location. After the second day, employees complained that the auditors were being disruptive because of the amount of their time the auditors consumed. To make matters worse, during the third day the lead auditor told the location manager that the scorings being given to safety and health management systems by the auditors were higher than usual and that the auditors would have to delve further into operations. Why? Because, the lead auditor said, it was expected that their report outlines management system shortcomings that need attention.
>
> Employees at the location became more irritated and complained because of the repetitive, valueless and duplicatory interferences in their work.
>
> One of the criticisms made by location management to the decision taken by headquarters personnel to send four auditors to their 370-employee location, in light of the fact that maintaining tight cost controls was required by the organization.

AUDITOR COMPETENCY

Throughout Z10.0 and 45001, there is an emphasis on identifying, prioritizing, and acting on occupational health and safety management process issues. Those issues are defined in Z10.0 as "hazards, risks, management system deficiencies, legal requirements and opportunities for improvement." (Definitions). Comparable definitions, but not entirely the same, appear in 45001.

To be able to identify and evaluate those "issues" as they exist in the operation being audited, safety professionals making the audit must have the necessary qualifications and competency. Managements have said in the past that auditors had little knowledge of the technical aspects of their hazards and risks and that the audit report they submitted was superficial and of little value. If safety audits are to be perceived as having value, the auditors must have the professional qualifications to make them.

Safety professionals who make audits need to consider how well they are prepared for the situation at hand and how best to approach the management personnel in the organization to be audited. Similarly, if persons external to an organization are engaged to make safety audits, the safety professionals who engage them should evaluate their credentials.

Fortunately, a group foresaw the need for auditors of health and safety management systems being competent. On a web page for the International Organization for Standardization (ISO), information can be found on:

> ISO/IEC TS 17021-10:2018—*Conformity assessment*—Requirements for bodies providing audit and certification of management systems audit and certification of management systems. Part 10: Competence requirements for auditing and certification of occupational health and safety management systems.

This is the Scope of this standard: ISO/IEC TS 17021-10:2018 specifies additional competence requirements for personnel involved in the audit and certification process for an occupational health and safety (OH&S) management system and complements the existing requirements of ISO/IEC 17021-1.

ONE SIZE DOES NOT FIT ALL

When drafting an audit guide and selecting the elements to be emphasized, much should be made of sector-specific hazards and risks—meaning, those inherent in the operations at the location to be audited.

All hazards are not equal: Neither are the risks deriving from them equal. In a chemical operation where the inherent fire and explosion hazards are significant, the processes in place and their effectiveness with respect to design and engineering, control of fire and explosion potential, occupational health exposures, training, inspection, management of change, and procurement require much greater attention than a warehouse where the only chemicals used are for cleaning purposes.

Similarly, provisions to avoid auto accidents are more significant in the operation of a distribution center than if driving by employees is only incidental to operations.

In some organizations, the same audit guide is used for all locations, and the audit system requires that numerical or alpha scorings be recorded for each element being evaluated. Weightings for the elements are the same, regardless of their significance at the location being audited. That practice is questionable.

This book emphasizes serious injury, illness, and fatality prevention. When the audit system requires that identical weightings be given to elements regardless of the nature of the operations, the greater import of a particular management system in the operation may be overlooked. Also, the additional probing necessary into that management system to identify those hazards that may be the causal factors for low probability/serious consequence events may be less-than-adequate.

Greater effectiveness can be achieved if audit guides are structured so that modifications can be made to suit the hazards and risks at a location. In our communication age, it is suggested that safety professionals who craft audit systems consider using a flexible computer-based model in which the descriptive content of the elements to be audited can be abbreviated or expanded, and their weightings varied to suit the exposures at individual locations. That would truly be a self-built audit system focused on the hazards and risks that are there.

GUIDELINES FOR AN AUDIT SYSTEM

Since it is proposed here that safety professionals not develop a one-size-fits-all audit system, a specifically recommended audit guide is not being presented in this chapter.

Nevertheless, to create or improve an audit guide, a very good reference base is the example audit plan in Addendum A, which is an abbreviated version of OSHA's VPP Site-Based Participation—Site Worksheet.

OSHA evaluators are to support their conclusions as they evaluate system elements by recording what they derive from interviews, observations, or documentation.

Requirements for the VPP safety management system are similar to many of the Z10.0 and 45001 provisions. There are differences. Some of the VPP requirements are not as specific as comparable provisions in Z10.0 and 45001.

And, the VPP requirements may include provisions that are not required by Z10.0 or 45001. Examples are for routine inspections and for preventative/predictive maintenance.

One other resource also comes from OSHA. Although it is ancient, it poses good questions for a safety professional to consider when developing or upgrading an audit system. It is The Program Evaluation Profile (PEP) system. On the Internet, OSHA says: NOTICE: This is an OSHA Archive Document, and may no longer represent OSHA Policy. It is presented here as historical content, for research and review purposes only.

Both of these publications have another feature that will interest some safety and health professionals. They are a resource on scoring systems. If a numerical or alpha scoring system is to be used, the right scoring system is the one with which the auditors and the personnel who review and act on audit reports are comfortable.

CONCLUSION

Auditing performance with respect to established operational goals is good business practice. The audit requirements in Z10.0 and 45001 are to meet that purpose. Professionally done, safety audits provide valuable information to decision-makers who desire to achieve superior safety and health results.

Safety professionals who propose that an organization meet the requirements of Z10.0 or 45001, expecting an audit in the future, start with a gap analysis. The result would be comparisons between the elements in the safety and health management systems in place with the provisions in Z10.0 or 45001. Many organizations do not have management systems in place that meet all the provisions in those standards.

For a very large percent of organizations, a gap analysis will reveal the shortcomings with respect to design reviews, management of change, risk assessments, a hierarchy of controls, and procurement practices.

After the gap analysis is made, safety and health professionals would assist management in formulating an action plan to fulfill the requirements.

REFERENCE

ANSI/ASSP Z10.0-2019. *Occupational Health and Safety Management Systems*. Park Ridge, IL: American Society of Safety Professionals, 2019.

ANSI/ASSP/ISO 45001-2018. *Occupational Health and Safety Management Systems—Requirements with Guidance for Use*. Park Ridge, IL: American Society of Safety Professionals, 2018.

ISO/IEC TS 17021-10:2018—Conformity assessment—Requirements for bodies providing audit and certification of management systems—Part 10: Competence requirements for auditing and certification of occupational health and safety management systems. Available at https://www.iso.org/standard/71102.html. Accessed June 3, 2019. n.d.

Kase, D.W. and K.J. Wiese. *Safety Auditing: A Management Tool*. Hoboken, NJ: John Wiley & Sons, 1990.

OSHA. The Program Evaluation Profile (PEP) System. Available at https://www.osha.gov/archive/SLTC/safetyhealth/pep.html. Accessed June 3, 2019. n.d.

OSHA's VPP Site-Based Participation Evaluation Report. Available at http://www.osha.gov/dcsp/vpp/vpp_report/site_based.html. Accessed June 3, 2019. n.d.

FURTHER READING

Manuele, F.A. *On the Practice of Safety*, Fourth Edition. Hoboken, NJ: John Wiley & Sons, 2013.

ADDENDUM A

OSHA'S VPP SITE-BASED PARTICIPATION SITE WORKSHEET

SECTION I: MANAGEMENT LEADERSHIP & EMPLOYEE INVOLVEMENT

A. Written Safety & Health Management System

A1. Are all the elements (such as Management Leadership and Employee Involvement, Worksite Analysis, Hazard Prevention and Control, and Safety and Health Training) and sub-elements of a basic safety and health management system part of a signed, written document? (For Federal Agencies, include 29 CFR 1960.) If not, please explain.

A2. Have all VPP elements and sub-elements been in place at least 1 year? If not, please identify those elements that have not been in place for at least 1 year.

A3. Is the written safety and health management system at least minimally effective to address the scope and complexity of worksite hazards? If not, please explain.

A4. Have any VPP documentation requirements been waived (as per FRN, VOL. 74, NO. 6, 01/09/09 page 936, IV, and A.4)? If so, please explain.

Advanced Safety Management: Focusing on Z10.0, 45001, and Serious Injury Prevention, Third Edition. Fred A. Manuele.
© 2020 John Wiley & Sons, Inc. Published 2020 by John Wiley & Sons, Inc.

B. Management Commitment & Leadership

B1. Does management overall demonstrate at least minimally effective, visible leadership with respect to the safety and health management system (as per FRN, VOL. 74, NO. 6, 01/09/09 page 936, IV. A.5. a-h)? Provide examples.

B2. How has the site communicated established policies and results-oriented goals and objectives for employee safety to employees?

B3. Do employees understand the goals and objectives for the safety and health management system?

B4. Are the safety and health management system goals and objectives meaningful and attainable? Provide examples supporting the meaningfulness and attainability (or lack thereof if answer is no) of the goal(s). (Attainability can either be unrealistic/realistic goals or poor/good implementation to achieve them.)

B5. How does the site measure its progress towards the safety and health management system goals and objectives? Provide examples.

C. Planning

C1. How does the site integrate planning for safety and health with its overall management planning process (for example, budget development, resource allocation, or training)?

C2. Is safety and health effectively integrated into the site's overall management planning process? If not, please explain.

C3. For site-based construction sites, is safety included in the planning phase of each project?

D. Authority and Line Accountability

D1. Does top management accept ultimate responsibility for safety and health? (Top management acknowledges ultimate responsibility even if some safety and health functions are delegated to others.) If not, please explain.

D2. How is the assignment of authority and responsibility documented and communicated (for example, organization charts, job descriptions, etc.)?

D3. Do the individuals assigned responsibility for safety and health have the authority to ensure that hazards are corrected or necessary changes to the safety and health management system are made? If not, please explain.

D4. How are managers, supervisors, and employees held accountable for meeting their responsibilities for workplace safety and health? (Are annual performance evaluations for managers and supervisors required?)

D5. Are adequate resources (equipment, budget, or experts) dedicated to ensuring workplace safety and health? Provide examples.

D6. Is access to experts (for example, Certified Industrial Hygienists, Certified Safety Professionals, Occupational Nurses, or Engineers), reasonably available, based upon the nature, conditions, complexity, and hazards of the site? If so, under what arrangements and how often are they used?

E. Contract Employees

E1. Does the site utilize contractors? Please explain.

E2. *Were there contractors/sub-contractors onsite at the time of the evaluation?*

E3. When selecting onsite contractors/sub-contractors, how does the site evaluate the contractor's safety and health management system and performance (including rates)?

E4. Are contractors and subcontractors required to maintain an effective safety and health management system and to comply with all applicable OSHA and company safety and health rules and regulations? If not, please explain.

E5. Does the site's contractor program cover the prompt correction and control of hazards in the event that the contractor/sub-contractor fails to correct or control such hazards? Provide examples.

E6. How does the site document and communicate oversight, coordination, and enforcement of safety and health expectations to contractors?

E7. Have the contract provisions specifying penalties for safety and health issues been enforced, when appropriate? If not, please explain.

E8. How does the site monitor the quality of the safety and health protection of its contract employees?

E9. Do contract provisions for contractors require the periodic review and analysis of injury and illness data? Provide examples.

E10. If the contractors' injury and illness rates are above the average for their industries, describe the site's procedures that ensure that all employees are provided effective protection on the worksite? If yes, please explain.

E11. Based on your answers to the above items, is the contract oversight minimally effective for the nature of the site? (Inadequate oversight is indicated by significant hazards created by the contractor, employees exposed to hazards, or a lack of host audits.) If not, please explain.

F. Employee Involvement

F1. How were employees selected to be interviewed by the VPP team?

F2. How many employees were interviewed formally? How many were interviewed informally?

F3. Do employees support the site's participation in the VPP?

F4. Do employees feel free to participate in the safety and health management system without fear of discrimination or reprisal? If so, please explain.

F5. Are employees meaningfully involved in the problem identification and resolution, or evaluation of the safety and health management system (beyond hazard reporting)? (As per FRN page 936 IV, A.6.) For site-based construction sites, does the company encourage strong labor-management communication in the form of supervisor and employee participation in toolbox safety meetings and training, safety audits, incident investigations, etc.?

F6. Are employees knowledgeable about the site's safety and health management system? If not, please explain.

F7. Are employees knowledgeable about the VPP? If not, please explain.

F8. Are the employees knowledgeable about OSHA rights and responsibilities? If not, please explain

F9. How were employees informed of the safety and health management system, VPP and OSHA rights and responsibilities? Please explain.

F10. Did management verify employee's comprehension of the site's safety and health management system, VPP and OSHA rights and responsibilities?

F11. Do employees have access to results of self- inspection, accident investigation, appropriate medical records, and personal sampling data upon request? If not, please explain.

G. Safety and Health Management System Evaluation

G1. Briefly describe the system in place for conducting an annual evaluation.

G2. Does the annual evaluation cover the aspects of the safety and health management system, including the elements described in the Federal Register? If not, please explain.

G3. Does the annual evaluation include written recommendations in a narrative format? If not, please explain.

G4. Is the annual evaluation an effective tool for assessing the success of the site's safety and health management system? Please explain.

G5. What evidence demonstrates that the site responded adequately to the recommendations made in the annual evaluation?

G6. Is the annual evaluation conducted by competent site, corporate or other trained personnel experienced in performing evaluations?

SECTION II: WORKSITE ANALYSIS

A. Baseline Hazard Analysis

A1. Has the site been at least minimally effective at identifying and documenting the common safety and health hazards associated with the site (such as those found in OSHA regulations, building standards, etc., and for which existing controls are well known)? If not, please explain.

A2. What methods are used in the baseline hazard analysis to identify health hazards? (Please include examples of instances when initial screening and full-shift sampling were used. See FRN page 937, B.2.b)

A3. Does the company rely on historical data to evaluate health hazards on the worksite? If so, did the company identify any operations that differed significantly from past experience and conduct additional analysis such as sampling or monitoring to ensure employee protection? If so, please describe.

A4. Does the site have a documented sampling strategy used to identify health hazards and assess employees' exposure (including duration, route, and frequency of exposure), and the number of exposed employees? If not, please explain.

A5. Do sampling, testing, and analysis follow nationally recognized procedures? If not, please explain.

A6. Does the site compare sampling results to the minimum exposure limits or are more restrictive exposure limits (PELs, TLVs, etc.) used? Please explain.

A7. Does the site identify hazards (including health) that need further analysis? If not, please explain. For site-based construction sites, does the hazard analysis include studies to identify potential employee hazards, phase analyses, task analyses, etc.?

A8. Does industrial hygiene sampling data, such as initial screening or full shift sampling data, indicate that records are being kept in logical order and include all sampling information (for example, sampling time, date, employee, job title, concentrated measures, and calculations)? If not, please explain the deficiencies and how they are being addressed.

A9. For site-based construction sites, are hazard analyses conducted to address safety and health for each phase of work?

B. Hazard Analysis of Significant Changes

B1. When purchasing new materials or equipment, or implementing new processes, what types of analyses are performed to determine impact on safety and health, and are these analyses adequate?

B2. When implementing/introducing non-routine tasks, materials or equipment, or modifying processes, what types of analyses are performed to determine impact on safety and health, and are these analyses adequate?

C. Hazard Analysis of Routine Activities

C1. Is there at least a minimally effective hazard analysis system in place for routine operations and activities?

C2. Does hazard identification and analysis address both safety and health hazards, if appropriate? If not, please explain.

SECTION II: WORKSITE ANALYSIS   **413**

C3. What hazard analysis technique(s) are employed for routine operations and activities (e.g., job hazard analysis, HAZ-OPS, fault trees)? Please explain.

C4. Are the results of the hazard analysis of routine activities adequately documented? If not, please explain.

C5. For site-based construction sites, are hazard analyses conducted to address safety and health hazards for specialty trade contractors during each phase of work?

D. Routine Inspections

D1. Does the site have a minimally effective system for performing safety and health inspections (i.e., a minimally effective system identifies hazards associated with normal operations)? If not, please explain.

D2. Are routine safety and health inspections conducted monthly, with the entire site covered at least quarterly (construction sites: entire site weekly.

D3. For site-based construction sites, are employees required to conduct inspections as often as necessary, but not less than weekly, of their workplace/area and of equipment?

D4. Does the site incorporate hazards identified through baseline hazard analysis, accident investigations, annual evaluations, etc., into routine inspections to prevent reoccurrence?

D5. Are employees conducting inspections adequately trained in hazard identification? If not, please explain.

D6. Is the routine inspection system written, including documentation of results indicating what needs to be corrected, by whom, and by when? If not, please explain.

D7. Did the VPP team find hazards that were not found/noted on the site's routine inspections? If so, please explain.

E. Hazard Reporting

E1. Is there a minimally effective means for employees to report hazards and have them addressed? If not, please explain.

E2. Does the hazard reporting system have an anonymous component?

E3. Does the site have a reliable system for employees to notify appropriate management personnel in writing about safety and health concerns? Please describe.

E4. Do the employees agree that they have an effective system for reporting safety and health concerns? If not, please explain.

F. Hazard Tracking

F1. Does a minimally effective hazard tracking system exist that result in hazards being controlled? If not, please explain.

F2. Does the hazard tracking system result in hazards being corrected and provide feedback to employees for hazards they have reported? If not, please explain.
F3. Does the hazard tracking system result in timely correction of hazards with interim protection established when needed? Please describe.
F4. Does the hazard tracking system address hazards found by employees, hazard analysis of routine and non-routine activities, inspections, and accident or incident investigations? If not, please explain.

G. Accident/Incident Investigations

G1. Is there a minimally effective system for conducting accident/incident investigations, including near-misses? If not, please explain.
G2. Is the accident/incident investigation policy and procedures documented and understood by all? If not, please explain.
G3. Is there a reporting system for near-misses that include tracking, etc.? If not, please explain.
G4. Are those conducting the investigations trained in accident/incident investigation techniques? Please explain what techniques are used, e.g., Fault-Tree, Root Cause, etc.
G5. Describe how investigators discover and document all the contributing factors that led to an accident/incident or a near-miss.
G6. Were any uncontrolled hazards discovered during the investigation previously addressed in any prior hazard analyses (e.g., baseline, self-inspection)? If yes, please explain.

H. Trend Analysis

H1. Does the site have a minimally effective means for identifying and assessing trends?
H2. Have there been any injury and/or illness trends over the last three years? If so, please explain.
H3. Did the team identify trends that should have been identified by the site? If so, please describe.
H4. If there have been injury and/or illness trends, what adequate courses of action have been taken? Please explain.
H5. Does the site assess trends utilizing data from hazard reports and/or accident/incident investigations to determine the potential for injuries and illnesses? If not, please explain.
H6. Are the results of trend analyses shared with employees and management and utilized to direct resources, prioritize hazard controls and modify goals to address trends? If not, please explain.

SECTION III: HAZARD PREVENTION AND CONTROL

A. Hazard Prevention and Control

A1. Does the site select at least minimally effective controls to prevent exposing employees to hazards?

A2. When the site selects hazard controls, does it follow the preferred hierarchy (engineering controls, administrative controls, work practice controls [e.g., lockout/tagout, bloodborne pathogens, and confined space programs], and personal protective equipment) to eliminate or control hazards? Please provide examples, such as how exposures to health hazards were controlled.

A3. Describe any administrative controls used at the site to limit employee exposure to hazards (for example, job rotation).

A4. Do the work practice controls and administrative controls adequately address those hazards not covered by engineering controls? If not, please explain.

A5. Are the work practice controls (e.g., lockout/tagout, bloodborne pathogens, and confined space programs) recommended by hazard analyses and implemented at the site? If not, explain.

A6. Are follow-up studies (where appropriate) conducted to ensure that hazard controls were adequate? If not, please explain.

A7. Are hazard controls documented and addressed in appropriate procedures, safety and health rules, inspections, training, etc.? Provide examples.

Disciplinary System

A8. Are there written employee safety procedures including a disciplinary system? Describe the disciplinary system?

A9. Has the disciplinary system been clearly communicated and enforced equally for both management and employees, when appropriate? If not, please explain

Emergency Procedures

A10. Does the site have minimally effective written procedures for emergencies?

A11. Did the site explain the frequency and types of emergency drills held (including at least an evacuation drill annually)?

A12. Is the emergency response plan updated as changes occur in the work areas e.g., evacuation routes or auditory systems?

A13. Did the site describe the system used to verify all employees' participation in at least one evacuation drill each year?

Preventative/Predictive Maintenance

A14. Does the site have a written preventative/predictive maintenance system? If not, please explain.

A15. Did the hazard identification and analysis (including manufacturers' recommendations) identify hazards that could result if equipment is not maintained properly? If not, please explain.

A16. Does the preventive maintenance system detect hazardous failures before they occur? If not, please explain. Is the preventive maintenance system adequate?

Personal Protective Equipment (PPE)

A17. How does the site select Personal Protective Equipment (PPE)?

A18. Did the site describe the PPE used at the site?

A19. Where PPE is required, do employees understand that it is required, why it is required, its limitations, how to use it, and how to maintain it? If not, please explain.

A20. Did the team observe employees using, storing, and maintaining PPE properly? If not, explain.

Process Safety Management (PSM)

A21. Is the site covered by the Process Safety Management standard (29 CFR 1910.119)? If yes, please answer questions A22-A25 below. Additionally, please complete either the onsite evaluation supplement A or B, and the onsite evaluation supplement C. If not, skip to section B.

A22. Which chemicals that trigger the Process Safety Management (PSM) standard are present?

A23. Which process(es) were followed from beginning to end and used to verify answers to the questions asked in the PSM application supplement, the PSM Questionnaire, and/or the Dynamic Inspection Priority Lists?

A24. Verify that contractor employees who perform maintenance, repair, turnaround, major renovation or specialty work on or adjacent to a covered process have received adequate training and demonstrate appropriate knowledge of hazards associated with PSM, such as non-routine tasks, process hazards, hot work, emergency evacuation procedures, etc.? Please explain.

A25. Is the PSM program adequate in that it addresses the elements of the PSM standard and the PSM directive? Please explain.

B. Occupational Health Care Program

B1. Describe the occupational health care program (including availability of physician services, first aid, and CPR/AED) and special programs such as audiograms or other medical tests used.

B2. How are licensed occupational health professionals used in the site's hazard identification and analysis, early recognition and treatment of illness and injury, and the system for limiting the severity of harm that might result from workplace illness or injury? Is this use appropriate?

B3. Is the occupational health program adequate for the size and location of the site, as well as the nature of hazards found here? If not, please explain.

C. Recordkeeping

C1. Are OSHA required recordkeeping forms being maintained properly in terms of accuracy, form completion, etc.? If not, please explain.

C2. Is the record keeper knowledgeable of 29 CFR 1904, OSHA's recordkeeping standard?

C3. What records were reviewed to determine compliance with the recordkeeping standard?

C4. Do the injury and illness rates accurately reflect work performed by contractors/sub-contractors at the site evaluated? Please explain.

C5. Was there any evidence of recordable injuries/illnesses not being reported due to management pressure, production concerns, incentive programs, etc.? If yes, please explain.

SECTION IV: SAFETY AND HEALTH TRAINING

A. Safety and Health Training

A1. What are the safety and health training requirements for managers, supervisors, employees, and contractors? Please explain.

A2. Is the training delivered by qualified instructors?

A3. Does the training provided to managers, supervisors, and non-supervisory employees (including contract employees) adequately address safety and health hazards?

A4. Does the company/site operate an effective safety and health orientation program for all employees including new hires? Please explain.

A5. How are the safety and health training needs for employees determined? Please explain.

A6. Does the site provide minimally effective training to educate supervisors and employees (including contract employees) regarding the known hazards of the site and their controls? If not, please explain.

A7. Are managers, supervisors, and non-supervisory employees (including contract employees) taught the safe work procedures to follow in order to protect themselves from hazards during initial job training and subsequent reinforcement training?

A8. Who is trained in hazard identification and analysis?

A9. Is training in hazard identification and analysis adequate for the conditions and hazards of the site? If not, please explain.

A10. Does management have a thorough understanding of the hazards of the site? Provide examples that demonstrate their understanding.

A11. Do managers, supervisors, and non-supervisory employees (including contract employees) and visitors on the site understand what to do in emergency situations? Please explain.

420 INCIDENT INVESTIGATION

Note 1: The level of documentation should be appropriate to the actual or potential severity of the incident, potential impact, potential learning, or other opportunities for improvement. Lessons learned should be communicated to relevant workers and other interested parties.

Note 2: When appropriate, OHS Incident Investigations should be similar to other business processes to identify contributing factors.

Note 3: The ultimate goal is to identify and correct hazards and system deficiencies before any injury or illness occurs. Incident investigations should be used to support operational and organizational learning by causal analysis to identifying system or other deficiencies for developing and implementing corrective action plans.

Note 4: Feedback loops and lessons learned from investigations should be fed into the leadership, planning, implementation, support, evaluation and corrective action, and management review processes.

COMMENTS REFLECTING ON THIS AUTHOR'S RESEARCH

As is said on incident investigation in Chapter 7, Innovations in Serious Injury, Illness, and Fatality Prevention:

- Research has shown that the quality of incident investigations in many companies has been less than stellar.
- This is an important subject which should be given emphasis in an operation risk management system.
- Opportunities to reduce risks and avoid injury or damage to people, property, or the environment are lost when incident investigations do not identify the reality of causal factors.

It was suggested that safety professionals analyze the incident investigation system in place for their effectiveness, asking—do they or don't they identify the reality of causal factors. When incident investigations are done properly, they are a good source from which determinations can be made about the adequacy of barriers, controls, and relative management systems. Doing so is a requirement of an effective risk management system.

Also, it was said in Chapter 7: *One of the purposes of this chapter is to convince safety professionals that their learned advice when incident investigations are made is seriously needed at all levels of management.* And that is a purpose of this chapter also.

To provide an information base for safety professionals who choose to take action to improve the incident investigation process, this chapter:

- Reviews the research done by this author.
- Explains why supervisors who complete incident investigations may be reluctant to record the reality of causal factors.

CHAPTER 19

INCIDENT INVESTIGATION

Incident investigation in 45001 is addressed in Clause 10 Improvement. Within Clause 10, section 10.2 is titled Incident, nonconformity and corrective action. Duplication of the entirety of Clause 10 as in 45001 appears in Chapter 18.

In this chapter, the entirety of the requirements for incident investigation in Z10.0 is shown, followed by comment on the results of this author's research on incident investigation.

Expectations of management on incident investigation in Z10.0 are pretty much the same as the requirements in 45001.

9.2 INCIDENT INVESTIGATION IN Z10.0

The organization shall establish a process to report, investigate, analyze, document, and communicate incidents in order to learn from the event, improve the management system, and address OHSMS nonconformances and other factors that may be causing or contributing to the occurrence of incidents. The investigations shall be performed by personnel demonstrating competence in investigation procedures, conducted in a timely manner and include worker participation.

Organizations shall ensure that barriers to reporting are removed.

Advanced Safety Management: Focusing on Z10.0, 45001, and Serious Injury Prevention,
Third Edition. Fred A. Manuele.
© 2020 John Wiley & Sons, Inc. Published 2020 by John Wiley & Sons, Inc.

- Discusses why supervisors and others may not be adequately qualified to make thorough incident investigations.
- Suggests that safety professionals review their own concepts of how accidents happen.
- Encourages that evaluations be made of the quality of incident investigation systems in place to determine needs and opportunities and courses of action for improvement.
- Promotes use of the Five Why Problem Solving System.
- Makes observations on selected resources.
- Again, tries to convince safety professionals that their learned advice when incident investigations are made is seriously needed at all levels of management.

RESULTS OF RESEARCH

In the studies of incident investigation reports made by this author, on a scale of 10 with 10 being best, companies were given scores ranging from 2 to 8, with an average of 5.7, and that average could be a stretch. These relatively poor scores were troubling and prompted inquiry into situations that may exist in the investigation process that might be obstacles to in-depth determinations of contributing factors and why this important safety management function is often done superficially.

In many situations, it was found that the content of incident investigation reports did not fit well with issued guidelines. This author ran a Five Why exercise to determine why there was such a huge gap between issued procedures on incident investigation and what actually takes place. As the Five Why exercise proceeded, it became apparent that our model is flawed on several counts.

CONTINUED FOCUS ON WORKER UNSAFE ACTS AS CAUSAL FACTORS

In organizations where there is a reluctance to explore systemic causal factors, the incident investigation stops after identifying the worker human error—the so-called unsafe act. Thus, a more thorough investigation that looks into the reality of the systemic causal factors is avoided. This is a huge, widespread problem.

In such situations, technical, organizational, and management systems and cultural causal factors for incidents are not identified. Wherever that occurs and avoidance of reality is deeply embedded into the system of expected performance, real opposition to culture changes should be expected.

As is shown in several chapters in this book, the occurrence of an unsafe act is an alert to the need for additional inquiry to determine the possible existence of design, engineering, procedural, operational, or cultural causal factors.

THE POSITION IN WHICH SUPERVISORS ARE PLACED

Typically, first-line supervisors are given the responsibility to initiate an incident investigation report. It is presumed that since they are closest to the work, they know more about the details of what has occurred and the existence of hazards and risks relative to the incident.

When supervisors complete incident investigation reports, they are being asked to write performance reviews on themselves and on the people to whom they report—all the way up to the board of directors.

It is understandable—supervisors would avoid expounding on their own shortcomings. The probability is close to zero that a supervisor will write:

> This accident occurred in my area of supervision and I take full responsibility for it. I overlooked .... I should have done .... .My boss did not forward the work order for repairs I sent him two months ago ....

It is not surprising that supervisors would be reluctant to write about shortcomings in the management systems for which the people to whom they report are responsible. If supervisors do write about such system shortcomings, adverse personnel relationships could result.

With respect to operators (first line employees) and incident causation, James Reason (1990) wrote in *Human Error* that:

> Rather than being the main instigator of an accident, operators tend to be the inheritors of system defects created by poor design, incorrect installation, faulty maintenance and bad management decisions. Their part is usually that of adding the final garnish to a lethal brew whose ingredients have already been long in the cooking. (p 173)

Supervisors are one step above line employees. They also work in a "lethal brew whose ingredients have already been long in the cooking". They have little or no influence on the original design of operations and work systems and are hampered in so far as being able to have major changes made in them.

In some organizations, the procedure is to have a team of two or three to investigate certain accidents—such as all OSHA recordables. Then, if the team consists of fellow supervisors, the team is expected to write a performance appraisal on the supervisor for the area in which the accident occurred and that supervisor's boss, and her boss—all the way up to the board of directors.

It is not difficult to understand that supervisors would be averse to criticizing another supervisor and of the management personnel above the supervisor's level. At every level of management above the line supervisor, it would also be normal to try to avoid being factually critical of themselves. Self-preservation dominates at all levels.

CULTURAL IMPLICATIONS THAT ENCOURAGE GOOD INCIDENT INVESTIGATIONS

Overcoming such obstacles requires a culture change. In some operations, senior management insists upon being informed of the factors contributing to accidents and that reports be factual. That is an element of its culture. An example follows.

In a company where management is fact based and sincere when they say that they want to know about the contributing factors for accidents, regardless of where the responsibility lies, a special investigation procedure is in place for serious injuries and fatalities.

Management recognized that it was difficult for leaders at all levels to complete factual investigation reports that may be critical of themselves. Thus, an independent facilitator serves as the investigation and discussion team leader. At least five knowledgeable people serve on the team. All members of the team know that a factual report is expected.

It is known that the CEO reads the reports, asks questions to assure that they are complete, and insists that the people who report to him resolve all the recommendations made to a proper conclusion. Thus, the CEO demonstrates by what she does that the organization's culture requires that facts be determined.

Incident investigation is done well when the safety culture established by management will not tolerate other than superior performance. In the studies made of the quality of investigations, some companies scored an 8 out of a possible 10. (More than one safety director accused me of being a hard marker.)

In those companies, the positive safety culture is driven by the senior executives, and in some instances, by the board of directors. At those levels, personnel are held accountable for results.

CULTURAL IMPLICATIONS THAT MAY IMPEDE GOOD INCIDENT INVESTIGATIONS

Throughout this book, the significance of an organization's culture and how it impacts favorably or unfavorably on safety-related decision-making has been emphasized. There is a relative and all-too-truthful paragraph in *Guidelines for Preventing Human Error in Process Safety* where comments are made on the "Cultural Aspects of Data Collection System Design". It also pertains to incident investigation. It follows.

> A company's culture can make or break even a well-designed data collection system. Essential requirements are minimal use of blame, freedom from fear of reprisals, and feedback which indicates that the information being generated is being used to make changes that will be beneficial to everybody.
>
> All three factors are vital for the success of a data collection system and are all, to a certain extent, under the control of management. (p 259)

424 INCIDENT INVESTIGATION

In relation to the foregoing, the title of R.B. Whittingham's (2004) book *The Blame Machine: Why Human Error Causes Accidents* is particularly appropriate. Whittingham says that his research shows that, in some organizations, a "blame culture" has evolved whereby the focus in their investigations is on individual human error and the corrective action stops at that level. That avoids seeking data on and improving the management systems that may have enabled the human error. (xii).

What Whittingham wrote is indicative of an inadequate safety culture. As an example of one aspect of a negative safety culture, consider the following scenario. It represents a culture of fear.

> An electrocution occurred. As required in that organization, the corporate safety director visited the location for a review of the investigation made at the location level. During discussion with the deceased employee's immediate supervisor, it became apparent that the supervisor knew of the design shortcomings in the lockout/tagout system, of which there were many at the location.
>
> When asked why the design shortcomings were not recorded as causal factors in the investigation report, the supervisor's response was: "Are you crazy? I would get fired if I did that. Correcting all these lockout/tagout problems will cost money and the boss doesn't want to hear about things like that".

This culture of fear arose from the system of expected performance that management created. The supervisor completed the investigation report in accord with what he believed management expected. He recorded the causal factor as "employee failed to follow the lockout/tagout procedure" and the investigation stopped there.

Overcoming such a culture of fear in the process of improving incident investigation processes, wherever and to what extent it exists, will require careful analysis, and much persuasive diplomacy. For an incident investigation system to be effective, management must demonstrate by what it does that it **wants to know** about the contributing causal factors.

Assume that the safety culture does not require effective incident investigations. Consider the following examples, limited to 10, of statements that could be made in investigation reports but may be perceived as self-incriminating. More importantly, they could be interpreted as being accusatory of management levels above the supervisor—and thus avoided.

1. We did not take the time to train this employee because we were too busy.
2. The injured employee mentioned the hazard to me but it was the kind of thing we have tolerated.
3. We have had work orders in maintenance for three months to fix the wiring on this equipment.
4. We did not do pre-job planning and hazardous situations came up that we did not expect.
5. It was the kind of a rush situation that often happens and we understand that sometimes the workers take short cuts and don't follow the SOP.

6. The work is overly stressful and risky. Hazards were not properly considered in the design process.
7. The equipment is being run beyond its normal life cycle, and the risks in operating it are high.
8. What we are asking our people to do is exhausting and they make mistakes.
9. We haven't had time to write a Standard Operating Procedure for this job.
10. The stuff that purchasing bought is cheaply made and it falls apart.

Not doing thorough incident investigations is normal in organizations where management is not fact based and does not promote and require that hazard and risk problems be identified and acted upon. Situations of that sort define safety culture problems.

If safety professionals promote improving the quality of incident investigation, the reality of the safety culture in place must be evaluated and accurately defined as an action plan for improvement is formulated.

HAVING COMPASSION FOR SUPERVISORS

As was stated previously, first line supervisors are given the responsibility to initiate an incident investigation report because it is assumed that they are close to the work, know the most about the hazards and risks and are the best qualified. That premise needs re-thinking.

Safety professionals should ask—how much training with respect to hazards and risks do supervisors get, does the training make them knowledgeably and technically qualified, and how often is training provided?

This author doubts that supervisors, after taking a one- or two-day course on incident investigation, become knowledgeable about hazards and risks and thus are well qualified to make good incident investigations.

A question that logically follows pertains to all personnel at all levels who do incident investigations. How often do they complete incident investigations and do the forms and procedure manuals provide adequate support? It is unusual for a supervisor to complete two or three incident investigations in a year. That may also be the case for members of teams that are given the responsibility to investigate an accident.

Also, consideration needs to be given to the time lapse between when supervisors and others attend a training session and when they complete an incident investigation report. It is generally accepted that knowledge obtained in a training session will not be retained without frequent use.

Supervisors, and others, should be provided with readily available reminder references, the content of which should be comparable to and an extension of Addenda A to this chapter. An additional resource for the development of a reference guide is Guidelines for Investigation of Logistics Incidents and Identifying Root Causes (2015). This is a 31-page document, on the Internet and downloadable without charge.

For emphasis, it is said again: *One of the purposes of this chapter is to convince safety professionals that their learned advice when incident investigations are made is seriously needed at all levels of management.*

ON THE WAY TO IMPROVEMENT: START WITH A SELF-EVALUATION OF ONE'S OWN BELIEFS AND THE CULTURE

Safety professionals who undertake to improve the quality of incident investigation should commence with the first step in problem solving—define the problem, which may be binary.

A Self-Evaluation of One's Beliefs on Causation

As safety professionals give advice when incident investigations are made, they are applying their adopted causation models. Their models relate to what they have learned and their beliefs concerning how accidents happen. They must recognize as fact that they are obligated to have their advice be based on a sound thought process that takes into consideration the reality of the hazards, the risks that derive from them, and the relative management system deficiencies that may be causal factors.

Chapter 20 is titled Incident Causation Models. That chapter and the three following chapters are good resources for the self-evaluations that many safety professionals should make with respect to the appropriateness of the causation model they have adopted. It is recognized that it is difficult, perhaps agonizing and uncomfortable, to think about one's own beliefs and test them against facts.

A Self-Evaluation of the Culture in Place

Safety professionals should make an evaluation of a sampling of completed incident investigation reports for the cultural implications they may reveal. In the studies this author made, the identification entries in incident investigation forms—such as name, department, location of the accident, shift, time, occupation, age, time in the job etc.—got relatively high scores for thoroughness of completion.

Thus, it is suggested that the evaluation concentrate on the incident descriptions, causal and contributing factor determination, and the corrective actions taken. For efficient time usage, a safety professional may want to have the evaluation include only incidents resulting in serious injury or illness and near miss incidents that could have had serious results in slightly different circumstances.

Several chapters in this book provide guidance on culture evaluation and serious injury, illness, and fatality prevention. A safety professional who undertakes such a study should keep in mind that its outcome is to be an analysis of the:

- activities in which serious injuries occur, for which concentrated prevention efforts will be beneficial
- quality of causal factor determination and corrective action taking

- culture that has been established over time with respect to good or not so good causal factor determination and corrective action taking
- organization levels that are to be influenced if improvements are to be made.

From that analysis, a plan of action would be drafted to favorably influence the system of expected performance—thus, the culture. It is a near absolute that the organization's safety culture with respect to the quality of incident investigation will not be changed without support from senior management.

So, the plan of action must be well crafted to convince the management of the value of making the changes proposed—avoiding injuries to employees, avoiding operations downtime, cost reduction, waste reduction (lean), personnel relations, and fulfilling community responsibility.

It is much, much easier to write all this than it will be for safety professionals to get it done. Culture changes are not easily accomplished. They require considerable time and patience to achieve small steps forward.

TEAMS

To conclude on this subject: It is strongly recommended that, if practicable, selected categories of incidents be investigated by well-chosen teams. Some of the team members should be personnel not directly related to the area in which the accident occurred.

In this author's studies of the quality of investigations, reports prepared by teams that were encouraged to be factual got the highest scores. That is an idea that safety professionals could promote.

OTHER SUBJECTS TO BE REVIEWED

As a part of the improvement endeavor, other evaluations should be made, such as—what is being taught about incident investigation, what guidance is given in procedure manuals, and whether the content and structure of the incident investigation form assist or hinder thorough investigations.

The following define real-world situations as discovered in the studies made. Consider how such situations would affect the quality of investigations.

- in courses taught on incident investigation, the instructor leads attendees to conclude that 80-90 percent of accidents are caused principally by the unsafe acts of workers
- instruction is plainly given that the corrective actions proposed in incident investigation forms should focus on improving worker behavior
- in the incident investigation procedure manual the same thought is conveyed and little guidance is given on causal factors at levels above the worker

- the first instruction in an incident investigation form, after a description of the incident is recorded, is – Identify the unsafe act committed by the worker.
- instructions for the computer-based data entry system with respect to accidents say – Enter the unsafe act code. The system allows the entry of only one causal factor code.

When there is a lack of understanding of the fundamentals of incident causation and there is no need to identify contributing causal factors, supervisors, upper levels of management, and safety professionals may put their signatures on forms, indicating approval, when the reality is that investigations are shallow and of little value. Making the additional reviews proposed in this chapter will help a safety professional define the extent of the problem and provide assistance in crafting a course of action for improvement.

THE 5 WHY PROBLEM SOLVING TECHNIQUE

As incident investigation procedures are improved, the goal is to have contributing causal factors determined and properly acted upon. As a beginning step where the incident investigation system needs such improvement, it is suggested that a problem-solving process be considered for which the training and administrative requirements are not extensive—that is The 5 Why Problem Solving Technique.

Skilled incident investigators may say that the "5 Why" process is inadequate because it does not promote the identification of causal factors resulting from decisions made at a senior executive level. That is not necessarily so. Usually, when inquiry gets to the fourth "why", considerations are at the management levels above the supervisor and may consider decisions made by the board of directors.

The origin of the "5 Why" process is attributed to Taiichi Ohno while he was at Toyota. He developed and promoted a practice of asking "why" five times to determine what caused a problem so that root causal factors can be identified, and effective countermeasures can be implemented. The "5 Why" process is applied in a large number of settings for a huge variety of problems.

Since the premise on which the 5 Why concept is based is uncomplicated, it can be easily adopted in the incident investigation process, as some safety professionals have done. Given an incident description, the discussion leader would ask "why" five times to get to the contributing causal factors and outline the necessary corrective actions.

This author is a promoter of The Five Why Problem Solving Technique because for many organizations, achieving competence in applying the Technique for incident investigations will be a major step forward. Chapter 21 in this book is titled The Five Why Problem Solving Technique. It contains several real-world applications of the Technique.

They say in Annex A for Z10.0 at E9.2 that there are many techniques for incident investigation, and they list a few tools used in business management systems. It was satisfying to note that the first technique in the list is The Five Why Problem Solving Technique.

WHAT THIS CHAPTER IS NOT

Since the literature giving guidance on incident investigation techniques is hugely abundant, comments are not being made here on such as: investigation criteria; immediate actions to be taken; fact determination; objectivity; interviewing witnesses; developing incident investigation teams; action plans, etc. The chapter titled "Designer Incident Investigation" in the fourth edition of *On The Practice of Safety* (2013) gives a detailed review of the methodology.

For safety professionals who choose to be educated on more sophisticated incident investigation methods, the following resources provide information on Barrier analysis; Change analysis; Event tree analyses; Failure mode and effects analysis; and Fishbone (Ishikawa) diagrams.

INCIDENT INVESTIGATION RESOURCES

Since the names of the authors and publishers for each of the resources listed here are shown, as well as the websites for some of them, they are not repeated in the Reference list at the end of this chapter. The first three resources listed are highly recommended for their content. Also, they are well worth the price. They are available on the Internet and can be downloaded, free.

"Root Cause Analysis Guidance Document, DOE-NE-STD-1004-92". Washington, DC: US Department of Energy, 1992. Accessed June 2, 2019

http://us.yhs4.search.yahoo.com/yhs/search?p=DOE-NE-STD-1004-92&hspart=att&hsimp=yhs-att_001&type=att_lego_portal_home

> This is a sixty-nine page highly informative document. It is an instructive read. Various incident investigation techniques are discussed in an "Overview of Occurrence Investigation". Thus, it is a resource on Events and Causal Factor Analysis; Change Analysis; Barrier Analysis; Management Oversight and Risk Tree (MORT) analytical logic diagram; Human Performance Evaluation; and Kepner-Tregoe Problem Solving and Decision Making.

MORT User's Manual. DOE, 1992. See the comments made beneath the entry for NRI MORT which follows. It is available on the Internet and downloadable for free. Accessed June 4, 2019 https://www.osti.gov/biblio/5254810/

NRI MORT User's Manual—NRI-I (2002), a Generic Edition For Use with the Management Oversight and Risk Tree Analytical Logic Diagram. This is a later edition of the previous resource shown. It is published by The Noordwijk Risk Initiative Foundation in The Netherlands.

In a discussion under the caption "What is MORT," these comments are made: "By virtue of public domain documentation, MORT has spawned several variants, many of them translations of the MORT User's Manual into other languages. The durability of MORT is a testament to its construction and content; it is a highly logical expression of the functions required for an organization to manage risks effectively." They say that this 2002 version of the MORT User's Manual aims to:

- Rephrase the questions in British English
- Improve guidance on the investigative application of MORT
- Restore 'freshness' to the 1992 MORT question set
- Simplify the system of transfers in the chart
- Remove DOE-specific references
- Help users tailor the question set to their own organizations

I believe they accomplished their purposes—to improve guidance on the investigation application, restore 'freshness,' and simplify the system. What they have done is fascinating. The 69-page document is available on the Internet. Accessed June 3, 2019 http://www.nri.eu.com/NRI1.pdf#search='NRI%20Mort%20User%27s%20Manual'©

I recommend that safety professionals who want to identify the reality of causal factors acquire an understanding of the thinking on which MORT is based. Chapter 22 in this book is titled MORT—The Management Oversight and Risk Tree

A review of the aforementioned documents will provide an inexpensive and valuable education. Now, to extend the resource list, two books on incident investigation and causal factor identification and analysis are listed. There are many other resources besides these.

Guidelines for Preventing Human Error in Process Safety. New York, NY: Center for Chemical Process Safety of the American Institute of Chemical Engineers, 1994

This is a highly recommended text. Chapter 6 is titled "Data Collection and Incident Analysis Methods." Elsewhere, comments are made on types of human error causal factors, their nature, and how to identify and analyze them.

Hendrick, Kingsley and Ludwig Benner, Jr. *Investigating Accidents With STEP.* New York, NY: Marcel Dekker, Inc., 1987

Hendrick and Benner have developed an incident investigation system called "Sequentially Timed Events Plotting." Several authors refer to this thought provoking system. This book is devoted entirely to the STEP system.

One other book that is highly recommended is *Managing the Risks of Organizational Accidents* by James Reason (1997). Chapter 23 in this book is titled Reason's Swiss Cheese Model

CONCLUSION

This author's studies of incident investigation prompt the conclusion that significant risk reduction can be achieved if investigations are done well. If incident investigations are thorough, the reality of the technical, organizational, methods of operation, and cultural causal factors will be revealed.

If safety professionals want to select leading indicators for safety management system improvement, they would have a good data source for that purpose if incident investigation reports realistically identify causal factors. This author now believes that the quality of incident investigation is one of the principle markers in evaluating an organization's safety culture.

Assume that a safety professional decides to take action to improve the quality of incident investigation. It is proposed that the following comments about incident investigation as excerpted from the August 2003 "Report of the Columbia Accident Investigation Board" be kept in mind as a base for reflection throughout the endeavor.

The Report pertains to the Columbia space ship disaster. Accessed June 3, 2019. http://www.nasa.gov/columbia/home/CAIB_Vol1.html

> Many accident investigations do not go far enough. They identify the technical cause of the accident, and then connect it to a variant of "operator error." But this is seldom the entire issue.
>
> When the determinations of the causal chain are limited to the technical flaw and individual failure, typically the actions taken to prevent a similar event in the future are also limited: fix the technical problem and replace or retrain the individual responsible. Putting these corrections in place leads to another mistake—the belief that the problem is solved.
>
> Too often, accident investigations blame a failure only on the last step in a complex process, when a more comprehensive understanding of that process could reveal that earlier steps might be equally or even more culpable. In this Board's opinion, unless the technical, organizational, and cultural recommendations made in this report are implemented, little will have been accomplished to lessen the chance that another accident will follow. (Volume 1 – Chapter 7 – p 177)

For emphasis, I paraphrase: If the cultural, technical, organizational, and methods of operation causal factors are not identified, analyzed, and resolved, little will be done to prevent recurrence of similar incidents.

REFERENCES

ANSI/ASSP Z10.0-2019. *Occupational Health and Safety Management Systems*. Park Ridge, IL: American Society of Safety Professionals, 2019.

ANSI/ASSP/ISO 45001-2018. *Occupational Health and Safety Management Systems—Requirements with Guidance for Use*. Park Ridge, IL: American Society of Safety Professionals, 2018.

Center for Chemical Process Safety. *Guidelines for Preventing Human Error in Process Safety*. New York: Center for Chemical Process Safety of the American Institute of Chemical Engineers, 1994.

European Association of Chemical Distributors. *Guidelines for Investigation of Logistics Incidents and Identifying Root Causes*. UK: European Association of Chemical Distributors, Available at https://www.ecta.com/resources/Documents/Best%20Practices%20Guidelines/guidelines_for_investigation_of_logistics_incidents_and_identifying_root_causes.pdf. Accessed June 3, 2019. 2015.

Manuele, F.A. *On the Practice of Safety*, Fourth Edition. Hoboken, NH: John Wiley & Sons, 2013.

Reason, J. *Human Error*. New York, NY: Cambridge University Press, First published in 1990. At least 12 additional reprintings, n.d.

Reason, J. *Managing the Risks of Organizational Accidents*. Burlington, VT: Ashgate Publishing, 1997.

Whittingham, R.B. *The Blame Machine: Why Human Error Causes Accidents*. Burlington, MA: Elsevier Butterworth-Heinemann, 2004.

FURTHER READING

The Five Why System. Available at http://www.moresteam.com/toolbox/5-why-analysis.cfm. https://www.mindtools.com/pages/article/newTMC_5W.htm. Accessed June 3, 2019.

ADDENDUM A

A REFERENCE FOR THE SELECTION OF CAUSAL FACTORS AND CORRECTIVE ACTIONS FOR INCIDENT INVESTIGATION PROCEDURES AND REPORTS

Designers of incident investigation systems should understand that the causal factors and corrective actions included within investigation forms or as separate informational documents must be appropriate for the operations being conducted. The material presented here should not be used without modification to suit needs.

Work Place Design Considerations

1. Hazards derive from basic design of facilities, hardware, equipment or tooling.
2. Hazardous materials need attention.
3. Layout or position of hardware or equipment presents hazards.
4. Environmental factors (heat, cold, noise, lighting, vibration, ventilation, et cetera) presented hazards.
5. Work space for operation, maintenance, or storage is insufficient.
6. Accessibility for maintenance work is hazardous.

Work Methods Considerations

1. Work methods are overly stressful.
2. Work methods are error-provocative.
3. Job is overly difficult, unpleasant, or dangerous.

Advanced Safety Management: Focusing on Z10.0, 45001, and Serious Injury Prevention, Third Edition. Fred A. Manuele.
© 2020 John Wiley & Sons, Inc. Published 2020 by John Wiley & Sons, Inc.

4. Job requires performance beyond what an operator can deliver.
5. Job induces fatigue.
6. Immediate work situation encouraged riskier actions than prescribed work methods.
7. Work flow is hazardous.
8. Positioning of employees in relation to equipment and materials is hazardous.

Job Procedure Particulars

1. No written or known job procedure.
2. Job procedures existed but did not address the hazards.
3. Job procedures existed but employees did not know of them.
4. Employee knew job procedures but deviated from them.
5. Deviation from job procedure not observed by supervision.
6. Employee not capable of doing this job (physically, work habits, or behaviorally).
7. Correct equipment, tools, or materials were not used.
8. Proper equipment, tools, or materials were not available.
9. Employee did not know where to obtain proper equipment, tools, or materials.
10. Employee used substitute equipment, tools, or materials.
11. Defective or worn out tools were used.

Hazardous Conditions

1. Hazardous condition had not been recognized.
2. Hazardous condition was recognized but not reported.
3. Hazardous condition was reported but not corrected.
4. Hazardous condition was recognized but employees were not informed of the appropriate interim job procedure.

Personal Protective Equipment

1. Proper personal protective equipment (PPE) not specified for job.
2. PPE specified for job but not available.
3. PPE specified for job, but employee did not know requirements.
4. PPE specified for job, but employees did not know how to use or maintain.
5. PPE not used properly.
6. PPE inadequate.

Management and Supervisory Aspects

1. General inspection program is ineffective.
2. Inspection procedure did not detect the hazards.
3. Training as respects identified hazards not provided, inadequate, or didn't take.
4. Maintenance with respect to identified hazards is inadequate.
5. Review not made of hazards and right methods before commencing work for a job done infrequently.
6. This job requires a job hazard/task/ergonomics analysis.
7. Supervisory responsibility and accountability not defined or understood.
8. Supervisors not adequately trained for assigned safety responsibility.
9. Emergency equipment not specified, not readily available, not used, or did not function properly.

Corrective Actions to Be Considered

1. Job study to be recommended: job hazard/task/ergonomics analysis needed.
2. Work methods to be revised to make them more compatible with worker capabilities and limitations.
3. Job procedures to be changed to reduce risk.
4. Changes are to be proposed in work space, equipment location or work flow.
5. Improvement is to be recommended for environmental conditions.
6. Proper tools to be provided along with information on obtaining them and their use.
7. Instruction to be given on the hazards of using improper or defective tools.
8. Job procedure to be written or amended.
9. Additional training to be given concerning hazard avoidance on this job.
10. Necessary employee counseling will be provided.
11. Disciplinary actions deemed necessary, and will be taken.
12. Action is to be recommended with respect to employee who cannot become suited to the work.
13. For infrequently performed jobs, it is to be reinforced that a pre-job review of hazards and procedures is to take place.
14. Particular physical hazards discovered will be eliminated.
15. Improvement in inspection procedures to be initiated or proposed.
16. Maintenance inadequacies are to be addressed.
17. Personal protective equipment shortcomings to be corrected.

CHAPTER 20

INCIDENT CAUSATION MODELS

Incident investigation is one of the most important elements in an operational risk management system. Nevertheless, research has shown that in many organizations what is done for incident investigation results in missed opportunities to reduce risks. Thus, they do not learn from their histories, and the causal factors remain dormant, with potential.

One of the reasons that incident investigations are not done well is that many safety practitioners have convinced management, based on their adopted causation model, that the principal causal factors for incidents are the unsafe acts of employees. This chapter pleads with safety practitioners that they:

- Do a self-analysis concerning how they think incidents happen.
- Adopt a causation model that encourages inquiry into the reality of causal factors.
- Initiate an educational activity in the organizations for which they give advice to have the causation model they have adopted to become the base for incident investigations.

If incident investigations are done properly, an organization identifies the relative hazards, risks, and possible deficiencies in management systems and can take corrective action to reduce the probability of similar incidents occurring in the future.

Advanced Safety Management: Focusing on Z10.0, 45001, and Serious Injury Prevention, Third Edition. Fred A. Manuele.
© 2020 John Wiley & Sons, Inc. Published 2020 by John Wiley & Sons, Inc.

REQUIREMENTS FOR CAUSATION MODELS

This author concludes from his research that an incident causation model must be inclusive of the following premises.

- An organization's culture is the primary determiner with respect to the avoidance, elimination, reduction, or control of hazards and whether acceptable risk levels are to be achieved and maintained.
- Management commitment or noncommitment to operations risk management is an extension of an organization's culture.
- Causal factors may be related to an organization's culture and management's commitment or noncommitment to operations risk management and may derive from decisions made at the top management level that result in insufficiencies in policies, standards, procedures, and the accountability system.
- A large majority of the problems in any operation are systemic. They derive from the decisions made by management that establish the operating socio-technical system—that is, the work place, the work methods, and the governing social atmosphere/climate/environment.
- Many, but not all, accidents resulting in serious injury or fatality are unique and singular events, having multiple, complex, and interacting causal factors that may have cultural, organizational, technical, or operational systems origins.
- A sound causation model for hazards-related incidents must take into consideration the entirety of the socio-technical system—applying a holistic approach to both the technical aspects and the social aspects of operations in which continuous variation is the norm. It must be understood that those aspects are interdependent and mutually inclusive.
- A causation model must:
  - Be practically usable in the organization that adopts it;
  - Not be overly complex in relation to needs; and
  - Focus on the reality of hazards, risks and identification of the possible deficiencies in the relative management systems.

To apply these principles, a macro view of the socio-technical system must be taken. That requires thinking large about operating systems as integral and inseparable parts of a whole and their interrelationships. It also requires seeking the deficiencies in management systems, some of which could derive from the organizational, cultural, technical, and social aspects of an organization.

SAFETY PROFESSIONALS—YOU HAVE ADOPTED A CAUSATION MODEL

In *The Field Guide to Understanding Human Error,* Sidney Dekker (2006) makes this astute and factual observation—one that must be considered as fact by safety professionals who give advice when incident investigations are made.

> Where you look for causes depends on how you believe accidents happen. Whether you know it or not, you apply an *accident model* to your analysis and understanding of failure. An accident model is a mutually agreed, and often unspoken, understanding of how accidents occur. (p. 81)

As safety professionals participate in determining causal and contributing factors for an incident, they are applying their adopted causation model. Their models relate to what they have learned and their beliefs concerning how accidents happen.

They must recognize as fact that they are obligated to have their advice be based on a sound thought process that takes into consideration the reality of the hazards, the risks that derive from them, and the relative management system deficiencies.

For effectiveness, an organization must have adopted a causation model that gets to the reality of risk-based causal factors. Stellar performance in incident investigation cannot be achieved otherwise.

In *Barriers and Accident Prevention,* Erik Hollnagel (2004) supports that premise. He writes extensively and persuasively about the need for an organization to have selected a causation model to serve as a base for investigation, communication, and taking corrective action. (p. 44)

Earlier, it was said that a large percent of safety professionals still apply a causation model that focuses on the so-called unsafe acts of the worker as the causal factors. Thus, there are many missed opportunities to improve on the reduction of injuries, illnesses, and damage to property or the environmental. Their thinking is Heinrichean, for which comment follows.

THE HEINRICHEAN CAUSATION MODEL

It is unfortunate that many safety practitioners have adopted a causation model based on the Heinrichean premise that occupational accidents are principally caused by unsafe acts of employees. That premise is repeated in all four editions of H.W. Heinrich's book *Industrial Accident Prevention*.

Heinrich said that among the direct and proximate causes of industrial accidents: 88% are unsafe acts of persons; 10% are unsafe mechanical or physical conditions; and 2% are unpreventable. (p. 43—1931: p. 22—1941: p. 19—1950: p. 22—1959).

440 INCIDENT CAUSATION MODELS

If safety practitioners believe that incidents are caused principally by worker unsafe acts, by their human errors, and that is what they have taught others, the problem is hugely huge because that concept may be imbedded in an organization's culture. It will be difficult to change. That perception—that 88% of accidents are caused by unsafe acts of workers—does not encompass what has been learned about the sources of human error and the complexity of incident causation.

Heinrich's statements were first published in 1931. Since then, numerous texts and articles have been published indicating that incident causation is much more complex than Heinrich indicated. Only a few of those publications are cited here.

Guidelines for Preventing Human Error in Process Safety by the Center for Chemical Process Safety

This book is highly recommended because it is one of the best references available on incident causation. Excerpts shown here are informative about how accidents happen, where human errors occur, who commits errors and at what level, the influence of organizational culture and where attention is needed to reduce the occurrence of human errors. These excerpts are risk-based and management system oriented.

- It is readily acknowledged that human errors at the operational level are a primary contributor to the failure of systems. It is often not recognized, however, that these errors frequently arise from failures at the management, design, or technical expert levels of the company. (p. xiii)
- Almost all major accident investigations in recent years have shown that human error was a significant causal factor at the level of design, operations, maintenance, or the management process. (p. 5)
- One central principle presented in this book is the need to consider the organizational factors that create the preconditions for errors, as well as the immediate causes. (p. 5)

Since "failures at the management, design, or technical expert levels of the company" effect the design of the workplace and the work methods—it is logical for safety practitioners to encourage the selection of causal factors and a causation model that relate to deficiencies in the operating system.

In the last bulleted item mentioned previously, reference is made to "organizational factors that create the preconditions for errors". It is a hard fact—if the design of the work place or the work methods produces an error-provocative environment, it is likely that errors will occur. But the error—the unsafe act—is not the principle causal factor. A causation model should encourage inquiry into the relative organizational factors.

Human Error by James Reason

In *Human Error*, Reason (1990) makes an observation that relates directly to the changes that have occurred in the thinking about unsafe acts/human errors.

Rather than being the main instigators of an accident, operators tend to be the inheritors of system defects created by poor design, incorrect installation, faulty maintenance and bad management decisions. Their part is usually that of adding the final garnish to a lethal brew whose ingredients have already been long in the cooking. (p. 173)

Reason (1997) is a prominent and often quoted writer. In *Organizational Accidents Revisited*, he made a comment that is relative to this chapter and for a discussion of causation and causation models. Reason comments on the commonality of thinking held by the people who have been influential in his field. He says "What they have in common is a rejection of unsafe acts as the focus of accident investigation". (p. 107) Chapter 23 in this book pertains to Reason's Swiss Cheese Model.

Techniques of Safety Management by Dan Petersen

In his seventh of The Ten Basic Principles of Safety, Petersen (1989) expressed his view concerning individual behavior being a derivative of the work environment created by management:

In most cases, unsafe behavior is normal behavior; it is the result of normal people reacting to their environment. Management's job is to change the environment that leads to the unsafe behavior. Unsafe behavior is the result of the environment that has been constructed by management. In that environment, it is completely logical and normal to act unsafely. (p. 35)

Assume that the environment that has been constructed results in logical and normal behavior that is considered unsafe. Then, incident investigators and their causation models should give emphasis to the decision-making out of which those environmental causal factors arose.

The Field Guide to Understanding Error by Sidney Dekker

Sidney Dekker's text was previously referenced in this chapter. Dekker is one of the most influential authors on human error. Excerpts from his book follow. His Field Guide has achieved a deserved and international prominence.

- Human error is not a cause of failure. Human error is the effect, or symptom, of deeper trouble. Human error is systematically connected to features of people's tools, tasks, and operating systems. Human error is not the conclusion of an investigation. It is the starting point. (p. 15)
- Sources of error are structural, not personal. If you want to understand human error, you have to dig into the system in which people work. (p. 17)

- Error has its roots in the system surrounding it; connecting systematically to mechanical, programmed, paper-based, procedural, organizational and other aspects to such an extent that the contributions from system and human error begin to blur. (p. 74)

This transition whereby the source of human error is considered systemic and largely relates to the processes created by management is opposite to focusing principally on worker unsafe acts as primary causal factors. Incident investigators, as they determine how accidents happen, must have a causation model that requires delving far beyond what the worker does or does not do and into the operating system.

ON CATEGORIES OF CAUSATION MODELS

Erik Hollnagel listed several major categories of causation models in his book *Barriers and Accident Prevention*. Sidney Dekker, in *The Field Guide to Understanding Human Error*, cites the same categories as did Hollnagel. Variations of what Hollnagel and Dekker wrote appear in other texts. Paraphrasing from Hollnagel's book follows.

Sequential Accident Models

Models in this category indicate that accidents occur as a result of a sequence of events that occur in a specific order. Heinrich's domino sequence was the first such model. As others and this author have done, Hollnagel observed that sequential models are inadequate when they focus on the unsafe act – the human error – of the worker as the causal factor for an incident. They are also inadequate when incidents result from multiple causal factors that may develop linearly or in parallel after an initiating event. (p. 47)

Epidemiological Accident Models

Epidemiological models relate to studies of how diseases develop and the searches made for their pathogens and causal factors. A depiction of this type of model would show incidents as deriving from a blend of causal factors, some active and some having been built up over time and existing in combination at the time the incident occurred. (p. 55)

Dekker made the following comments on epidemiological models in *The Field Guide to Understanding Human Error*.

The epidemiological model sees accidents as an effect of the combination of:

- Active errors or unsafe acts, committed by those at the sharp end of a system, and

- Latent errors, or conditions buried inside the organization that lie dormant for a long time but can be triggered in a particular set of circumstances. (p. 88)

Hollnagel says that, in the application of the epidemiological model, consideration would be given to:

- Performance deviations
- Deviations from safe practice
- Environmental conditions (technical and social aspects)
- Absence or ineffectiveness of barriers that could have prevented the accident and latent hazardous conditions or practices that have accumulated over time. (p. 54)

It is significant that Dekker comments beneficially on epidemiological models as follows because epidemiological models are sufficient for many incident investigations.

Even though they may linearize and oversimplify, the epidemiological model has been helpful in portraying the resulting imperfect organizational structure that let the accident happen. (p. 89)

Systems Accident Models

Application of systems accident models requires considering individual systems as interrelated and inseparable parts of a whole. Significantly, in his discussion of systems models, Hollnagel recognized the value of a form of linear plotting. For an incident having causal factors that develop in parallel after the initiating event, more than one linear plotting may be necessary, each being related causally to each other. Hollnagel writes:

Events can still be ordered post hoc either temporally or in terms of causal relations. But in the systemic model each event may be preceded by several events (temporally or causally), as well as be followed by several events. (p. 59)

Applying a systems approach requires macro thinking rather than micro thinking and seeking the possible hazards, risks, and the relative deficiencies in management systems.

An example of micro thinking is taking the position that unsafe acts of employees are the principal causal factors for occupational accidents, stopping the investigation after a worker unsafe act is identified, and focusing corrective actions on improving the work practices of the worker rather than delving into possible systems causal factors.

BARRIERS AND CONTROLS AND CAUSATION MODELS

In *System Safety for the 21st Century,* Richard A. Stephans (2004) gave a broad definition of barriers and commented on what a barrier analysis should accomplish.

> Barrier – Anything used to control, prevent, or impede energy flows. Types of barriers include physical, equipment design, warning devices, procedures and work processes, knowledge and skills, and supervision. Barriers may be control or safety barriers or act as both. (p. 357)
>
> Barrier analysis includes determining whether or not a barrier was provided and used and whether the barrier failed. If a barrier was provided but not used, the analysis then seeks to identify the task performance error which caused the barrier not to be used. This process is to evaluate barriers on the energy source, between the energy source and the persons and/or objects to be protected, or on the target person or object. The same process is also used if the barrier evaluated consists of separating the potential targets from potentially harmful energy flows or environments by time or space. (p. 227)

In Erik Hollnagel's book *Barriers and Accident Prevention* (2004), Chapter 3 is titled "Barrier Functions and Barrier Systems" and Chapter 4 is on "Understanding the Role of Barriers in Accidents". Much is made by Hollnagel of the need for and the appropriateness of barriers in accident prevention. This excerpt from Hollnagel relates directly to accident causation and analysis and to what a causation model should include.

> Barriers are important for the understanding and prevention of accidents in two different, but related, ways. Firstly, the very fact that an accident has taken place usually means that one or more barriers have failed – either because they did not serve their purpose adequately or because they were missing or dysfunctional. The search for barriers that failed must be an important part of accident analysis. (p. 68)

In the NRI MORT User's Manual (2009), descriptions of barriers and controls are given, as follows. MORT is the acronym for the *Management Oversight and Risk Tree.*

> Barriers are means of separation present solely for protection purposes. Controls are means of channeling energy or substances to do work (and provide protection as a by-product). Controls also limit the exposure to targets [people, property, or the environment]. MORT analysis will tease out the procedural and upstream issues. (p. xx)

Mention is made in forgoing citations from Stephan and the NRI MORT User's Manual to preventing or impeding energy flows. Those citations relate to Dr. William Haddon's energy release and control theory.

Dr. Haddon was the first Director of the National Highway Safety Bureau. His stated concept is that unwanted transfers of energy can be harmful (and wasteful) and that a systematic approach in the design and operating processes should be taken to limiting their possibility. A part of his approach consists of providing physical or procedural barriers to prevent contact by persons or property and to direct energy flows into wanted channels. Haddon's concepts are soundly based. They are also good design references.

Haddon (1970) stated in his paper "On the Escape of Tigers: An Ecologic Note" that "the concern here is the reduction of damage produced by energy transfer." But he also said that "the type of categorization here is similar to those useful for dealing systematically with other environmental problems and their ecology." Excerpts follow from "On the Escape of Tigers: An Ecologic Note" as the article appeared in the American Journal of Public Health. Note that all of the strategies have to do with facility design or work methods design.

A major class of ecologic phenomena involves the transfer of energy in such ways and amounts, and at such rapid rates, that inanimate or animate structures are damaged.

Several strategies, in one mix or another, are available for reducing the human and economic losses that make this class of phenomena of social concern. In their logical sequence, they are as follows:

1. Prevent the marshaling of the form of energy.
2. Reduce the amount of energy marshaled.
3. Prevent the release of the energy.
4. Modify the rate or spatial distribution of release of the energy from its source.
5. Separate, in space or time, the energy being released from that which is susceptible to harm or damage.
6. Separate, by interposing a material barrier, the energy released from that which is susceptible to harm or damage.
7. Modify appropriately the contact surface, subsurface, or basic structure, as in eliminating, rounding, and softening corners, edges, and points with which people can, and therefore sooner or later do, come in contact.
8. Strengthen the structure, living or non-living, that might otherwise be damaged by the energy transfer.
9. Move rapidly in detection and evaluation of damage that has occurred or is occurring, and counter its continuation or extension.
10. After the emergency period following the damaging energy exchange, stabilize the process. (p. 2229)

All hazards are not addressed by the unwanted energy release concept. Examples are: the potential for asphyxiation from entering a confined space filled with gas; or inhalation of asbestos fibers. But all hazards are encompassed within a goal that is to avoid *both* unwanted energy releases *and* exposures to hazardous environments.

To prevent unwanted energy flows or exposures to hazardous environments, a causation model must determine whether appropriate barriers existed or did not exist and whether the barriers that existed failed or were dysfunctional.

Haddon's energy release and control theory was influential in the later development of barrier theory and of certain causation models.

SELECTED RESOURCES ON CAUSATION MODELS

Unfortunately, there is an overabundance of incident causation models. But that may have resulted from investigators knowing of the importance of incident investigation and their desire to have a sound base for what they do. That can be a problem. For illustration, comment is made here on two collections of descriptions of causation models. There are several others.

Enter *Models of Causation: Safety* (2012) into a search engine to access a paper published by the Safety Institute of Australia Ltd. Comments are made in that paper on 13 accident causation models. It is downloadable without charge and is a recommended resource.

Enter *Methods for accident investigation* into the same search engine to locate a publication issued by NTNU—The Norwegian University of Science and Technology (n.d.). This paper can also be downloaded without cost and is also a recommended resource. Extensive comments are made in the Norwegian University publication on 14 incident investigation techniques. They vary muchly from those described in the Safety Institute of Australia publication.

In addition to the two resources just previously mentioned, comment is made here on additional resources that were selected because they are truly educational and available without charge.

MORT—The Management Oversight and Risk Tree

What MORT is and how it can be used is addressed in the Abstract for the third edition of the MORT User's Manual issued by the Department of Energy (DOE) in 1992. Excerpts follow.

> MORT is a comprehensive analytical procedure that provides a disciplined method for determining the causes and contributing factors of major accidents. Alternatively, it serves as a tool to evaluate the quality of an existing program. (p. iii)

Note the term "major accidents" in the foregoing citation. While MORT provides a valuable informational base on possible causal factors, it is not suggested that safety professionals propose its adoption by their clients as the incident investigation model to be used unless their clients have a frequency of "major incidents."

Developing skills for the application of MORT requires study and continuous application to retain knowledge of the technique. The use of MORT is time consuming and expensive. Fortunately, major events that would justify establishing a MORT incident investigation team do not occur often.

Nevertheless, having knowledge of MORT principles and methods helps understand how incidents happen. If I was teaching at a university in a safety and environmental science degree program, I would expose students to MORT for the thought base it provides on how accidents happen.

MORT is unusual in that it promotes inquiry in much greater detail into upstream decision-making for the identification of causal factors than most other incident investigation models. In the application of MORT, a review would be made of the reality of an organization's culture and how culture deficiencies can be the sources of causal factors.

An interesting and later version of a MORT User's Manual was published in 2009 (second edition) by The Noordwijk Risk Initiative Foundation in The Netherlands. It also is an important resource. They say that their second edition is to improve guidance in the investigation process, restore freshness, and simplify the system. What they have done is fascinating. I believe they accomplished their purposes. Chapter 22 in this book is on MORT.

Accident and Operational Safety Analysis: Volume I: Accident Analysis Techniques

DOE HANDBOOK issued by the U.S. Department of Energy (2012).

> *This is an excellent resource that provides the functional and technical basis for an understanding of accident prevention and investigation principles and practice.* (p. i)

Guidance for Safety Investigation of Accidents

This publication was issued by ESReDA Working Group (n.d.) on Accident Investigation. They say in their preface that incident investigation is improving and that the purpose of their publication is to reflect the state-of-the-art and to address future challenges. They also say that they provide practical guidelines for users. (p. iii)

Comparison of Some Selected Methods for Accident Investigation

Interesting opinions are expressed in this publication on selected causation models. Other authors and publications give differing evaluations. This publication was issued by The Norwegian University of Technology and Science (NTNU)/SINTEF Industrial Management is educational (n.d.).

SUGGESTED CAUSATION MODELS FOR CONSIDERATION BY SAFETY PROFESSIONALS

Comments on specific causation models here are purposely brief. If a safety professional wanted to become proficient in the use of a model or models, it is suggested that the necessary study time be applied. There are many variations in the pictorial depictions of causation models—as a journey into the Internet will reveal. Intentionally, they are not duplicated here

- Adoption of The Five Why Problem Solving Technique is highly recommended as the initiating step in those organizations toward improvement. Chapter 21 is devoted to The Five Why System (n.d.).

  On occasion, the causal factors for an incident are considered to be more complicated, and an additional or replacement causation model should be used. Comments on a few of those causation models follow.

- In *Risk Assessment: A Practical Guide for Assessing Operational Risks*. Popov, Lyon, and Hollcroft (2016) devote a chapter to Bow-Tie Risk Assessment Methodology as they do for several other risk assessment methods. Although described as a tool for risk assessments, the Bow-Tie method can also be used for incident investigation. There are several papers on the Internet that pertain to the Bow-Tie method and incident investigation.

  The use of the Bow-Tie analysis methodology results in clear communication on the adequacy or inadequacy of barriers for the incident that occurred. Being barrier-based, it follows James Reason's "Swiss Cheese" Model of Defenses. Others have similarly observed.

- Chapter 23 in this book is devoted to Reason's Swiss Cheese Model—another highly recommended incident investigation technique. Reason's model illustrates how inadequacies may occur with respect to barriers and that under specific circumstances those inadequacies can be in alignment, such as the holes in Swiss cheese theoretically can align. Thus, the potentials of hazards may penetrate the barriers and an incident may occur.

- One of the causation models discussed in *Models of Causation: Safety* is titled Accident Evolution and Barrier Function (AEB) Model. Comment on it is made here because of how it depicts an incident—as a series of interactions between human and technical systems. Also, it is another model related to Reason's thinking, and it relies on the effectiveness of barriers. In an incident investigation, this model requires inquiry into the entirety of the socio-technical system. It also promotes a review of the adequacy of barrier installation and usage. Those are themes that this author supports.

  What follows paraphrases from an article written by Ola Svenson (2005), Stockholm University, Department of Psychology, SE-106 91 Stockholm, Sweden.

  The Accident Analysis and Barrier Function (AEB) Method considers an incident as a series of interactions between human and technical systems. In the human and technical interrelations, errors can occur leading to an accident. In principle, barriers could arrest the further development of the error.

  A barrier function can be performed by one or several *barrier function systems*. To illustrate, a mechanical system, a computer system or another operator can all perform a given barrier function to stop an operator from making

an error. An AEB analysis would be done in two steps. A linear depiction of the incident would be followed by a barrier function analysis.
- By how many names is the next causation model known – at least by the <u>Ishikawa Diagram, The Fishbone Diagram and The Cause and Effect Diagram. In a sense,</u> The Fishbone diagram, is really a cause and effect causation model. In its use, the intent is to determine the likely causes and sub-causes for an undesired outcome. It looks like the bones of a fish. (Go into the Internet to see the many variations of a fishbone diagram.)

  It is not unusual for a safety professional to enter categories on the principal bones, such as Management, Environment, Supervision, Methods/Processes, Barriers, Machinery and Equipment, People Qualifications and Staffing, Materials and Measurement. Discussions are to identify the possible causal factors under the principal categories.
- Enter <u>Tripod Beta Methodology</u> into a search engine and several pages of references and advertisements (software and training) will appear. Tripod Beta is mentioned here because it:
  - is an extensive application of Reason's Swiss Cheese Model.
  - requires inquiry both to determine the human factor contributions to an incident and to identify the organizational weaknesses that allowed the incident to occur.

Enter the following into a computer to obtain "Guidance on using Tripod Beta in the investigation and analysis of incidents, accidents and business losses."

While that publication is informative, it includes a Caveat, the gist of which is: Before one tries to apply the Tripod Beta method, adequate education should be obtained. (2015, p. 2)

CONCLUSION

There is a serious need for many of those engaged in the practice of safety to upgrade their views about how accidents happen. Much has been learned about the sources of human error and a bit of that history, and resources have been included in this chapter. Safety professionals, it will be to your advantage and that of the organizations to which you give advice if you analyze the accident causation model that you use and adopt a model that reflects on current thinking.

It is not suggested that once you adopt a more relevant model that it will be easy to achieve the culture change that will be necessary in some organizations to utilize that model. A planned course of action, having the support and direction of management, will be necessary.

It is understood that to discard concepts that have been at the base of a safety practitioner's work and to accept and act on other concepts may be very difficult to do.

In his book *The Standardization of Error,* Vilhjamur Stefansson (1928) comments on the difficulty to be expected when attempts are made to dislodge beliefs that are embedded in a system even though they are proven to be untrue. That premise applies not only to safety practitioners but also to employees at all levels who have been taught what are now known to be outmoded concepts. That includes senior management and members of boards of directors.

In a discussion of the implications of the causation models used in the practice of safety, the writers of *Models of Causation: Safety* also made the following statements. They are noteworthy.

> Thus, in developing their mental model, occupational health and safety professionals should be aware of a range of models of causation and be able to critically evaluate the model for application to their practice. In applying a particular model, a safety professional also needs to be able to differentiate between what actually occurs in the workplace with what should happen. (p. 20)

For emphasis: Do not propose a causation model that is overly complex in relation to the needs of an organization.

REFERENCES

Dekker, S. *The Field Guide to Understanding Human Error*. Burlington, VT: Ashgate Publishing Company. 2006.

DOE HANDBOOK: Accident and Operational Safety Analysis—Volume I: Accident Analysis Techniques. DOE-HDBK-1208-2012, U.S. Department of Energy, Washington, D.C. 20585, July 2012.

Comparison of Some Selected Methods for Accident Investigation. The Norwegian University of Technology and Science (NTNU)/SINTEF Industrial Management, N-7465 Trondheim, Norway. Available at https://dvikan.no/ntnu-studentserver/reports/Comparison%20of%20some%20selected%20methods%20for%20accident%20investigation.pdf. Accessed June 3, 2019. n.d.

Guidance for Safety Investigation of Accidents. ESReDA Working Group on Accident Investigation. Available at https://esreda.org/wp-content/uploads/2016/07/ESReDA_GLSIA_Final_June_2009_For_Download.pdf. Accessed June 3, 2019. n.d.

Haddon, W. "On the Escape of Tigers: An Ecologic Note". *Journal of Public Health*, December 1970.

Heinrich, HW. *Industrial Accident Prevention*, Fourth Edition. New York, NY: McGraw-Hill Book Company, 1959.

Hollnagel, E. *Barriers and Accident Prevention*. Burlington, VT: Ashgate Publishing, 2004.

Methods for Accident Investigation. *NTNU*—Norwegian University of Science and Technology. Available at http://www.learnfromaccidents.com.gridhosted.co.uk/images/uploads/Norwegian_university_of_science_and_technology_Method_for_accident_investigation.pdf. Accessed June 2, 2019.

Models of Causation: Safety. Safety Institute of Australia, Ltd., 2012. Available at http://www.ohsbok.org.au/wp-content/uploads/2013/12/32-Models-of-causation-Safety.pdf. Accessed June 3, 2019.

MORT User's Manual. DOE. On the Internet. Downloadable free. 1992. Available at https://www.osti.gov/biblio/5254810/. Accessed June 2, 2019.

NRI MORT User's Manual. Downloadable without charge. Enter "MORT User's Manual" into a search engine and this publication appears. 2009. Available at http://nri.eu.com/NRI1.pdf. Accessed June 3, 2019.

Petersen, D. *Techniques for Safety Management*. Park Ridge, IL: American Society of Safety Professionals, 1989.

Popov, G., Lyon, B., and B. Hollcroft. *Risk Assessment: A Practical Guide to Assessing Operational Risks*. Hoboken, NJ: John Wiley, 2016.

Reason, J. *Human Error*. New York, NY: Cambridge University Press, 1990.

Reason, J. *Managing the Risks of Organizational Accidents*. Burlington, VT: Ashgate Publishing Company, 1997.

Stephans, R.A. *System Safety for the 21st Century*. Hoboken, NJ: John Wiley & Sons, 2004. (This is an updated version of *System Safety 2000* by Joe Stephenson).

Stefansson, V. *The Standardization of Error*. London: K. Paul, Trench, Truber & Co. Ltd, 1928.

Svenson, O. *Accident Analysis and Barrier Function (AEB) Method, Manual for Incident Analysis*. Stockholm, Sweden: Stockholm University, Department of Psychology, 2005.

The Five Why System. Available at http://www.moresteam.com/toolbox/5-why-analysis.cfm. Accessed June 3, 2019. n.d.

Tripod Beta. *Guidance on Using Tripod Beta in the Investigation and Analysis of Incidents, Accidents and Business Losses*. London, England: Energy Institute, 2015. Available at https://publishing.energyinst.org/__data/assets/pdf_file/0004/235183/TRIPOD-BETA-v.5.01-04.03.15-primer.pdf. Accessed June 3, 2019.

FURTHER READING

ANSI/ASSP Z10.0-2019. *Occupational Health and Safety Management Systems*. Park Ridge, IL: American Society of Safety Professionals, 2019.

ANSI/ASSP/ISO 45001-2018. *Occupational Health and Safety Management Systems—Requirements with Guidance for Use*. Park Ridge, IL: American Society of Safety Professionals, 2018.

Fishbone Quality Resources. About Fishbone Diagrams. Available at https://asq.org/quality-resources/fishbone. Accessed June 3, 2019. n.d.

Fishbone: MORE Steam.com. Fishbone Tutorial. Available at https://www.moresteam.com/toolbox/fishbone-diagram.cfm. Accessed June 3, 2019. n.d.

Fishbone Study Circle. Fishbone (Cause and Effect or Ishikawa) Diagram, 2018. Available at https://pmstudycircle.com/2014/07/fishbone-cause-and-effect-or-ishikawa-diagram/. Accessed June 3, 2019.

Lyon, B.K. and G. Popov. *Risk Management Tools for Safety Professionals*. Park Ridge, IL: The American Society of Safety Professionals, 2018.

Manuele, F.A. *Heinrich Revisited: Truisms or Myths*, Second Edition. Itasca, IL: National Safety Council, 2014.

Manuele, F.A. "Incident investigation: our methods are flawed." *Professional Safety*, October 2014.

Manuele, F.A. *On The Practice of Safety*, Fourth Edition. New York: John Wiley & Sons, 2013.

Manuele, F.A. "Reviewing Heinrich: dislodging two myths from the practice of safety." *Professional Safety*, 2011.

Manuele, F.A. "Root-cause analysis: uncovering the hows and whys of incidents." *Professional Safety*, May 2016.

Manuele, F.A. "Serious injuries & fatalities – a call for a new focus on their prevention." *Professional Safety*, December 2008.

Understanding How to Use The 5-Whys for Root Cause Analysis. Lifetime Reliability—Solutions. Available at http://www.lifetime-reliability.com/tutorials/lean-management-methods/How_to_Use_the_5-Whys_for_Root_Cause_Analysis.pdf. Accessed June 3, 2019. n.d.

CHAPTER 21

THE FIVE WHY PROBLEM-SOLVING TECHNIQUE

As has been said previously, the quality of incident investigations, even in some very large organizations, is significantly less than stellar. Yet incident investigation is a vital element within an operational risk management system. For a first step forward to improve on incident investigation, I promote adoption of the Five Why problem-solving technique, with emphasis. Why do so?

1. There is a substantial need for a causation model that is relatively easy to learn and apply. Application of The Five Why System does not require: extensive and costly training and retraining, the use of advanced mathematics, or overly complex methods.
2. If successfully adopted, The Five Why System would:
   - Be a major step forward to improve the quality of investigations;
   - Lead to the reality of systemic, risk-based causal factors;
   - Assist in noting the absence, inappropriateness, misuse or nonuse of barriers that should have been in place and properly used; and
   - Provide risk-related data to determine when risk assessments should be made.
3. Because of its structure, The Five Why technique promotes:
   - Involvement of individuals in the investigation group who are close to the work;

Advanced Safety Management: Focusing on Z10.0, 45001, and Serious Injury Prevention, Third Edition. Fred A. Manuele.
© 2020 John Wiley & Sons, Inc. Published 2020 by John Wiley & Sons, Inc.

- Critical and systematic thinking; and meaningful discussion to lead to agreed-upon risk-related causal factors and the preventive actions to be taken.
4. Use of The Five Why technique helps determine whether the causal factor chains are of such complexity that other incident investigation systems should be used.

Some safety practitioners may conclude that the incident investigation system in place attains results that are superior to what can be achieved through application of the Five Why technique. This advice is given, based on this author's experience: Analyze risk-related causal factor identification to determine whether superiority is actually achieved.

Creation of the Five Why Technique is attributed to Taiichi Ohno and was widely used within the Toyota Motor Corporation. Its intent is to provide a problem-solving methodology to determine cause and effect relationships and to identify root causal factors. Usage of the technique is extensive for countless problem-solving purposes in a great variety of organizations.

Some safety practitioners have found that applying the technique has resulted in moving forward from an inadequate incident investigation level to a system that focuses on the reality of risk-related causal factors.

As a case in point—in one very large company, it was agreed that if the Five Why system was promoted and they were able to give themselves a B+ grade on the quality of investigations in two years, they would have taken huge steps forward. Although general guidance is available on how to use the technique, the outcome depends on the knowledge and persistence of the users.

FOR WHAT IS THE FIVE WHY TECHNIQUE TO BE USED AND WHEN?

This technique can be used for any hazard and risk-related incident: an employee injury or illness, an accident involving a non-employee, product quality and safety, property damage such as resulting from a fire, an incident resulting in business down time, damage to properties other than the organization's properties, and an environmental incident.

It would be highly unusual for the Five Why technique to be used for such as a paper cut or a scratch from a staple that was not properly set. In a few organizations, the term *Significant Incident*, or something similar, is used to designate the types of incidents for which immediate upward reporting is necessary and a thorough investigation is to be made. A composite follows of the categories of *Significant Incidents* that safety practitioners have set forth in their guidelines. It is offered as a reference.

Significant Incidents

- An OSHA reportable Incident.
- Hospitalization of an employee for more than three days.

- An incident resulting in injury to three or more employees.
- A fatality.
- An incident, such as a fire, resulting in property damage in excess of US$10,000.
- A product loss valued in excess of US$20,000.
- An environmental incident (air, water, land) that must be reported to a governmental authority.
- An incident that:
  - required the evacuation of a building or job site,
  - required an emergency shutdown of operations,
  - could incite public interest, and
  - created a crisis or significant emergency.
- A near-miss incident that did not result in harm or damage, but could have had serious consequences under other circumstances.
- An incident that will provide a lesson learned for other locations.

PROCEDURES FOR USE OF THE FIVE WHY TECHNIQUE

A general outline follows on how the Five Why technique can be managed. Adjustments should be made in the process by safety professionals as they see fit to achieve effectiveness. But, without a buy-in by senior management, the technique will not be effectively applied.

1. Encourage management at a proper level to appoint an incident review group with the understanding that success is more likely when The Five Why process is used if the participants have real-world experience with the situation in which the incident occurred. It is encouraged that operations personnel be participants.
2. Provide all members with a thorough description of the incident.
3. Either the incident review group leader or a safety professional serving as a consultant to the group would commence by asking why could this incident happen and continue asking why until the group agrees on the causal factors—the issues.
4. Whoever the discussion leader, that person should try to have comments made by participants related to factual data, to steer discussions beyond symptoms to get to causal factors and particularly to avoid placing blame.
5. Consider using a mechanism to record the agreed upon responses to the why questions (see the following composite of a work sheet, purposely simple).
6. If discussions reveal that there was more than one path of activities or nonactivities that contributed to the occurrence, each path would be discussed.
7. Be cognizant of the time to end discussions—when pertinent comments are no longer forthcoming.

8. When the causal factors—the issues—are identified, assignments must be made with respect the preventive actions to be taken—persons responsible and time for completion.
9. Action on the recommendations should be followed to a logical conclusion.

A colleague who has adopted The Five Why system says that he has taught incident investigators to inject a "how could that happen" occasionally into the Five Why discussion. That could add to the thinking and considerations of causal factors.

TRAINING

Prior to application of the technique, training should be given to those who are to be involved in incident investigations concerning: hazards, risks, and possible deficiencies in management systems, hazard identification and analysis, and risk assessment. That training would be followed by instruction on The Five Why problem-solving technique.

Videos may help in the training to be given. If "Five Why Videos" is entered into a search engine, the outcome would indicate that a large number of organizations have training videos available. Some are good and would be of value as an introduction to The Five Why technique.

Also, some vendors of training materials are now offering training programs on The Five Why technique and sell software systems that, they say, are to assist in establishing and conducting the system. They can also be found through a voyage into the Internet.

A safety director who contributed material to the research undertaken on this subject said the following about his training of personnel and his application of The Five Why System.

> I have trained supervisors, shift managers, department managers and facility managers in the use of The Five Why System for accident investigations. I taught them the difference between fact finding and fault finding. They understand that documenting a failure on their part does not necessarily mean that they are lousy supervisors and doing so will help us identify system problems that we must correct. I review every investigation report. Any time I feel they have stopped asking "Why" too soon, I assist them with additional investigation to assure the root cause(s) are identified and appropriate corrective actions are developed and implemented.

OPPOSITION

Some incident investigators, particularly those who promote sales of their own systems, may say that the Five Why process is inadequate because it does not promote identification of causal factors resulting from decisions made at a senior management level. That is not so. Usually, when inquiry gets to the third or fourth Why, considerations are at the management levels above the supervisor and could consider decisions as far up as those made by the board of directors.

Nevertheless, safety professionals should keep in mind that achieving success in the use of the Five Why system will be a hugely huge step forward in many organizations for the quality of incident investigations.

CULTURE CHANGE NECESSARY

Improving the quality of incident investigations in most organizations will require changes in their culture, and safety professionals must be aware of the enormity of that task. Literature is abundant on the difficulty of making changes in any system that is deeply embedded within an organization's processes.

Safety professionals must recognize that the culture change necessary can be accomplished only with top management support. Also, they must have the knowledge of methods to successfully manage change. Being a culture change agent is discussed in Chapter 3 in this book.

FIVE EXAMPLES OF APPLICATIONS OF THE FIVE WHY TECHNIQUE

Examples given here pertain to real-world operational situations. They outline the thought and inquiry process used in application of the Five Why technique.

EXAMPLE 1

The written description of the incident says that a tool-carrying wheeled cart tipped over while an employee was trying to move it. She was seriously injured.

1. Why did the cart tip over?
   The diameter of the casters is too small and the carts are tippy. This has happened several times but there was no injury.
2. Why weren't the previous incidents reported?
   We didn't recognize the extent of the hazard and that a serious injury could occur when the cart tipped over.
3. Why is the diameter of the casters too small?
   They were made that way in the fabrication shop.
4. Why did the fabrication shop make carts with casters that are too small?
   They followed the dimensions given to them by engineering.
5. Why did engineering give fabrication dimensions for casters that have been proven to be too small?
   Engineering did not consider the hazards and risks that would result from using small casters.

6. Why did engineering not consider those hazards and risks?
   It never occurred to the designers that use of small casters would create hazardous situations.
   Causal/contributing factors: Hazard was not recognized by operations personnel; design of the casters resulted in hazardous situations.

Conclusion I [the department manager] have made engineering aware of the design problem. In that meeting, emphasis was given to the need to focus on hazards and risks in the design process. Also, engineering was asked to study the matter and has given new design parameters to fabrication: the caster diameter is to be tripled. On a high priority basis, fabrication is to replace all casters on similar carts. A 30-day completion date for that work was set.

I have also alerted supervisors to the problem in areas where carts of that design are used. They have been advised that when deciding to report or not report an incident that did not result in injury, they are to be extra cautious. And, they have been advised to gather all personnel who use the carts and instruct them that larger casters are to be placed on tool carts and that, until that is done, moving the carts is to be a two-person effort. I have asked our safety director to alert her associates at other locations of this situation and how we are handling it.

EXAMPLE 2

Operations personnel report concern over injury potential because of conditions that develop when a metal forming machine stops when the overload trip actuates. This is an example of how the Five Why technique can be used to resolve hazard/risk situations before an accident occurs. The safety director met with the supervisor who is directly responsible for the work.

1. Why are you concerned?
   The electrical overload trip actuates very often when we use this new forming machine. It gets risky when it stops in mid-cycle and the work that has to be done to clear the partially formed metal adds risks that our employees think are more than they should have to bear. Occasionally, that's O.K. Often is too much.
2. Why does the overload trip actuate?
   This is a new problem for us. We rarely had the overload trip actuate. It started after a new order for metal was received. We are told that the purchasing department thought that it got a very good deal from a metals distributor but it turns out that what was delivered to us did not meet our specifications. This metal is not as malleable and workable and the metal former struggles in the forming process. So, the overload trip actuates. Maintenance is furious with us because we have to call on them as often as we do.
3. Why can't the amperage for the overload trip be increased for this batch of metal?
   Our engineers say they don't want greater power fed into this machine.

4. Why do you have to call on maintenance so often?
   The rule here is that no overload trip is to be reset without a review of why it tripped and clearance from maintenance.
5. Why haven't you recommended to your operating manager that he arrange a get together with the engineer and the maintenance manager to decide on what needs to be done to resolve the overload trip problem for this batch of metal?
   That's not easy for me to do at my level. But it would be good if you could find a way to get that done.
   Possible causal/contributing factors: Over-exertion; machine actuating when cleaning the partially formed metal; fall potential; partially formed metal being hazardous in the handling process.
   Temporary resolution of this risk-related problem involved discussions with: the purchasing department as to future purchases; operations; engineering; and maintenance. It is often the case that risk reduction actions require participation by several interrelated functions. And, of interest, the supervisor does not feel free to discuss a hazardous situation with the person to whom he reports.

EXAMPLE 3

Machine broke down; a repairman who was fixing it did not turn off the electric power and was electrocuted. Initially the causal factor was recorded as the unsafe act of the repairman who did not follow the lockout/tagout procedure.

1. Why would the repairman take such a shortcut?
   He was under considerable pressure from production people to get the machine back in operation.
2. Why would production people put that much pressure on him?
   This machine is vital to the overall process and production had fallen way behind.
   Some production people were idle and doing nothing.
3. Why did the repairman not merely take the few steps that would be necessary and follow the lockout/tagout procedure?
   We checked that out and found that the lockout/tagout station wasn't within just a few steps.
4. How far away was the lock/out station?
   Over 300 ft.
5. Why was the lock/out station so far away?
   That's the way the system was designed. We have a lot of situations where the lockout/tagout station is not nearby. They have been discussed but decisions were not made to get them fixed.
   Causal factors/contributing factors: The distance to the lockout/tagout station made it inconvenient to go there under the circumstances; the significance of

the risks of not having lockout/tagout stations immediately available was not recognized; and the pressure applied by production to get the repairs finished. How could this situation be resolved?

Senior management is all shook up because of this fatality happening on their watch. Engineering and maintenance have been tasked to prepare a list of all the lockout/tagout situations that need attention and to prioritize them. We have been told that the work will get done.

EXAMPLE 4

What follows was the entirety of entries in the incident investigation report for:

- Incident description. Machine operator fell and broke his hip.
- Causal factors. Oil on the floor.
- Corrective actions. Cleaned the floor.

Further inquiry was made.

1. Why was there oil on the floor?
   A gasket leaked.
2. Why did the gasket leak?
   Bearings are worn on this machine and when it is stressed, it vibrates a lot and the vibration loosened the joint.
3. Why is the machine stressed?
   When production is at full pace, which is often, this machine just barely meets the demand.
4. Why haven't the bearings been replaced?
   We sent a work order to maintenance on two occasions with no response.
5. Why hasn't maintenance responded?
   We have been through two expense reductions and maintenance is short of staff. They prioritize work orders and ours have not reached sufficient priority status.
6. Why hasn't the machine been replaced with one that can handle production at full pace?
   That has been discussed at our department meetings but we haven't been able to get approval.

Causal/contributing factors include the leaking gasket; the worn bearings; the inability of maintenance to respond to work orders on a timely basis; and running a machine beyond its capacity.

How could this situation be resolved?

Management has been alerted to what happened at this machine and the potential problems we could have with other machines that are not being properly maintained. Additions are made to the maintenance staff. Our department head submitted a request for approval to replace the machine with

- It is important that the first inquiry step asks "Why" and not a "What" or a person or a diversionary symptom.
- When "because" is given as a causal factor, it is necessary to continue the discussion to explore the "why" of the "because".
- Focusing on what an individual did or did not do and ending inquiry at that point will not lead to the identification of causal factors—the discussion should be continued into the "why" process.
- It often occurs that the discussion leads to a statement indicating that the employee didn't follow the prescribed procedure. In such situations, it is appropriate to continuously ask "why" the employee did what was done. It might be found that the procedure—work as imagined—is not a good fit with what is needed to do the work. Also, inquiry should be made to determine what the employee did increased the risk, was neutral, or reduced the risk. Similar questions should be asked about productivity.
- If the investigation is for a type of incident that has occurred previously, questioning should explore the adequacy of the corrective actions taken for the previous incident and whether they were sufficient.
- Not all problems have a single root cause. To explore multiple root causes, the Five Why method should be repeated asking a different sequence of questions for each line of inquiry.
- Occasionally, application of the Five Why technique to interrelated systems identifies actions that should be taken in several operational entities to resolve a problem.
- It should be understood that the results from using a Five Why methodology may be the starting point for the use of a more sophisticated causal factor determination system when there is complexity.
- Discussions should not be ended until the investigation group is satisfied that the causal factors have been identified.
- Remedial actions proposed must be actionable and reduce the risk to an acceptable level. Also, responsibility for taking action and in what time should be established.

CONCLUSION

Safety professionals should make an assessment of the quality of incident investigation in the organizations to which they give counseling. More than likely, they will observe that there are inadequacies in risk-related causal factor identification.

They should then develop a convincing proposal to decision-makers on the benefits to be obtained from improving the system, giving examples of incident investigations that did not identify the reality of the risks and causal factors. They should also present their views on the actions necessary to initiate the Five Why technique.

In their proposals, they should comment on the fact that the system in place for incident investigation and the beliefs about how accidents happen may be deeply embedded and that a culture change may be necessary for which leadership from senior management is imperative.

RESOURCES

Literature on the Five Why system is much more extensive than it was a few years ago. But a search through Amazon and Google found only two books, mini-books, on the subject. They are listed in the following. It is understandable that lengthy books have not been written on The Five Why Technique. It is not that complicated and does not require an extensive explanatory text.

Selected resources for this chapter are taken from the Internet. Many other similar publications are available.

Liker, J.K. *Root Cause Using 5 Whys*. A mini book available through Amazon.

Pamoga, P.A. *Getting to the Root Cause through the 5 Why Methodology*. A mini-book available through Amazon

Asian Development Bank—The Five Whys Technique. Available at https://www.adb.org/sites/default/files/publication/27641/five-whys-technique.pdf. Accessed June 3, 2019.

Problem Solving, Root Cause & The Five Whys. How To Do It Better. Business Processes. Available at http://cart.critical-thinking.com/five-whys-how-to-do-it-better. Accessed June 3, 2019.

5-Why Analysis.MoreStream. Available at http://www.moresteam.com/toolbox/5-why-analysis.cfm. Accessed June 3, 2019.

Understanding How to Use The 5-Whys for Root Cause Analysis. Lifetime Reliability—Solutions. Available at http://www.lifetime-reliability.com/tutorials/lean-management-methods/How_to_Use_the_5-Whys_for_Root_Cause_Analysis.pdf. Accessed June 3, 2019.

one of larger capacity. Workers compensation claims cost for the cracked hip was estimated at about US$800,000. Replacement of the machine received a higher priority and the pump was replaced.

EXAMPLE 5

An untrained employee was trying to replace a dead light bulb that was slightly over 15 ft. above the floor. Ladder shifted and collapsed and the employee fell to the floor—a fatality.

1. Why did this accident occur?
   No step ladder was available that could reach high enough for the employee to reach the light bulb. So, the supervisor decided to place the foot pads on the side rails of the ladder on two inverted part's boxes. As the employee worked and twisted to unscrew the dead light bulb while standing on the first step of the ladder, the side rails of the ladder shifted and he fell to a concrete floor.
2. Why did the supervisor make such a decision?
   Even though the environmental, safety, and health department had outlined the procedures for the use of ladders, including the training necessary, it was generally accepted that it was O.K. not to follow the procedures for some jobs. There is only one ladder in this place that could reach high enough so that a worker could stand, say no higher than the second step, and reach the dead light bulb, and it was being used in another place.
3. Why was there only one ladder available for work at this height?
   Management was considering a different way to access the light bulbs because they had to be replaced often. So, the responses to requests for the purchase of additional ladders were negative with the explanation that other methods were being considered.
4. Why did the supervisor make such a risky arrangement to replace the dead light bulb and let an untrained employee take the risks that he did?
   This supervisor was good at ignoring the rules. Even though he went through the ladder use training program, he insisted that he knew better. As an example, he made fun of the procedure that required a fall arrest system for work at these heights. And, since the supervisor did not think much of the ladder use training program, he did not think it urgent to schedule the trainee for the class.
5. Why was violation of the rules by the supervisor condoned?
   Since it had become the accepted practice for supervisors to ignore safe practice rules, it is now realized that what he did was not out of the ordinary.
6. Why was violation of safe practice rules condoned and how could similar accidents be prevented?
   Senior management (editorial note—senior management was represented in the accident review group) did not recognize that its acceptance of the violation of safe practice rules produced an atmosphere in which getting things

done was more important than the excessive risks to which employees were exposed. It was concluded that all management personnel, including first-line supervisors, shared responsibility for this accident and that a major change was needed with respect to priorities to prevent the occurrence of similar accidents.

Causal factors/contributing factors: The overarching causal factor for this incident was that management created a culture in which it was acceptable for supervisors to take unacceptable risks; failure to decide on the acquisition of mechanical or hydraulic equipment to access frequent work at such heights; using an inadequate step ladder arrangement for the job and placing its side rails on part's boxes; assigning an untrained employee to do a job for which he was unaware of the risks and who, under supervision, violated several safe practice rules pertaining to the use of ladders.

Actions taken: Management realized that it should reverse the culture in which unacceptable risks were condoned—a culture that it had created; as its first step to demonstrate its intent, the supervisor was terminated (editorial note—action questionable); senior management issued a communication in which operational causal factors were cited as examples of what would no longer be tolerated; a meeting of supervisors and department heads was held in which management's reversal of its position was discussed and responsibilities for accomplishment of the changes necessary were established; equipment was purchased to access dead light bulbs without using step ladders.

OBSERVATIONS

Having analyzed incident reports in which the Five Why System was used, these cautions are offered to safety practitioners who give counsel on incident investigation:

- Management commitment to identifying the reality of causal factors is an absolute necessity for success.
- Expect that repetition of Five Why exercises will be necessary to get the idea across. Doing so in group meetings at several levels, but particularly at the top management level, is a good idea.
- Be sure that management is informed and prepared to act on the increase in recommendations made for improvements in the workplace as causal factors are identified and as skill is developed in applying the Five Why process.
- Since the causal factors for an incident may relate to hazards, risks, or deficiencies in the management systems for which participants in the incident investigation process are responsible, attempts to avoid responsibility or diverting responsibility to others may be encountered.
- Expect that participants will express preconceived ideas about how accidents happen, particularly if they have been lead to believe that the unsafe acts of workers are the causal factors, and be prepared to steer comments into a Five Why sequence.

CHAPTER 22

MORT—THE MANAGEMENT OVERSIGHT AND RISK TREE

If I taught at a university in a safety and environmental science degree program, students would be exposed to The Management Oversight and Risk Tree (MORT) for the thought base it provides on how accidents happen and what is needed for their prevention.

What MORT is and how it can be used is addressed in the Abstract for the third edition of the MORT User's Manual issued by the U.S. Department of Energy (DOE).

> MORT is a comprehensive analytical procedure that provides a disciplined method for determining the causes and contributing factors of major accidents. Alternatively, it serves as a tool to evaluate the quality of an existing program (p. iii).

Developing skills for the application of MORT requires study and continuous application to develop and retain knowledge of the technique and the skills necessary in its usage. It is not the intent here to have safety professionals become skilled in the application of MORT. But they would add to their knowledge and capabilities by developing an awareness of its content.

UNUSUAL ASPECTS

MORT is unusual in that it specifically promotes inquiry into management's upstream decision-making for the identification of sources of causal factors in

---

Advanced Safety Management: Focusing on Z10.0, 45001, and Serious Injury Prevention, Third Edition. Fred A. Manuele.
© 2020 John Wiley & Sons, Inc. Published 2020 by John Wiley & Sons, Inc.

much greater detail than most other incident investigation models. Why cite this difference?

Significant causal factors may derive from decisions made at an organization's governing body and senior management levels. Incident investigations that do not delve into possible causal factors that relate to such deficiencies are not inclusive. A statement made in the DOE Handbook on Accident and Operational Safety—Volume 1: Accident Analysis Techniques clearly makes that case.

> Investigations require delving into the basic organizational processes: designing, constructing, operating, maintaining, communicating, selecting and training, supervising, and managing that contain the kinds of latent conditions most likely to constitute a threat to the safety of the system. The inherent nature of organizational culture and strategic decision-making means latent conditions are inevitable. (p. 1-2).

MORT is also unusual because of its premise that where there are deficiencies in operating controls there will be related deficiencies in management decision-making. Thus, the management branch of MORT is dominant and determines whether there are adequate barriers and controls in operations—the specific control branch. (DOE Version - p. 11).

OTHER REASONS WHY MORT IS AN EDUCATIONAL RESOURCE

- MORT is truly comprehensive concerning possible causal factors, as will be seen.
- It places great significance on having the necessary barriers and controls in place to prevent incidents and to reduce their impact if incidents occur.
- Design processes and the risk assessment processes are emphasized.
- It includes provisions for the necessary training and support for supervision.
- Haddon's (1970) unwanted energy release theory is basic in its application.

A REVISION OF MORT

An interesting and later version of a MORT User's Manual was published by The Noordwijk Risk Initiative Foundation in The Netherlands. Its title is NRI Mort User's Manual for use with the Management Oversight & Risk Tree analytical logic diagram (NRI nd). A second edition was published in 2009. They say that it improves guidance in the investigation process, restores freshness, and simplifies the investigation system.

This publication recognizes that the MORT method had its beginning in work done for the U.S. Government in an activity undertaken by W.G. Johnson. (p. viii). William Johnson's work that resulted in the development of MORT was done in the 1970s for the U.S. Atomic Energy Commission, the predecessor to the U.S. Department of Energy. His book *Mort Safety Assurance Systems,* published in 1980, is the ultimate text on MORT. It has demonstrated staying power.

A review of the DOE and the NRI versions of MORT indicate that each has attributes that the other does not have. Both are recommended for their educational value.

DEFINING BARRIERS AND CONTROLS

MORT places great significance on having the necessary barriers and controls in place to prevent accidents and to reduce their impact if they occur. What are barriers and controls?

In Erik Hollnagel's (2004) book *Barriers and Accident Prevention*, Chapter 3 is titled "Barrier Functions And Barrier Systems" and Chapter 4 discusses "Understanding The Role Of Barriers In Accidents". Hollnagel's coverage of barriers is extensive and informative. His writings are good references on barriers and are recommended. Hollnagel says:

> An accident can be described as one or more barriers that have failed, even though the failure of a barrier only rarely is a cause in itself. A barrier is, generally speaking, an obstacle, an obstruction, or a hindrance that may either:
>
> 1. Prevent an event from taking place, or
> 2. Thwart or lessen the impact of the consequences if it happens nonetheless. (p. 68).

Hollnagel makes an important point when he implies that barriers have two purposes: To prevent an incident from occurring and to limit the extent of harm or damage should an incident occur. To paraphrase Hollnagel, an accident happens because of the failures of one or more barriers. Think about that.

When Hollnagel says that "failure of a barrier only rarely is a cause in itself" the implication is that the failure probably results from deficiencies in the relative management systems.

In *System Safety for the 21st Century*, Richard A. Stephans (2004) gave a rather broad definition of barriers. And his definition is considered appropriate.

> Barrier—Anything used to control, prevent, or impede energy flows. Types of barriers include physical, equipment design, warning devices, procedures and work processes, knowledge and skills, and supervision. Barriers may be control or safety barriers or act as both (p. 357).

Lyon and Popov, in *Risk Assessment Tools for Safety Professionals*, comment on both the preventive and mitigative aspects of barriers in their definition as did Hollnagel. Repeating for emphasis—barriers and controls serve to avoid incidents and also to reduce the severity of outcomes if they occur.

> Barrier: Physical or procedural control measures that are put in place to prevent or reduce likelihood of risk exposure (preventive) and/or reduce severity of impact/consequences (mitigative) resulting from a hazardous event. (p. 313).

468 MORT—THE MANAGEMENT OVERSIGHT AND RISK TREE

Analyzing for the adequacy or inadequacy of barriers and controls is significant in applying the MORT system. Barriers are not exclusively physical. They may include having a qualified staff, training, supervision, appropriate procedures, maintenance, and communications.

THE MAIN BRANCHES OF MORT

Even though the procedures for applying MORT require in-depth inquiry to determine whether deficiencies in management systems were causal factors for incidents, depictions of the main elements of MORT do not give such deficiencies the emphasis needed. To overcome that shortcoming, an updated and simplified version of the main branches of MORT has been developed. It appears here as Figure 22.1.

Discussions of subjects follow in the order in which they appear in the Main Branches of MORT. But they do not represent the detail into which an application of MORT can take a safety professional. There are over 1,500 subsets to the Main Branches of MORT.

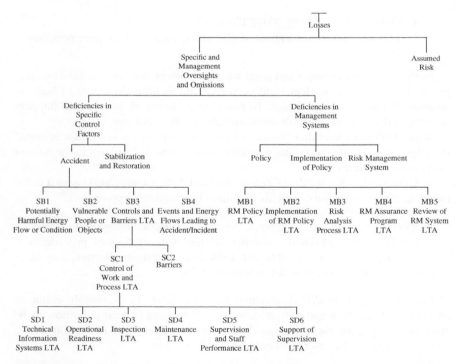

FIGURE 22.1 The Main Branches of the Management Oversight and Risk Tree (MORT) Updated and simplified

Throughout a MORT exercise, the purpose is to determine whether barriers and controls were sufficient or "Less than adequate"—which is close to a synonym for "Risk was unacceptable."

Using the DOE version of MORT as a principal guide, examples are now given of questions that would be asked during a MORT investigation, starting with Losses and Assumed Risks.

LOSSES AND ASSUMED RISKS

It is assumed that an incident occurred which is of such a nature as to require creation of an investigation team. Although the term Assumed Risks also appears at the top of the MORT chart, the subject is reviewed as a part of the analysis of the Risk Management System as is shown later in this chapter.

SPECIFIC & MANAGEMENT OVERSIGHTS & OMISSIONS

Note that the major caption Specific & Management Oversights & Omissions is followed by two sub-captions. They are Deficiencies in Management Systems and Deficiencies in Specific Controls. A significant observation made in the DOE MORT User's Manual is cited here.

> MORT suggests that an accident is usually multifactorial in nature. It occurs because of lack of adequate barriers and/or controls upon the unwanted energy transfer associated with the incident. It is usually proceeded by sequences of planning errors and operational errors that produce failures to adjust to changes in human factors or environmental factors (p. 7).

Safety professionals who give advice on incident investigation should take seriously the foregoing statement. It says that accidents occur "because of lack of adequate barriers and/or controls". It also says that "an accident is usually multifactorial in nature". Most often, there are multiple causal factors that result from deficiencies in management systems and in specific control systems.

Such premises are important in achieving an understanding of the multiplicity of causal factors and how accidents happen. As an incident investigation proceeds, investigators would determine whether there are deficiencies in the specific control activity as represented by barrier and control inadequacies and whether the deficiencies exist because of shortcomings in management decision-making.

DEFICIENCIES IN MANAGEMENT SYSTEMS

Policy and Implementation

For these purposes, a policy is a general statement about the organization's position with respect to safety. Investigators would determine whether the policy was sufficient and whether deficiencies in management's implementation of its policy were contributing causal factors.

Risk Management System

Investigators, as they review the Risk Management System, would determine whether the systems in place—resulting from decisions made at an executive management level—are sufficient with respect to:

- Barriers and controls
- Achieving and maintaining acceptable risk levels
- Assuring that there is a continuing information flow to decision-makers with respect to the status of those systems.

Several subjects are to be reviewed when considering the Risk Management System as to adequacy and the possible relation to sufficient or insufficient barriers and controls and possible incident causal factors. They follow.

Risk Management Policy and Its Implementation

Subjects to be addressed are

- Has management established systems and methods to achieve a high level of performance with respect to the policy and goals?
- Have managerial, supervisory, and line responsibilities been clearly established?
- Are training and retraining systems for management and supervisory personnel appropriate to the needs?
- Is budgeting sufficient?
- Have the accountability requirements been clearly established and administered?
- Has management demonstrated leadership and commitment, vigorously?

Risk Analysis Process

Evaluations would be made of the following in this section.

- Is the risk assessment process properly conceptualized, defined, and executed?
- Does the risk assessment system provide management with the information needed to take risk reduction decisions and to decide when residual risks are acceptable?
- Are appropriate standards and procedures established for the general design process (prevention through design)?

- Are the barriers and controls necessary included in the design processes?
- Have energy controls been specifically included in the design process and in operation specifications?
- Are human factors considerations given appropriate attention in the design of the workplace and work methods?
- Has the needed consideration been given to all identified hazards and risks with respect to task procedures and the work environment?
- Has a specific change analysis system been established that promotes consideration of hazards and risks when changes are made?
- Have reliability and quality assurance standards been established and applied?

Risk Management Assurance Program and Review of Risk Management System

- Are reviews and evaluations made periodically to identify possible shortcomings in environmental, safety, and quality assurance programs?
- Do such reviews include the effectiveness of the individual elements in the risk management system?
- Are audits occasionally made of existing risk management systems by an independent entity?

Review of Assumed Risks

- Has it been established that specific risk assessments would have been made for assumed risks?
- Has it been established that judgments as to acceptability are to be taken by a person who has the authority to do so?
- Is it understood that the risk assessment process, including documentation of the decision-making, is to be recorded and does management adhere to that premise?
- If risks had been assumed that relate to the incident, was an additional review conducted of the risk assessment decision-making?
- Is there indication in the assumed risk decision-making that disproportionate weight was given to the costs to reduce the risk in relation to the risk reduction to be achieved?

> Editorial note: This item relates to the definition of acceptable risk and ALARP, for which the following definitions appear in the American National Standard for *Prevention through Design: Guidelines for Addressing Occupational Hazards and Risks in Design and Redesign Processes*. Its designation is ANSI/ASSE Z590.3-2011 (R2016).
>
> Acceptable risk is defined as that risk for which the probability of an incident or exposure occurring and the severity of harm or damage that

may result are as low as reasonably practicable (ALARP) in the setting being considered (p. 12).

ALARP. That level of risk which can be further lowered only by an increase in resource expenditure that is disproportionate in relation to the resulting decrease in risk (p. 12).

It occurs that in the risk assessment process, excessive weight may be given to the cost of risk reduction in relation to the amount of risk reduction to be achieved. That results in condoning insufficient barriers and controls and unacceptable risks, which have been related to incidents that resulted in serious injuries and fatalities.

Just as significant—as management gives excessive weight to the cost of risk reduction, that becomes a part of the organization's culture.

It is the usual practice to accept some risks without formal risk assessments even though the consequences can be severe, but not controllable—such as a hurricane, cyclone, or an earthquake. Nevertheless, the potential for such events occurring requires consideration when establishing the emergency response system and outlining the procedures to be followed.

Accident

There is a relationship between "Accident" and the term "Losses" that appears at the top of the MORT chart. In the DOE version of MORT, an accident is defined as:

An unwanted transfer of energy or environmental condition because of lack or inadequate barriers and/or controls, producing injury to persons and/or damage to property or the process. (p. 14).

That definition relates directly to Haddon's Energy Release Theory. Haddon's concepts are soundly based. They are also good design references. Haddon was among the first to propose that barriers be placed between the source of energy and the person or property that could be harmed or damaged. Comments on his approach have been recorded previously in this book.

Stabilization and Restoration

Principally, the purpose here is to evaluate how effective was the emergency response system, with emphasis on the care of injured employees. If there are shortcomings, they would be identified, and recommendations would be made to management for consideration.

Stabilization and restoration also pertain to how quickly disturbed operations were again productive. (There are extensive emergency response plan review outlines in both the DOE and the NRI MORT User's Manuals.)

Potentially Harmful Energy Flow or Environmental Condition

In this section, consideration is given to the harmful energy/environmental conditions that might relate to the incident and to evaluate the barrier and control measures in place. If the incident description indicates that there were two or more energy flows, the analysts would prepare separate energy diagrams and evaluate the possible associated barrier and control inadequacies.

Nonfunctional energy and functional energy are considered. Nonfunctional energy is an energy flow that was not a functional part of the system or product being produced. Functional energy is that energy flow that was designed into the operating system or the work methods.

Inquiry would be made about the adequacy of the barriers and controls in place—design and engineering or administrative—to prevent harmful energy flows to impact on people or the physical environment.

Vulnerable People or Objects

To add to the understanding of the nature of the incident, inquiry is to be made to determine the number and work categories of the personnel exposed to the harmful energy release or the hazardous environment, and of the potential damage to property. This is a sort of an exposure evaluation. Recordings would be made of whether the personnel or property were functional or nonfunctional parts of the operating system.

Investigators would determine whether people or property were in the right place and the right time, whether the barriers and controls in place were adequate, or possibly impracticable, or whether the system established included or did not include provisions for and opportunities for personnel to evade the harmful energy flow or hazardous conditions.

This section could relate to the emergency response plan and whether it was current, well understood, and whether drills had been conducted.

CONTROLS AND BARRIERS

Understandably, this section of the MORT User's Manual is extensive. In the DOE version, it fills 21 pages: and 27 pages in the NRI publication. Brief comments follow on each of the principal sections. Their purpose is to determine whether deficiencies in controls and barriers for possible unwanted energy transfers were causal factors for the incident being investigated.

There are two subsections in this part of MORT—Barriers and Control of work processes. They overlap significantly. Very brief comments follow on each of the subjects to be reviewed during an incident investigation.

Barriers

This guidance is given in the DOE Manual.

> The term 'barrier' has the connotation of physical intervention; however, the barrier may be a 'paper barrier'. Separation by time or space in particular may be accomplished by written procedure or some other type of administrative control. (p. 21).

Subjects to Be Considered

- Were the barriers sufficient in relation to the operational hazards and risks?
- Where the outcome of an initiating event can be result in serious injuries and fatalities or extensive damage to property or the environment, were multiple barriers provided, and were they sufficient?
- Were there precursors indicating that a damaging and unwanted energy flow might occur (near misses), and if yes, was supervisory attention deficient?
- Did the barriers fail?
- Were barriers provided not used?
- Does the non-use of provided barriers indicate that there were task performance errors or lack of supervision?
- Were the necessary personal protective device barriers provided and used?
- Was providing barriers impossible because of the process circumstances?
- Were arrangements made to have appropriate barriers of time or space which separated the energy source from the persons of objects that might be injured or damaged?
- Were there deficiencies in the maintenance of barriers?

CONTROL OF WORK AND PROCESSES

A determination would be made of the suitability of barriers and control systems in place for the work processes in the area in which the incident occurred.

Technical Information Systems

- Was the information provided sufficient with respect to the technological aspects of the work processes?
- Were there sufficient numbers of trained personnel?
- Was technical leadership and guidance sufficient?
- Was the exchange or transmittal of knowledge sufficient in relation to the potential unwanted energy transfer?
- If there had been planned or unplanned changes in the processes, were the people making the change made aware of the risks and the possible need for a risk assessment?

Operational Readiness

For this section, inquiry would be made in relation to the causal factors for the incident to determine whether the work process and the facility were ready to be used as intended.

- Was the process and the facility operationally ready?
- Were the criteria and procedure used to check the operational readiness sufficient and were the personnel who determined operational readiness properly trained and qualified?
- Were the outcomes of the "upstream elements" in the safety management system sufficient for the processes undertaken—such as system design, work methods design, training, hazard identification and analysis and risk assessment, setting operating limits and performance specification, and adequacy of barriers and controls?
- Was there sufficient interface between operation personnel and maintenance and testing personnel?

Inspection and Maintenance

Although there are separate identifications for inspection and maintenance in the chart for the main branches of MORT, the criteria to be evaluated in both the DOE manual and the NRI manual are the same. These are samples of questions to be asked in relation to the causal factors for the incident and whether barriers and controls were sufficient.

- Was the planning broad enough to include all areas that should be inspected or maintained?
- Was provision made in the design process for ease of access for maintenance and inspection?
- Was the scheduling for inspection and maintenance sufficient in relation to the inherent risks in the processes?
- Was the schedule followed?
- Were the inspection and maintenance personnel well qualified for tasks undertaken?
- Were the assigned inspection and maintenance tasks performed in accord with established procedures?
- Did the causal factors for the incident relate to the inspection or maintenance tasks performed?

Supervision and Staff Performance

As will be seen by the extent of the subjects to be evaluated for this section, in-depth inquiry to be made into the quality and adequacy of supervision and the performance by staff personnel. In the manuals, several pages are devoted to supervision and staff performance.

Recall that supervision is a form of barrier and that supervision can be a form of control. Questions similar to the following apply.

- Were supervisors properly trained in relation to the inherent hazards and risks in the operation?
- Was enough time allotted to supervisors to thoroughly examine the work processes in relation to the causal factors for the incident?
- Did gaps exist in the communications when transfers of shift responsibility occurred that relate to the incident?
- Was the supervisor sufficiently energetic in detecting hazards and having corrective action taken?
- Did the supervisor fail to make the expected inspection of operations?
- In the work arraignment, was enough time allotted to make an inspection and detect the hazards?
- Was the supervisor attentive to hazards that may have arisen when changes were made in the operations and was appropriate action taken?
- Did the supervisor encourage the workforce to recognize hazards and was appropriate action taken with respect to those hazards?
- If the supervisor was aware of a hazard and postponed the necessary corrective action, thereby accepting the risk, did that occur because of error in supervisory judgment, or the authority to stop operations had not been given, or because of production pressures?
- Had a precedent been established indicating that the supervisor was expected to condone an unacceptable risk and not take action in such circumstances?
- If the incident involved two or more departments, was there sufficient and unambiguous communication and co-ordination between the departments?

Task Performance Errors

In both the DOE and NRI manuals, an interesting observation in comparable terminology is made that has to do with causal factor determination. This is what is said in the NRI manual.

> There are few 'unsafe acts' in the sense of blameworthy frontline employee failure. Assignment of 'unsafe act' responsibility to a frontline employee should not be made unless or until the following preventive steps have been shown to be adequate:
>
> - Risk analysis.
> - Management of supervisory detection.
> - Review of procedures for working safely.
> - Human factors review. (p. 18).

Task performance errors are those that contributed to the incident as causal factors and may also relate to the insufficiency of barriers and controls or their usage.

- Was the task assigned to employees properly scoped with clearly established steps and objectives?
- Do the contributing factors for the incident relate to shortcomings in the way the task was assigned by the supervisor to a staff member?
- Was a risk assessment made for a work task involving a high potential for error, injury, damage, or for encountering an unwanted energy flow?
- Were the recommendations developed through the risk assessment applied?
- Was the task identified as one for which there was low risk potential and was that a reasonable assessment?
- Was the work force given a pretask briefing prior to commencing the work and did it contain sufficient knowledge about the hazards and risks?
- Did supervisors give active attention to hazard-related suggestions made by the work staff?
- Were proper worksite risk controls placed on the work processes, facility, equipment, and personnel?
- Did the way the work was done conform to the standard operating procedure, and if it did not, was the risk increased, neutral, or reduced? (Deviance is a norm.)
- Did the deviance from established procedures result from a poorly designed workplace or work methods?
- Was a person assigned to the task who did not meet the standards for the task?
- Was the training given to the task performer sufficient?
- Did the supervisor note the deviation that increased the risk and was reinstruction given?
- Did the supervisor observe worker personnel problems such as an emotional upset, drug use, or excessive use of alcohol?
- Did a worker willfully and habitually break the rules and did the risks increase because of his actions? If so, was appropriate action taken?
- Was employee motivation lacking and did the shortcoming add to risk or condone risky performance?

Support of Supervision

It is highly probable that no other operational risk management system contains a specific provision that is to determine whether upper management provided the type of support and guidance so that all lower level supervisors could adequately control risks.

As previously noted, barriers and controls would have been considered, desirably, in the initial design and subsequent redesign processes, which are largely under the control of upper management. Inquirers would seek information with respect to the causal factors related to the following.

- Were supervisors given the needed training for the technical aspects of their assignments?
- Were supervisors given the necessary training for general supervision and responding to the needs and quirks of the work staff?
- Was the necessary training updated as changes occurred?
- Was the necessity of having, maintaining, and usage of barriers and control emphasized in the training given to supervisors?
- Was there an accessible, open line of communication which encouraged transmitting needed information between the executive staff and lower level management and supervisors?
- Did supervisors have access to technical staff at a senior management level?
- Were the available resources used effectively and to the advantage of lower level managers and supervisors?
- Was senior management effectively responsive to risks referred to them by lower level management and supervisors?

CONCLUSION

While the excerpts taken from the MORT User's Manual may seem extensive, they are but a sampling of the actual content of the manual. Often, the excerpts were revised and combined for simplicity. This author recommends that safety professionals obtain both the DOE and the NRI manuals since they provide an excellent educational resource and can be downloaded from the Internet.

A statement that appears at the beginning of this paper is repeated here because it is stressed that having knowledge of MORT assists in understanding how accidents happen and the barriers and controls that should be in place for an effective socio-technical operational risk management system.

> If I taught at a university in a safety and environmental science degree program, students would be exposed to The Management Oversight and Risk Tree (MORT) for the thought base it provides on how accidents happen and what is needed for their prevention.

REFERENCES

ANSI/ASSE Z590.3-2011 (2016). *American National Standard for Prevention through Design: Guidelines for Addressing Occupational Hazards and Risks in Design and Redesign Processes*. Park Ridge, IL: The American Society of Safety Professionals, 2016.

DOE. MORT User's Manual. On the Internet. Downloadable free. Available at https://www.osti.gov/servlets/purl/5254810. Accessed Jun 3, 2019. 1992.

Haddon, W. "On the escape of tigers: an ecologic note". *Journal of Public Health*, 1970 60(12): 2229–2234.

Hollnagel, E. *Barriers and Accident Prevention*. Burlington, VT: Ashgate Publishing, 2004.

Johnson, W. *Mort Safety Assurance Systems*. New York: Marcel Decker, 1973.

NRI MORT User's Manual. On the Internet. Downloadable without charge. Enter 'MORT User's Manual' into a search engine and this publication appears. Available at http://nri.eu.com/NRI1.pdf. Accessed June 3, 2019. 2009.

Stephans, R.A. *System Safety for the 21st Century*. Hoboken, NJ: John Wiley & Sons, 2004. (This is an updated version of *System Safety 2000* by Joe Stephenson).

U. S. Department of Energy. *Handbook—Accident and Operational Safety Analysis—Volume 1: Accident Analysis Techniques*. Washington, DC: Department of Energy, 2012. It is also downloadable free. Enter the title into a search engine. Available at https://sites.ntc.doe.gov/partners/elearningcommunication/sba130de/Course%20Documents/EFCOG/pubs/AA%20Handbook%20Master%2012-3-15%20V41%20FINAL.PDF. Accessed June 3, 2019.

CHAPTER 23

JAMES REASON'S SWISS CHEESE MODEL

In recent years, no individual has had more influence on how people think about incident causation than James Reason. In *Managing the Risks of Organizational Accidents,* Reason (1997) wrote about two ways that systems could be disrupted. He recognized that active errors, what have been called unsafe acts, could be committed by operators and be the initiator for an incident. But he also emphasized that what he labeled latent conditions could be resident in an operation for some time.

Reason emphasized that latent conditions are present in all organizations; result from decisions made by "manufacturers, designers and organizational managers"; and could be the more significant causal factors for accidents. His views on the origins on latent conditions have had a major impact on the thinking of many safety professionals. (p. 10)

My interpretation of what Reason wrote about latent conditions fits particularly well with the citations that were taken from *Guidelines for Preventing Human Error in Process Safety* (1994) and appear in Chapter 8 on "Human Error Avoidance" in this book. Authors of the *Guidelines* book recognized that failures at the management, design, or technical expert levels of the company could result in adverse situations (p. xiii) and that organizational factors could create the preconditions for errors, as well as being the immediate causes of incidents. (p. 5)

Advanced Safety Management: Focusing on Z10.0, 45001, and Serious Injury Prevention, Third Edition. Fred A. Manuele.
© 2020 John Wiley & Sons, Inc. Published 2020 by John Wiley & Sons, Inc.

HOW EXTENSIVE IS THE USE OF THE SWISS CHEESE MODEL?

Only a few comments in the literature were selected to demonstrate the significance of and the range of adoption of variations of Reason's causation model. Keywords or phrases in these citations are: de facto template; the dominant safety metaphor of our time; revolutionized common views on accident causation; forces investigators to address latent failures within the causal sequence; can impact performance at all levels; and principle behind layered security and defense in depth.

> **Source**: *Models of Causation: Safety*, a 2012 publication
> In some industries the Swiss Cheese model is still considered best practice 22 years after its introduction. (p. 20)
> **Source**: University of New South Wales, Faculty of Science, Australia (n.d.)
> When it comes to understanding incidents and accidents, James Reason's 'Swiss cheese model' has become the de facto template. (p. 1)
> **Source**: CGE Risk Management Solutions (n.d.)
> The Swiss cheese model has proven to be one of the dominant safety metaphors of our time. Over the years, multiple barrier-based methods have been developed using this metaphor, often with slightly different goals and interpretations of the original Swiss cheese model. (p. 1)
> **Source**: The Human Factors Analysis and Classification System—HFACS (n.d.), U.S. Department of Transportation, Federal Aviation Administration. This is but one governmental body that has adopted Reason's thinking. Note particularly in the following that "the organization itself can impact performance at all levels."
> - In many ways, Reason's "Swiss cheese" model of accident causation has revolutionized common views of accident causation. (p. 2)
> - However, what makes the "Swiss cheese" model particularly useful in accident investigation, is that it forces investigators to address latent failures within the causal sequence of events as well. (p. 2)
> - Reason's model didn't stop at the supervisory level either; the organization itself can impact performance at all levels. (p. 2)
>
> **Source**: Wikipedia (n.d.)
> The Swiss Cheese model of accident causation is used in risk analysis and risk management, including aviation safety, engineering, healthcare, emergency service organizations, and as the principle behind layered security, as used in computer security and defense in depth. (p. 1)

REASON'S SWISS CHEESE MODEL: ORIGIN AND DEPICTION

Although the foundation on which Reason developed what has become known as the Swiss cheese model exists in his book *Human Error*, there is no mention of the model in that work. Much of Reason's thinking as expressed in *Human Error* is duplicated in *Managing the Risks of Organizational Accidents,* which was first published in 1997.

In this latter book, Reason used the term the Swiss Cheese Model in a brief discussion of what would be the ideal world for defenses in depth and the real world in which defenses in depth may be inadequate. (p. 10)

But the name he gave his model is The Accident Trajectory, which has been replaced by what has become popularly known as the Swiss cheese model. (Some writers say that the designation "Swiss Cheese Model" was created by users.)

Beneath his model, Reason said that it depicted "An accident trajectory passing through corresponding holes in the layers of defenses, barriers and safeguards. The holes can be created by active and latent failures." (p. 12)

As noted, Reason's text refers to barriers but his model for The Accident Trajectory does not. It does include the phrase "Defenses in Depth" which could be interpreted as inclusive of barriers. Because of the emphasis now given by some safety professionals to having appropriate barriers and controls (I am one of them), my depiction of a variation of the Swiss cheese model specifically mentions barriers. It follows in Figure 23.1.

Reason used the term "holes" symbolically and as a metaphor. He did not imply that there were actual holes in management systems. What the holes represent are inadequacies in barriers and controls that, if adequate and appropriately used, may have prevented an incident from occurring. Such inadequacies result from deficiencies in the management systems that relate to them. Relationship of holes to each other is not fixed: It is fluid and subject to change as are all operational situations.

As an incident occurs, so indicated Reason, there would be a confluence of deficiencies, initiated by either a latent condition or an active error.

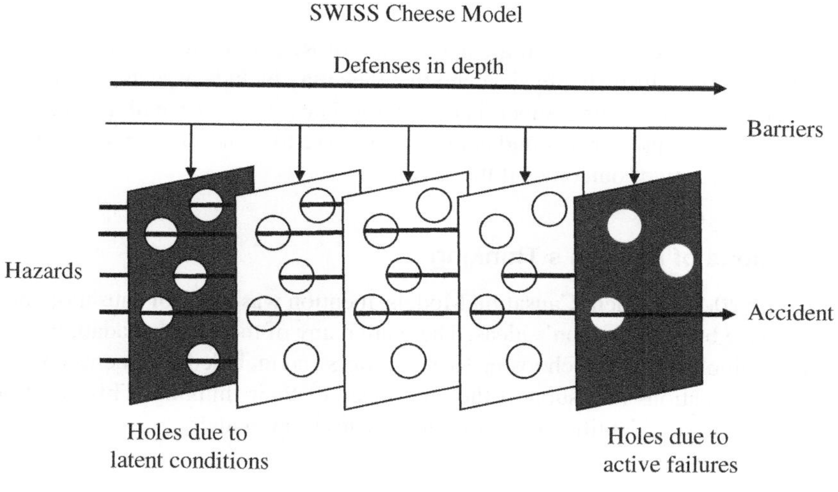

SWISS Cheese Model

FIGURE 23.1 A variation of Reason's Swiss Cheese Model.

WHAT IS A BARRIER?

For application of Reason's Swiss Cheese Model, having appropriate and properly used barriers is vital. Extensive comments are made on barriers in Chapter 20, on Causation Models. Only highlights of those comments appear here.

Reference was made to Erik Hollnagel's (2004) book *Barriers and Accident Prevention*, in which he discusses "Barrier Functions and Barrier Systems" and "Understanding the Role of Barriers in Accidents." His writings are good references on barriers and are recommended.

Hollnagel referred to two particular resources on barriers: William Haddon's (1970) unwanted energy release theory, which some 50 plus years ago promoted the use of barriers to protect people and property; and MORT—The Management Oversight and Risk Tree. As can be noted in Chapter 22 on MORT, having appropriate barriers and using them effectively is vital in the application of MORT.

Two definitions of barriers given in Chapter 22 are duplicated here: the definition given by Richard A. Stephans (2004) in *System Safety for the 21st Century* because of its breadth; and the definition given by Lyon and Popov (2018), in *Risk Assessment Tools for Safety Professionals*, because it mentions both the preventive and mitigative aspects of barriers.

> Barriers—Stephans: Anything used to control, prevent, or impede energy flows. Types of barriers include physical, equipment design, warning devices, procedures and work processes, knowledge and skills, and supervision. Barriers may be control or safety barriers or act as both. (p. 357)
>
> Barrier—Lyon and Popov: Physical or procedural control measures that are put in place to prevent or reduce likelihood of risk exposure (preventive) and/or reduce severity of impact/consequences (mitigative) resulting from a hazardous event. (p. 313)

It is important to note that in the application of Reason's Swiss Cheese Model: barriers are not exclusively physical and that they may include aspects such as having a qualified staff, training, supervision, appropriate procedures, maintenance, and communications; and barriers and controls can serve to avoid incidents and also to reduce the severity of outcomes if they occur.

Adaptations of Reason's Thinking

In Chapter 20, on Incident Causation Models, mention was made of causation models that were built on Reason's ideas. There are many of them. Some adaptors have added terminology of their choosing to the barriers and included comments on them in their explanations that support the legitimacy of their thinking. This is a brief selection of barrier identifications from those causation models.

Senior Management	Failed Defenses	System Design
Organization Factors	System/Hardware	Absent Defenses
Supervision	Training	Maintenance
Worker Behavior	Culture	Unsafe Acts
Working Conditions	Errors and Variations	Procedures

HOW ACCIDENTS HAPPEN

This author's thinking, as follows, relates to his research and experience, Reason's explanation of The Accident Trajectory, and to Reason's writings that preceded and followed comments on the Trajectory.

An accident occurs because of a confluence of causal factors that arise when there are deficiencies in successive barriers and controls that allow hazards to reach their potential and the actuation of that potential may have harmful effects on people, property, or the environment. When the barriers and controls:

- As of a given moment are non-existent, inadequate or improperly used;
- All of the elements necessary for an accident to occur may develop;
- The "holes" become in alignment;
- There are active or latent causal factors, or both;
- The incident process begins with an initiating event;
- There are unwanted energy flows or exposures to harmful substances;
- Multiple interacting events may occur sequentially or in parallel;
- Harm or damage results, or could have resulted in slightly different circumstances.

TO WHAT TYPES OF OPERATIONS DOES REASON'S THINKING APPLY?

What Reason says in *Human Error* about the basics of his theory **applies to all entities of every sort**.

> A basic premise of this framework is that systems accidents have their primary origins in fallible decisions made by designers and high-level (corporate or plant) managerial decision makers. (p. 203)

It is fact that causal factors in every type of operation may derive from management decisions when safety policies, standards, procedures, and the accountability system or their implementation are inadequate.

That is a key point in Chapter 2—Culture, Management Leadership and Worker Participation. In that chapter, the first three items in *A Socio-Technical Model for an Operational Risk Management System* were duplicated as they are here. They represent the decision-making that results in having or not having adequate barriers and controls and latent conditions. They are:

- The governing body and senior management establish a culture that requires having barriers and controls in place to achieve and maintain acceptable risk levels for people, property, and the environment.
- Management leadership, commitment, involvement, and the accountability system establish that the performance level to be achieved is in accord with the culture established.
- To achieve acceptable risk levels, management establishes (appropriate) policies, standards, procedures, and processes.

An organization's governing body and senior management establish and own the culture. What management does or does not do and the effectiveness of the accountability system determine what the culture is to be and whether operating risks are acceptable. Figure 23.2 depicts this author's version of *A Socio-Technical Model for an Operational Risk Management System*.

It should be obvious that I am an enthusiastic supporter of Reason's thinking and has been influenced muchly what he wrote.

Criticisms on the Use of Reason/s Swiss Cheese Model

Critics of Reason's causation model have indicated that in its use the significance of active failures (unsafe acts) is subordinated. That's not an appropriate reading. In Reason's model, "Holes due to active failures" are a main factor. It could be argued that for most situations Reason would take the position that active failures often trigger the process that becomes an accident.

Reason implied in *Human Error* that employees are exposed to that which management creates through its decision-making and, thus, shortcomings in the work system should be the focus of incident investigations. He wrote:

> Rather than being the main instigators of an accident, operators tend to be the inheritors of system defects created by poor design, incorrect installation, faulty maintenance and bad management decisions. Their part is usually that of adding the final garnish to a lethal brew whose ingredients have already been long in the cooking. (p. 173)

Reason also noted that unsafe acts are now seen more as consequences than as principal causes, but he did not diminish the significance of human errors—active failures—as possibly being the principle causal factors for incidents. But it would be a misapplication if a focus on active failures did not also give appropriate consideration to what Reason referred to as "poor design, incorrect installation, faulty

The governing board and senior management establish a culture that requires having barriers and controls in place to achieve and maintain acceptable risk levels for people, property, and the environment

Management leadership, commitment, involvement, and the accountability system, establish that the performance level to be achieved is in accord with the culture established

To achieve acceptable risk levels, management establishes policies, standards, procedures, and processes with respect to:

- Risk assessment
- Prevention through design
- Providing adequate resources
  - Competency and adequacy of staff
- Operating procedures
  - Organization of work
- Training and motivation
- Maintenance for system integrity
- Management of change/pre-job planning
  - Pre-start review
- Procurement — hazard and risk specifications
  - Relationships with suppliers
- Emergency planning and management
- Risk-related processes
  - Employee participation
  - Information — communication
  - Permits
  - Inspections
  - Incident investigation and analysis
  - Providing personal protective equipment
- Contractors — on premises
- Conformance/compliance assurance reviews

Periodic performance reports are prepared pertaining to the adequacy of systems and controls to maintain acceptable risk levels.

Figure 23.2 A Socio-Technical Model for an Operational Risk Management System.

maintenance and bad management decisions" that resulted in latent conditions or practices.

Other critics complain that Reason does not specify the processes that create the holes in defenses. But the holes are metaphoric and are not described as physical holes. They may also complain that Reason's literature does not fully describe how active failures and latent conditions interact. That is true. But a study of incident reports indicates that they do interact.

Reason does not define barriers. Correct, but others adequately define barriers, as in this chapter and in Chapter 20 on Causation Models.

Critics suggest a lack of clarity concerning the category in which Reason's model falls—in an epidemiological category or a systems category (see Chapter 20 for definitions). It does not make any difference. For instance, in Models of *Causation: Safety,* Reason's model is first mentioned in comments on epidemiological models (p. 10) and discussed as a systems model, including a modified depiction of his model. (p. 12)

Safety professionals—getting caught up in model categories is a waste of time. What is important is validity of thinking and applicability.

Criticism must be considered and examined for its substance. But sometimes critics are negative as they promote their own thinking and models.

REASON'S OBSERVATIONS ON CHANGING VIEWS ABOUT ACTIVE FAILURES

In *Organizational Accidents Revisited,* Reason wrote that in some organizations a conceptual change had occurred in the process of identifying causal factors and that the focus was moved from the unsafe acts of employees as being the principle causal factors to recognition of systemic causal factors. That book was published in 2016.

One could hope that what Reason wrote about the conceptual change he had witnessed applied broadly and not only to a few organizations. Nevertheless, what Reason observed is progress. (p. 10)

In practice, according to this author's research, it is not yet the accepted and universal practice for investigators to inquire beyond unsafe acts of employees for causal factors and to consider the possible deficiencies in management systems that may relate to causation.

What does all this mean? In the practice of safety, the emphasis is moving, although slowly, from trying to change the behavior of workers—which for quite some time was the emphasis of many safety practitioners—to reducing operational risks through delving into the systems in which people work. And Reason's writing has influenced this transition.

PREVENTION THROUGH DESIGN IMPLICATIONS

In *Managing the Risks of Organizational Accidents,* Reason wrote that latent conditions could result from poor design, poor maintenance, etc. (p. 12). In *Human Error,* he wrote that operators tend to be the inheritors of system defects created by poor design et al. In the same book and on the same page, Reason said that the greater good would be achieved if management and incident investigators concentrated on limiting the development of latent conditions. (p. 173)

Because of his experience, this author has been a promoter of safety professionals becoming more involved in the design processes. In so doing, they could assist in limiting the development of latent conditions that result when hazards and risk are not inadequately considered in the design of the workplace and of work methods. Interest in prevention through design among safety professionals is emerging, although slowly. The American Society of Safety Professionals is now giving courses on Prevention Through Design.

And we now have an American National Standard on the subject, the title of which is *Guidelines for Addressing Occupational Hazards and Risks in Design and Redesign Processes*. Its designation is ANSI/ASSE Z590.3 (2016). Initiation of the activity out of which the standard was developed came from the National Institute for Occupational Safety and Health (NIOSH).

In 2008, NIOSH announced that one of its major initiatives was to "Develop and approve a broad, generic voluntary consensus standard on Prevention through Design that is aligned with international design activities and practice."

Support for such a standard was obtained from the American Society of Safety Engineers. In September 2011, the standard was approved by the American National Standards Institute (ANSI) and reapproved in 2016.

This is the definition of Prevention through Design in the standard. It is occupational safety related, as is the National Institute for Occupational Safety and Health (NIOSH).

> Prevention through Design. Addressing occupational safety and health needs in the design and redesign process to prevent or minimize the work-related hazards and risks, associated with the construction, manufacture, use, maintenance, retrofitting, and disposal of facilities, processes, materials, and equipment. (p. 13)

As this author wrote in the fourth edition of *On the Practice of Safety* (2013), there are sound reasons in support of hazards and risks being addressed in the design process to avoid the development of latent hazardous conditions.

> Hazards and risks are most effectively and economically avoided, eliminated, or controlled in the design and redesign processes. (p. 388)

Experience has shown that it is less costly to address hazards and risks in the original design process than it is to have to retrofit during operations because hazards and risks were not adequately addressed.

This author proposes that very early in the incident investigation endeavor a specific inquiry be made to determine whether there are design implications as causal factors both with respect to the work processes and the work methods. Investigations are not complete if design causal factors are not recognized in the process of determining how accidents happen.

CONCLUSION

Studies of the quality of incident investigations indicate that accomplishments may be less than stellar. Nevertheless, examples of outstanding investigations are occasionally received. It is said in this book that an effective incident investigation must consider the hazards, their accompanying risks, and the possible deficiencies in management systems that could derive from the cultural, organizational, technical, or operational aspects of an entity. That reflects directly on Reason's thinking.

A report in which references to shortcomings in the "cultural, organizational, technical, or operational aspects of an entity" was received by this author. It pertained to the investigation of an incident for a significant event that resulted in extensive property and environmental damage.

These subjects were cited as deficient and contributing causal factors, for which it is suggested that Reason would be pleased. These are excerpts from that report.

- Absence of management systems, controls and standards.
- Inadequate culture of accountability.
- Gaps of which the Board needed to be aware.
- Performance expectations were not clearly established.
- Management allowed such conditions and practices to exist and continue to occur.
- Hazardous operations were accepted and unaddressed.
- Lack of standardization.
- Lack of prevention priority and resources.
- Budget process was an impediment.
- Management was fragmented.
- Design inadequacies.
- Inconsistent oversight.
- Poor communications.

This report was among the most insightful of all the reports that this author has reviewed. It establishes—it can be done. It relates well to Reason's concepts on the accumulation of latent hazardous conditions and practices, over time.

Reason's influence throughout the world has been immense. This author suggests that safety professionals study his premises and relate them to the causation models that they use. Reason's ideas have been so broadly accepted because they represent sound thinking.

REFERENCES

ANSI/ASSE Z590.3. *Guidelines for Addressing Occupational Hazards and Risks in Design and Redesign Processes.* 2011 (R2016) Park Ridge, IL: American Society of Safety Professionals, 2016.

Barrier Effectiveness, CGE Risk Management Solutions. Available at https://www.cgerisk.com/knowledge-base/risk-assessment/barrier-effectiveness. Accessed June 3, 2019 (n.d.).

Guidelines for Preventing Human Error in Process Safety. New York: Center for Chemical Process Safety of the American Institute of Chemical Engineers, 1994.

HFACS, U.S. Department of Transportation, Federal Aviation Administration. The Human Factors Analysis and Classification System. Available at https://www.nifc.gov/fireInfo/fireInfo_documents/humanfactors_classAnly.pdf. Accessed June 3, 2019.

Haddon, W. "On the escape of tigers: an ecologic note". *Journal of Public Health*, December, 1970.

Hollnagel, E. *Barriers and Accident Prevention.* Burlington, VT: Ashgate Publishing, 2004.

Lyon, B.K. and G. Popov. *Risk Management Tools for Safety Professionals.* Park Ridge, IL: The American Society of Safety Professionals, 2018.

Manuele, F.A. *On the Practice of Safety*, Fourth Edition. Hoboken, NJ: John Wiley & Sons, 2013.

Models of Causation: Safety. Safety Institute of Australia, Ltd., 2012. Available at http://www.ohsbok.org.au/wp-content/uploads/2013/12/32-Models-of-causation-Safety.pdf. Accessed June 3, 2019.

Reason, J. *Managing the Risks of Organizational Accidents.* Burlington, VT: Ashgate Publishing, 1997.

Reason, J. *Organizational Accidents Revisited.* Burlington, VT: Ashgate Publishing, 2016.

Stephans, R.A. *System Safety for the 21st Century.* Hoboken, NJ: John Wiley & Sons, 2004 (This is an updated version of *System Safety 2000*, by Joe Stephenson.)

University of New South Wales, Faculty of Science, Australia. Available at https://www.google.com/search?q=Reason's+swiss+cheese+model+-+University+of+New+South+Wales&tbm=isch&source=hp&sa=X&ved=2ahUKEwigheim4M3iAhUPnq0KHVwFAHYQ7Al6BAgFEA0&biw=1024&bih=362#imgrc=yebcCwsoUhpN3M. Accessed June 3, 2019.

Wikipedia. Available at https://en.wikipedia.org/wiki/Swiss_cheese_model. Accessed June 2, 2019 (n.d.).

FURTHER READING

ANSI/ASSP Z10.0-2019. *Occupational Health and Safety Management Systems.* Park Ridge, IL: American Society of Safety Professionals, 2019.

ANSI/ASSP/ISO 45001-2018. *Occupational Health and Safety Management Systems—Requirements with Guidance for Use.* Park Ridge, IL: American Society of Safety Professionals, 2018.

Dekker, S. *The Field Guide to Understanding Human Error*. Burlington, VT: Ashgate Publishing, 2006.

MORT User's Manual. DOE, 1992. Available at https://www.osti.gov/servlets/purl/5254810. Accessed June 3, 2019.

NRI MORT User's Manual 2009 is available on the Internet. Downloadable without charge. Enter "MORT User's Manual" into a search engine and this publication appears. Available at http://nri.eu.com/NRI1.pdf. Accessed June 3, 2019.

Reason, J. *Human Error*. Cambridge, UK: Cambridge University Press, 1990.

Reason, J. and A. Hobbs. *Managing Maintenance Error: A Practical Guide*. Burlington, VT: Ashgate Publishing, 2003.

CHAPTER 24

ON SYSTEM SAFETY

Identifying and analyzing hazards and making risk assessments as early as practicable in the design and redesign processes, and additionally as needed, are the bases on which system safety is built. For system safety initiatives, the outcome is to be having acceptable risk levels.

Consider these selected sections in Z10.0 as they relate to hazards, risks, risk assessments, the design process, and acceptable risks. These citations are abbreviated substantially. There are close to identical provisions in 45001.

- Section 6.2 – Assessment and Prioritization – the process shall assess the level of risk for identified hazards; establish priorities based on factors such as the level of risk; and identify factors related to system deficiencies that lead to hazards and risks.
- Section 8.2 – Identification of OHSMS Issues: Processes shall be in place for the identification of hazards.
- Section 8.3 – Risk Assessment: The organization shall establish and implement a risk assessment process(es) appropriate to the nature of hazards and level of risk.
- Section 8.4 – Hierarchy of Controls: The organization shall establish a process for achieving feasible risk reduction based upon a preferred and given order of controls.

Advanced Safety Management: Focusing on Z10.0, 45001, and Serious Injury Prevention,
Third Edition. Fred A. Manuele.
© 2020 John Wiley & Sons, Inc. Published 2020 by John Wiley & Sons, Inc.

- Section 8.5 – Design Review and Management of Change: The organization shall establish a process to identify and take appropriate steps to prevent or otherwise control hazards at the design and redesign stages.

There is a direct relationship between system safety concepts and the overall practice of safety and the processes necessary to implement Z10.0 and 45001. Terminology for Section 8.5 above indicating that processes are to be established "to identify and take appropriate steps to prevent or otherwise control hazards at the design and redesign stages" could be a very brief description of system safety.

This author believes that generalists in the practice of safety will improve the quality of their performance by acquiring knowledge of applied system safety concepts and practices. It is not suggested that safety generalists must become supraspecialists in system safety, although trends indicate that they will be expected to apply at least the fundamentals of system safety.

To influence safety generalists to acquire knowledge of and apply system safety concepts, this chapter:

- Relates the generalist's practice of safety to applied system safety concepts.
- Gives a history of the origin, development, and application of system safety methods.
- Reviews several definitions of system safety.
- Outlines The System Safety Idea in terms applicable to the generalist's practice of safety.
- Encourages safety generalists to acquire knowledge and skills in system safety.

RELATING THE GENERALIST PRACTICE OF SAFETY TO SYSTEM SAFETY

ANSI/ASSE Z590.2 (R2012) is the designation for the American National Standard titled Criteria for Establishing the Scope and Function of the Professional Safety Position. Its secretariat is the American Society of Safety Professionals. Item 3 on page 6 is captioned Scope and Functions of the Professional Safety Position. They say that those functions are to:

A. Anticipate, identify and evaluate hazardous conditions and practices
B. Develop hazard control designs, methods, procedures and programs
C. Implement, administer and advise others on hazard controls and hazard control programs
D. Measure, audit and evaluate the effectiveness of hazard controls and hazard control programs

According to item A, the safety professional is to "anticipate hazards." And item B indicates that safety professionals are to "develop hazard control designs." Provisions in Z10.0 and 45001 require the same.

To be in a position to anticipate hazards, one must be involved in the design process. To effectively participate in the design process, a safety professional must be skilled with respect to hazard analysis and risk assessment techniques. Influencing the design process and using hazard analysis and risk assessment techniques to achieve acceptable risk levels are fundamental in system safety.

Enterprising generalists in safety will become proficient with respect to the hazard identification and control aspects and the design aspects of the "Scope and Function of a Safety Professional." That will be to their advantage as they give counseling to clients to achieve acceptable risks with respect to the protection of people, property, and the environment.

AFFECTING THE DESIGN AND REDESIGN PROCESSES

System safety professionals make a great deal of designing things right the first time and being participants throughout the design and redesign processes. Richard A. Stephans, the author of *System Safety for the 21st Century*, expresses that view well.

Safety Is Productive
Safety is achieved by doing things right the first time, every time. If things are done right the first time, every time, we not only have a safe operation but also and extremely efficient, productive, cost-effective operation (p 12).

Safety Requires Upstream Effort
The safety of an operation is determined long before the people, procedures, and plant and hardware come together at the work site to perform a given task (p 13).

For products, facilities, equipment, and processes, and for their subsequent alteration, the time and place to economically and effectively avoid, eliminate, reduce, or control hazards is in the design or redesign processes. Participating in those processes presents opportunities for upstream involvement by safety professionals, using system safety concepts.

Also, there has been an extended recognition that applying design and engineering solutions is the preferred course of action in operational risk management. That extended recognition derives from several sources, among which is the involvement of safety professionals in:

- Applied ergonomics
- Giving counsel to meet European requirements whereby risk assessments are to be made on goods that go into workplaces in the European Community countries
- Applying the requirements of guidelines and standards that: propose or require risk assessments; and present an ordered sequence of measures to be taken in a hierarchy of controls to achieve acceptable risk levels. Examples are:

- ANSI/ASSP Z10.0-2019. *Occupational Health and Safety Management Systems*
- MIL-STD-882E-2012. *Department of Defense Standard Practice for System Safety*
- ANSI-ASSE Z590.3-2011. *Prevention through Design: Guidelines for Addressing Occupational Hazards and Risks in Design and Redesign Processes*
- ANSI B11.0-2015. *Safety of Machinery—General Safety Requirements and Risk Assessments*
- BS OHSAS 18001:2007. *Occupational health and safety management systems—requirements*
- *Guidance On The Principles Of Safe Design For Work*. Canberra, Australia.
- *Machine Safety: Prevention of mechanical hazards*. (2009). Quebec, Canada
- Risk Assessment. The European Union, 2008
- CSA Z1002-12. *Occupational health and safety—Hazard identification and elimination and risk assessment and control*. Toronto.
- EN ISO 12100-2010. *Safety of Machinery. General principles for Design. Risk assessment and Risk reduction*. Geneva, Switzerland.
- Meeting the requirements for hazards analysis in OSHA's standard for Process Safety Management of Highly Hazardous Chemicals and for EPA risk management program requirements.

Of all the foregoing references, the Z590.3 standard gets closest to applied System Safety. It has been referred to as System Safety light.

It is given that making hazard identifications and analysis and risk assessments has become a more prominent part of safety professionals' responsibilities.

DEFINING SYSTEM SAFETY

Unfortunately, the term system safety does not convey a clear meaning of the practice as it is applied. Published definitions of system safety are of some help in understanding the concept, but they do not communicate clearly.

To give indications of the differences in the definitions of system safety, and to move this discussion forward, five sources are cited.

In MIL-STD-882E-2012, the *Department of Defense Standard Practice for System Safety*, system safety is defined as:

> The application of engineering and management principles, criteria, and techniques to achieve acceptable risk within the constraints of operational effectiveness and suitability, time, and cost throughout all phases of the system life-cycle (p 8).

This outline of "The System Safety Idea" encompasses most of the definitions previously given and goes beyond several. Safety generalists should ask—How closely does "The System Safety Idea" come to the results expected in applying the provisions in Z10.0 and 45001? Do safety professionals serve themselves well by becoming knowledgeable in system safety?

System safety commences with hazard identification and analysis and risk assessment. So do all hazards and risk-based activities, whatever they are called. This author is confident that application of system safety concepts in the business and industrial setting will result in significant reductions in injuries and illnesses, damage to property, and environmental incidents.

For emphasis, the intent of Section 8.5 in Z10.0 is repeated here: The organization shall establish a process to identify and take appropriate steps to prevent or otherwise control hazards at the design and redesign stages.

HAZARD IDENTIFICATION AND ANALYSIS AND RISK ASSESSMENT TECHNIQUES

It is not surprising that many safety generalists are turned away from system safety when they encounter the number of analytical techniques that have been developed, the complexity of some of them, and the skill necessary to apply them. Earlier in this chapter, it was made clear that safety generalists need not become supraspecialists in system safety. Nevertheless, the trends indicate that they will be expected to be skilled in applying basic system safety methods.

How many analytical systems are there? The second edition of the System Safety Analysis Handbook fills 626 pages and contains a compilation of 101 analysis techniques and methodologies. That Handbook serves as a desk reference for the accomplished system safety professionals who may have to resolve highly complex or infrequently encountered situations.

Three national standards that constitute a set should be of interest to safety generalists who want to become familiar with system safety techniques. The American Society of Safety Professionals is the secretariat.

> ANSI/ASSE Z690.1-2011: *Vocabulary for Risk Management (National Adoption of ISO Guide 73:2009)*. This standard provides definitions of terms that, the originators hope, will be used in other standards.
>
> ANSI/ASSE Z690.2-2011: *Risk Management Principles and Guidelines (National Adoption of ISO 31,000:2009)*. The intent of this standard is to provide a broad ranged primer on risk management systems that could be applied in any type of organization. The need to make risk assessments is introduced in section 5.4: Risk Assessment.
>
> ANSI/ASSE Z 690.3-2011: *Risk Assessment Techniques (National Adoption of IEC/ISO 31,010:2009)*. For safety generalists who want an education in risk assessment concepts and methods and a ready reference, this standard is worth

acquiring. It commences with a 15-page dissertation on risk assessment concepts and methods. In five pages, Appendix A provides brief comparisons of 31 risk assessment techniques. Reviews of the 31 techniques—Overview, Use, Inputs, Process, Strengths and Limitations—are provided in Annex B which covers 79 pages.

ANSI/ASSE Z 690.3-2011 is a valuable resource. A list of the 31 risk assessment techniques follows. Some could be applied only by experienced system safety professionals. But knowledge of a few of them will serve a huge percentage of the needs of a safety generalist.

B01	Brainstorming	B02	Structured or Semi-Structured Interviews
B03	Delphi	B04	Checklists
B05	Preliminary Hazard Analysis	B06	Hazard and Operability Studies
B07	Hazard Analysis and Control Points	B08	Environmental Risk Assessment
B09	Structure—What if Analysis	B10	Scenario Analysis
B11	Business Impact Analysis	B12	Root Cause Analysis
B13	Failure Mode Effect Analysis	B14	Fault Tree Analysis
B15	Event Tree Analysis	B16	Cause and Consequence Analysis
B17	Cause-and-Effect Analysis	B18	Layer Protection Analysis
B19	Decision Tree	B20	Human Reliability Analysis
B21	Bow Tie Analysis	B22	Reliability Centered Maintenance
B23	Sneak Circuit Analysis	B24	Markov Analysis
B25	Monte Carlo Simulation	B26	Bayesian Statistics and Bayes Nets
B27	FN Curves	B28	Risk Indices
B29	Consequence/Probability Matrix	B30	Cost Benefit Analysis
B31	Multi-Criteria Decision Analysis		

In ANSI/ASSE Z590.3-2011, the *Prevention through Design standard, Addendum G comments on eight hazard analysis and risk assessment techniques*. They are:

- Preliminary Hazard Analysis
- What-If Analysis
- Checklist Analysis
- What-If Checklist Analysis

- Hazard and Operability Analysis
- Failure Mode and Effects Analysis
- Fault Tree Analysis
- Management Oversight and Risk Tree (MORT).

It was also said in Z590.3 that as a practical matter, having knowledge of three risk assessment concepts will be sufficient to address most, but not all, risk situations. They are: Preliminary Hazard Analysis and Risk Assessment; The What-If/Checklist Analysis methods; and Failure Mode and Effects Analysis. (p 23)

Having knowledge and capability with respect to the aforementioned three standards will be sufficient to deal with a huge majority of the needs of Z10.0 and 45001. Additional references on risk assessment techniques appear later in this chapter under the caption Recommended Reading.

RISK ASSESSMENT MATRICES

It would be highly unusual for a text or standard on system safety not to include risk assessment matrices and provide examples. There are many, many variations of matrices and the definitions of the terms used in them vary greatly. The matrix used should be the one to which management and users in an organization decide is best for them. It is strongly recommended that a suitable matrix be chosen because of its value in risk decision-making.

A risk assessment matrix provides a method to categorize combinations of probability of occurrence and severity of harm, thus establishing risk levels. Z10.0 and 45001 require that priorities be established in the application of their requirements.

A matrix helps in communicating with decision-makers on risk reduction actions to be taken. Also, risk assessment matrices assist in comparing and prioritizing risks, and in effectively allocating mitigation resources. A matrix that this author has found useful is presented here as Figure 24.1.

All personnel involved in the risk assessment processes must understand the definitions used for occurrence probability and severity and for risk levels in the risk assessment matrix chosen. This is vitally important. For there to be effective communication between those involved in a risk assessment, all must understand the intent of the terms used.

THE HIERARCHY OF CONTROLS

It was said in the "The System Safety Idea" that if risk reduction was necessary after a risk assessment, the steps to follow are those in the Hierarchy of Controls. That subject is covered in Chapter 11 in this book.

In Z10.0, under the sub-caption "Hierarchy of Controls" (Section 8.4), this is its opening sentence and its hierarchy of controls.

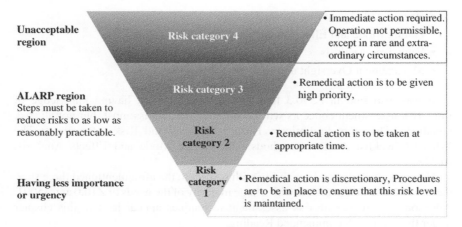

FIGURE 24.1 ALARP

The organization shall establish a process for achieving an acceptable level of risk reduction based upon the following preferred order of controls:

- Elimination;
- Substitution of less hazardous materials, processes, operations, or equipment;
- Engineering controls;
- Warnings;
- Administrative controls, and
- Personal protective equipment.

There are three lengthy notes following the Z10.0 hierarchy. Paraphrasing, the notes say that in the risk reduction process, organizations should consider:

- reliability, effectiveness, worker acceptance and the possible need for multiple controls and feedback loops to monitor control effectiveness;
- the possibility of reorganizing the work in the decision-making process;
- evaluating the effectiveness of controls periodically since most controls erode over time.

Then Z10.0 provides guidance on the subjects that should be considered when the hierarchy of controls is to be applied, as follows.

Application of this hierarchy of controls to achieve an acceptable level of risk shall take into account:

- The nature and extent of the risks being controlled;
- The degree of risk reduction desired;

- The requirements of applicable local, federal, and state statutes, standards and regulations;
- Recognized best practices in industry;
- The effectiveness, reliability, and durability of the control being considered
- Human factors (includes ergonomics)
- Available technology;
- Cost-effectiveness, and
- Internal organization standards.
- Strategies to eliminate or mitigate potential health exposures including those not originating from work activities.

At Section 8.1.2 in 45001, under Eliminating hazards and reducing OH&S risks, the "following hierarchy of controls" consists of five elements, all of which are included in the hierarchy shown in Z10.

Decision-makers should understand that, with respect to the six levels of control shown in Z10's hierarchy of controls, the ameliorating actions described in the first through the third control levels are more effective because they:

- are preventive actions that eliminate or reduce risk by design, elimination, substitution, and engineering measures
- rely the least on the human behavior—the performance of personnel
- are less defeatable by unit managers, supervisors, or workers.

Actions described in the fourth, fifth, and sixth levels are contingent actions and rely greatly on the performance of personnel for their effectiveness. Inherently, they are less reliable.

In application of the Hierarchy of Controls, the expectation is that consideration will be given to each of the steps in a descending order and that reasonable attempts will be made to avoid, eliminate, reduce, or control hazards, and their associated risks through steps higher in the hierarchy before lower steps are considered. A lower step in the hierarchy of controls is not to be chosen until practical applications of the preceding level or levels are considered. It is understood that for many risk situations, a combination of the risk management methods shown in the hierarchy of controls is necessary to achieve acceptable risk levels.

In applying the hierarchy, the outcome should be an acceptable risk level. In achieving that goal, the following should be taken into consideration.

- avoiding, eliminating, or reducing the probability of a hazard-related incident or exposure occurring.
- reducing the severity of harm or damage that may result if an incident or exposure occurs.
- the feasibility and effectiveness of the risk reduction measures to be taken, and their costs, in relation to the amount of risk reduction to be achieved.
- all of the requirements of Z10.0 and 45001.

A SLIGHTLY DIFFERENT PRESENTATION FOR A HIERARCHY OF CONTROLS

Item 2 in "The System Safety Idea" reads as follows: System safety personnel recognize that hazards are most effectively and economically anticipated, avoided, or controlled in the initial design or redesign processes. Citing Richard A. Stephans again as in *System Safety for the 21st Century*:

Safety Is Productive
Safety is achieved by doing things right the first time, every time. If things are done right the first time, every time, we not only have a safe operation but also and extremely efficient, productive, cost-effective operation (p 12).

As previously quoted definitions of system safety have shown, system safety is largely a design and redesign process. Content on a hierarchy of control in Chapter 13—Prevention Through Design—are duplicated here because that standard relates to the design processes.

Top management shall achieve acceptable risk levels by adopting, implementing, and maintaining a process to avoid, eliminate, reduce, and control hazards and risks. The process shall be based on the hierarchy of controls outlined in this standard, which is:

a. Risk avoidance
b. Eliminate
c. Substitution
d. Engineering controls
e. Warning systems
f. Administrative controls
g. Personal protective equipment

This is a "prevention" standard. Research was done to develop a variation in the hierarchy of controls to adapt to "prevention." Elimination, which is the first action to be taken in many hierarchies of controls, did not seem to fit with meaning of prevention.

- Elimination means: removal; purging; taking away; to get rid of something. To eliminate, there has to be something in place to remove.
- Avoidance means: to prevent something from happening; keeping away from; averting.

Designers start with a blank sheet of paper, or an empty screen in a CAD system. Designers have opportunities to avoid hazards in all design stages: conceptual, preliminary, and final. In the early design phases, there are no hazards, yet, to be eliminated, reduced, or controlled. So, avoidance is a better match for a hierarchy of controls for a prevention standard.

Whatever standard or guideline safety professionals use, the intent in applying a hierarchy of controls is to achieve acceptable risk levels and reduce the potential for injuries, illnesses, and fatalities occurring.

WHY SYSTEM SAFETY CONCEPTS HAVE NOT BEEN WIDELY ADOPTED

At least one other author expected a more widespread adoption of system safety concepts beyond the use by the military, aerospace personnel, and nuclear facility designers. He also had to recognize that it wasn't happening. In *The Loss Rate Concept In Safety Engineering*, R.L. Browning (1980) wrote this.

> As every loss event results from the interactions of elements in a system, it follows that all safety is "systems safety." The safety community instinctively welcomed the systems concept when it appeared during the stagnating performance of the mid-1960s, as evidenced by the ensuing freshet of symposia and literature. For a time, it was thought that this seemingly novel approach could reestablish the continuing improvement that the public had become accustomed to; however, this anticipation has not been fulfilled.
>
> Now, some three decades later, although systems techniques continue to find application and development in exotic programs (aerospace, nuclear power) and in the academic community, they are seldom met in the domain of traditional industrial and general safety (p 12).

Although there were countless seminars and a proliferation of papers on system safety, the generalist in the practice of safety seldom adopted system safety concepts. In response to his own question—Why this rejection?—Browning expressed the view that system safety literature and seminars on system safety may have turned off generalist safety professionals because of the "exotica" they usually presented. This author believes that to be so.

Nevertheless, Browning went on to build *The Loss Rate Concept In Safety Engineering* on system safety concepts. He also gave this encouragement:

> We have found through practical experience that industrial and general safety can be engineered at a level considerably below that required by the exotics, using the mathematical capabilities possessed by average technically minded persons, together with readily available input data (p 13).

There is a reality in Browning's observations: System safety literature at the time he wrote his book was loaded with governmental jargon, and it repelled the uninitiated. It made more of the complex hazard analysis and risk assessment techniques requiring extensive knowledge of mathematics and probability theory than it did of concepts and purposes.

Some system safety literature did, and still does, give the appearance of exotica. That has changed. Texts on system safety that are truly primers and slanted toward the neophyte have been written.

PROMOTING THE USE OF SYSTEM SAFETY CONCEPTS

With the hope of generating a further interest by generalist safety professionals in system safety concepts, it is suggested that they concentrate on those basics through which gains can be made in an occupational, environmental, or product design setting and avoid being repulsed by the more exotic hazard/risk assessment methodologies. Ted Ferry said it well in the Preface he wrote for Richard Stephans' book *System Safety for the 21st Century*:

> Professional credentials or experience in "system safety" are not required to appreciate the potential value of the systems approach and system safety techniques to general safety and health practice (p xiii).

To paraphrase Browning—all hazard-related incidents result from interactions of elements in a system. Therefore, all safety is system safety. Therein lies an important idea.

In *Safety Management*, John V. Grimaldi and Rollin H. Simonds (1989) wrote that:

> A reference to system analysis may merely imply an orderly examination of an established system or subsystem (p 287).

Applying system safety as "an orderly examination of an established system or subsystem" to identify, analyze, avoid, eliminate, reduce, or control hazards can be successful in the less complex situations, without using elaborate analytical methods.

Repeating for emphasis—applying the fundamentals of system safety can meet a huge percentage of the provisions of Z10.0 and 45001. For safety generalists who take an interest in system safety concepts, the following reading list is offered from which selections can be made.

RECOMMENDED READING

Clifton Ericson's book is what the title says it is, a *System Safety Primer*. This paperback, published in 2011, is only a 140-page read. Yet, it covers the system safety subject very well as a primer. It is an easy and recommended read.

System Safety for the 21st Century is an update by Richard Stephans of *System Safety 2000* by Joe Stephenson. Stephans followed the advice given to "keep it as a primer." This book begins with a history of and the fundamentals of system safety. Then, the author moves into system safety program planning and management, and system safety analysis techniques. About half of the book is devoted to those techniques. Stephans says:
This book is specifically written for:

- Safety professionals, including people in industrial and occupational safety, system safety, environmental safety, industrial hygiene, health, occupational medicine, fire protection, reliability, maintainability, and quality assurance.
- Engineers, especially design engineers and architects
- Managers and planners
- Students and faculty in safety, engineering, and management (p xv)

A safety generalist would find this book to be a good and not too difficult read.

Basic Guide To System Safety was written by Jeffrey W. Vincoli. These two sentences are taken from the Preface. "It should be noted from the beginning that it is not the intention of the Basic Guide to System Safety to provide any level of expertise beyond that of novice. Those practitioners who desire complete knowledge of the subject will not be satisfied with the information contained on these pages."

Vincoli also says: "The primary focus of this text shall be the advantage of utilizing system safety concepts and techniques as they apply to the general safety arena. In fact, the industrial workplace can be viewed as a natural extension of the past growth experience of the system safety discipline" (p 5).

Vincoli fulfilled his purpose. He has written a basic book on system safety that will serve the novice well.

MIL-STD-882E, the *Standard Practice for System Safety* issued by the Department of Defense, serves well as a primer. It is available on the Internet and can be downloaded without charge, at https://www.system-safety.org/Documents/MIL-STD-882E.pdf.

The Federal Aviation Administration's System Safety Handbook is also on the Internet as a free download. This is really a training manual. There are 17 Chapters and 10 Appendices—all individually downloadable as a separate PDF file. Enter Federal Aviation Administration System Safety Handbook into a search Engine, or go to https://www.faa.gov/regulations_policies/handbooks_manuals/aviation/risk_management/ss_handbook/.

The Loss Rate Concept In Safety Engineering, by R.L. Browning (1980), is a little but good book to which this author has referred several times. Browning believes that one can apply system safety concepts in an industrial setting without necessarily delving into exotic mathematics. He builds on "The Energy Cause Concept" and works through qualitative and quantitative analytical systems.

System Safety Engineering and Management, Second Edition, by Harold E. Roland and Brian Moriarty is a good but more involved book. It provides an extensive review of the concepts of system safety and their methods of application. An overview of a system safety program is given. The descriptions of several analytical techniques are valuable. For the application of some of them, quite a bit of knowledge about mathematics is necessary.

CONCLUSION

For this chapter, the principle intent is to establish that fundamental system safety concepts can be applied by generalists in the practice of safety to meet the provisions in Z10.0 and 45001; outline "The System Safety Idea"; and encourage generalists who have not adopted system safety concepts to commence to do so.

This author sincerely believes that generalists in the practice of safety can learn from system safety successes and be more effective in their work through adopting system safety concepts. Their application in the occupational, environmental, and product safety settings would result in significant reductions in incidents having adverse effects.

In summation: The entirety of purpose of those responsible for safety, regardless of their titles, is to manage their endeavors with respect to hazards so that the risks deriving from those hazards are acceptable.

REFERENCES

Browning, R.L. *The Loss Rate Concept In Safety Engineering*. New York: Marcel Dekker, 1980.

BS OHSAS 18001:2007. *Occupational Health and Safety Management Systems—Requirements*. London, UK: BSI Group, 2007.

Clements, P.L. and J.S. Rodney. *System Safety and Risk Management, NIOSH Instructional Module, A Guide for Engineering Educators*. Cincinnati, OH: National Institute for Occupational Safety and Health, 1998.

EN ISO 12100-2010. *Safety of Machinery. General Principles for Design. Risk Assessment and Risk Reduction*. Geneva, Switzerland: International Organization for Standardization, 2010.

Ericson, C.A. II. *System Safety Primer*. Available through Internet book sellers. Self-published, 2011.

Grimaldi, J.V. and R.H. Simonds. *Safety Management*. Homewood, IL: Irwin, 1989.

irsst. *Machine Safety: Prevention of Mechanical Hazards*. Quebec, Canada: The Institute for Research for Safety and Security at Work and The Commission for Safety and Security at Work in Quebec. Available at https://www.irsst.qc.ca/en/publications-tools/publication/i/100361/n/machine-safety-prevention-of-mechanical-hazards-fixed-guards-and-safety-distances-rg-597. Accessed June 3, 3019. 2009.

MIL-STD-882E. Department of Defense—Standard Practice System Safety, It is available on the Internet and can be downloaded. Available at https://www.system-safety.org/Documents/MIL-STD-882E.pdf. Accessed June 3, 2019. 2012.

Roland, H.E. and B.Moriarty. *System Safety Engineering and Management*, Second Edition. Hoboken, NJ: John Wiley & Sons, 1990.

Stephans, R.A. *System Safety in the 21st Century*. Hoboken, NJ: John Wiley & Sons, 2004.

FURTHER READING

ANSI/ASSP Z10.0-2019. *Occupational Health and Safety Management Systems*. Park Ridge, IL: American Society of Safety Professionals, 2019.

ANSI/ASSP/ISO 45001-2018. *Occupational Health and Safety Management Systems— Requirements with Guidance for Use*. Park Ridge, IL: American Society of Safety Professionals, 2018.

ANSI/ASSE Z690.1-2011. *Vocabulary for Risk Management*. Park Ridge, IL: American Society of Safety Professionals, 2011.

ANSI/ASSE Z690.2-2011. *Risk Management Principles and Guidelines*. Park Ridge, IL: American Society of Safety Professionals, 2011.

ANSI/ASSE Z690.3. *Risk Assessment Techniques*. Park Ridge, IL: American Society of Safety Professionals, 2011.

ANSI/ASSE Z590.2 (R2012). *The American National Standard titled* Criteria for Establishing the Scope and Function of the Professional Safety Position. Park Ridge, IL: The American Society of Safety Professionals, R2012.

ANSI/ASSE Z590.3-2011(R2016). *Prevention through Design: Guidelines for Addressing Occupational Hazards and Risks in Design and Redesign Processes*. Park Ridge, IL: American Society of Safety Professionals, R2016.

Australian Safety and Compensation Council. *Guidance On The Principles Of Safe Design For Work*. Canberra, Australia: Australian Safety and Compensation Council, an entity of the Australian Government, 2006.

CSA Z1002-12. *Occupational Health and Safety—Hazard Identification and Elimination and Risk Assessment and Control*. Toronto, Canada: Canadian Standards Association, 2012.

Environmental Management Systems: An Implementation Guide for Small and Medium Sized Organizations, Second Edition. Available at http://www.fedcenter.gov/_kd/Items/actions.cfm?action=Show&item_id=598&destination=ShowItem. Accessed June 3, 2019. n.d.

Federal Aviation Administration. *System Safety Handbook*. Enter title into a search Engine. Or go to: https://www.faa.gov/regulations_policies/handbooks_manuals/aviation/risk_management/ss_handbook/. Accessed June 3, 2019. December 39, 2000.

OSHA. OSHA's Rule for Process Safety Management of Highly Hazardous Chemicals, *29 CFR 1910.119*. Washington, DC: OSHA. 1992.

The International System Safety Society. *System Safety Analysis Handbook*. (For which Warner Talso and Richard A. Stephans provided stewardship). Unionville, VA: The International System Safety Society, 1999.

Vincoli, J.W. *Basic Guide To System Safety*. Hoboken, NJ: John Wiley & Sons, 1993.

CHAPTER 25

ACHIEVING ACCEPTABLE RISK LEVELS: THE OPERATIONAL GOAL

In the second edition of this book, published in 2014, the case was made that the term "acceptable risk" was more frequently used in standards and guidelines throughout the world. And several examples of that usage were given. In succeeding years, the use of the term has expanded considerably.

A journey into the Internet will show that a goodly number of additional entities in a variety of endeavors have adopted the term "acceptable risk" and made known their definitions. They differ somewhat.

Nevertheless, the premise that achieving zero risk levels in operations is not possible is broadly accepted. In many fields, then, there has been a search for an answer to the question that many have asked—how much risk is acceptable? For those involved in hazard-related entities, this author's answer to that question is provided here.

It was also said in the second edition that a substantial percentage of the personnel who have safety, health, and environmental responsibilities are reluctant to use the term. Evidence of that reluctance often arises in discussions on the development of new or revised standards or technical reports. Possibly, that aversion derives from:

- A lack of awareness of the factors that determine risk—probability of occurrence and severity of outcome of an incident or exposure.
- Aversion to the premise that some risks are acceptable.
- Not being familiar with the literature that convincingly demonstrates that, if a facility exists or an operation proceeds, attaining a zero-risk level is impossible.

Advanced Safety Management: Focusing on Z10.0, 45001, and Serious Injury Prevention,
Third Edition. Fred A. Manuele.
© 2020 John Wiley & Sons, Inc. Published 2020 by John Wiley & Sons, Inc.

- Concern over the subjective judgments made in the risk assessments process and the uncertainties that almost always exist.
- The unavailability of in-depth statistical probability and severity data that would allow precise and numerically accurate risk assessments.
- Insufficient knowledge and experience from which an understanding could be achieved that operational risks are necessarily accepted every day.

However, in recent years, the concept of acceptable risk has been interwoven into internationally applied standards and guidelines for a very broad range of equipment, products, processes, and systems. That has occurred as it has become recognized by many participants in the development of those standards and guidelines that:

- Risk acceptance decisions are made constantly in real-world applications.
- Society benefits if those decisions achieve acceptable risk levels.
- Knowledge has emerged on how to manage with respect to those risks.

This chapter provides a primer from which an understanding of risk and the concept of acceptable risk can be attained. For that purpose:

- A far-reaching premise is presented that is fundamental in dealing with risk.
- Several examples of the use of the term acceptable risk as taken from the applicable literature are given.
- Discussions address the impossibility of achieving zero risk levels.
- The inadequacy of using "minimum risk" as a replacement term for "acceptable risk" is explored.
- A Risk Assessment Matrix for use in determining acceptable risk levels is offered as an example.
- The ALARP concept is discussed (as low as reasonably practicable) with an example of how the concept is applied in achieving an acceptable risk level.

FUNDAMENTAL PREMISE

An all-encompassing premise follows that is basic to the work of all personnel who give counsel to avoid injury and illness, and property and environmental damage.

> The entirety of purpose of those responsible for safety, regardless of their titles, is to manage their endeavors with respect to hazards so that the risks deriving from those hazards are acceptable.

That premise is supported by this hypothesis. If there are no hazards, if there is no potential for harm, risks of injury or damage can not arise. The focus of the practice of safety is on hazards and the risks that derive from them to achieve acceptable risk levels.

PROGRESSION WITH RESPECT TO USE OF THE TERM ACCEPTABLE RISK

The term acceptable risk is becoming the norm. A list of selected references, intentionally lengthy, is presented here to show how broadly the concept of acceptable risk has been adopted.

- In *Of Acceptable Risk: Science and the Determination of Safety* (1976), Lowrance wrote that:

    A thing is safe if it risks are judged to be acceptable. (p. 8).

- This next citation, from a court decision made in 1980, developed significant importance because it has given long term guidance with respect to Department of Labor policy, but particularly for OSHA, and to the work done at the National Institute for Occupational Safety and Health (NIOSH).

    The Supreme Court's benzene decision of 1980 states that "before he can promulgate any permanent health or safety standard, the Secretary [of Labor] is required to make a threshold finding that a place of employment is unsafe--in the sense that significant risks are present and can be eliminated or lessened by a change in practices" (IUD v. API, 448 U.S. at 655). The Court broadly describes the range of risks OSHA might determine to be significant: It is the Agency's responsibility to determine in the first instance what it considers to be a "significant" risk. *Some risks are plainly acceptable and others are plainly unacceptable.* [Emphasis added].

    If, for example, the odds are one in a billion that a person will die from cancer by taking a drink of chlorinated water, the risk clearly could not be considered significant. On the other hand, if the odds are one in a thousand that regular inhalation of gasoline vapors that are 2% benzene will be fatal, a reasonable person might well consider the risk significant and take the appropriate steps to decrease or eliminate it. The Court further stated, 'The requirement that a "significant" risk be identified is not a mathematical straitjacket. Although the Agency has no duty to calculate the exact probability of harm, it does have an obligation to find that a significant risk is present before it can characterize a place of employment as "unsafe"' and proceed to promulgate a regulation.

- ISO/IEC Guide 51: 1999, *Safety aspects—Guidelines for their inclusion in standards,* was first published in 1990. This paper was issued by the International Organization for Standardization (ISO) and the International Electrotechnical Commission (IEC) to provide standardized terms and definitions to be used in standards for "any safety aspect related to people, property, or the environment." A second edition was issued in 1999, from which the following definitions are cited. (Work on a third edition is in progress.)

    - *Safety*: freedom from unacceptable risk. (Section 3.1)
    - *Tolerable risk*: risk which is accepted in a given context based on the current values of society. (Section 3.7)

- In the World Health Organization (WHO) publication *Water Quality: Guidelines, Standards and Health* (2001), Chapter 10 is titled "Water Quality: Guidelines, Standards and Health." The following guidelines are given in determining acceptable risk. A risk is acceptable when:
  - It falls below an arbitrary defined probability.
  - It falls below some level that is already tolerated.
  - It falls below an arbitrary defined attributable fraction of total disease burden in the community.
  - The cost of reducing the risk would exceed the costs saved.
  - The cost of reducing the risk would exceed the costs saved when the 'costs of suffering' are also factored in.
  - The opportunity costs would be better spent on other, more pressing, public health problems.
  - Public health professionals say it is acceptable.
  - The general public say it is acceptable (or more likely, do not say it is not).
  - Politicians say it is acceptable.
- In a March 2003 entry into the Internet, OSHA set forth its requirements for organizations to obtain certification under its Voluntary Protection Programs (VPP). An excerpt follows.

  Worksite Analysis. A hazard identification and analysis system must be implemented to systemastically identify basic and unforeseen safety and health hazards, evaluate their risks, and prioritize and recommend methods to eliminate or control hazards to an acceptable level of risk.

- The following appears in ANSI/ASSE Z244.1 (2003) *American National Standard for Control of Hazardous Energy: Lockout/Tagout and Alternative Methods,* (Reaffirmed, January 2009).

  A.2 Acceptable level of risk: If the evaluation in A.1.6 determines the risk to be acceptable, then the process is completed....

- This definition appears in the United Nations publication UN ISDR, *Terminology: Basic Terms of Disaster Risk Reduction,* 2009, p. 1:

  Acceptable Risk: The level of loss a society or community considers acceptable given existing social, economic, political, cultural, technical and environmental conditions.

- In the Sci-Tech Dictionary (2009; Answers, on the Internet), a definition of acceptable risk as used in geology is given.

  Acceptable Risk: (geophysics) In seismology, that level of earthquake effects which is judged to be of sufficiently low social and economic consequence, and which is useful for determining design requirements in structures or for taking certain actions.

PROGRESSION WITH RESPECT TO USE OF THE TERM ACCEPTABLE RISK **515**

- This is the definition of acceptable risk as shown in the Australian/New Zealand AS/NZS 4360: 2004 *Risk management standard*.

  1.3.16 Risk acceptance: An informed decision to accept the consequences and the likelihood of a particular risk.

- The Office of Hazards Materials Safety in the Pipeline and Hazardous Materials Safety Administration (PHMSA) in the U.S. Department of Transportation, has issued *Risk Management Definitions* (2005) from which the following definition is taken.

  Acceptable Risk: An acceptable level of risk for regulations and special permits is established by consideration of risk, cost/benefit and public comments. Relative or comparative risk analysis is most often used where quantitative risk analysis is not practical or justified. Public participation is important in a risk analysis process, not only for enhancing the public's understanding of the risks associated with hazardous materials transportation, but also for insuring that the point of view of all major segments of the population-at-risk is included in the analyses process.

  Risk and cost/benefit analysis are important tools in informing the public about the actual risk and cost as opposed to the perceived risk and cost involved in an activity. Through such a public process PHMSA [Pipeline and Hazardous Materials Safety Administration] establishes hazard classification, hazard communication, packaging, and operational control standards.

- In the 2007 revision of the British Standards Institution publication, BS OHSAS 18001:2007, *Occupational health and safety: Management systems—Requirements,* a significant change was made as follows.

  The term "tolerable risk" has been replaced by the term "acceptable risk." (Section 3.1)

- In the International Electrotechnical Commission document, IEC 60601-1-9, International Standard for *Environmentally Conscious Design of Medical Equipment*, published in July 2007, the Introduction says:

  The standard includes the evaluation of whether risks are acceptable (risk evaluation).

- This reference is a highly recommended document issued in 2009 by the Institute for Research for Safety and Security at Work and The Commission for Safety and Security at Work in Quebec, Canada. The title is "Machine Safety: Prevention of mechanical hazards." In the Introduction, they say—

  When machine-related hazards ... cannot be eliminated through inherently safe design, they must then be reduced to an acceptable level.

- In ANSI B11.0-2010. *American National Standard: Safety of Machinery— General Requirements and Risk Assessment*, a broad ranging standard on machinery and risk assessment, the following is written.

  Acceptable risk: A risk level achieved after protective measures have been applied. It is a risk level that is accepted for a given task (hazardous situation) or hazard. For the purpose of this standard the terms "acceptable risk" and "tolerable risk" are considered to be synonymous.

- This definition is taken from ANSI/PMMI B155.1 (2011), *American National Standard for Safety Requirements for Packaging Machinery and Packaging-Related Converting Machinery*.

  Acceptable risk: risk that is accepted for a given task or hazard. For the purpose of this standard the terms "acceptable risk" and "tolerable risk" are considered synonymous.

- In ANSI/ASSE Z590.3-2011 (2016), titled *Prevention Through Design: Guidelines for Addressing Occupational Hazards and Risks in the Design and Redesign Processes*, the following appear.

Definitions
Acceptable Risk: That risk for which the probability of an incident or exposure occurring and the severity of harm or damage that may result are as low as reasonably practicable (ALARP) in the setting being considered.

ALARP: As low as reasonably practicable. That level of risk which can be further lowered only by an increase in resource expenditure that is disproportionate in relation to the resulting decrease in risk. (p. 12). [This author would now amend this definition as follows: That low level of risk which can be further reduced only by an expenditure that is disproportionate in relation to the resulting decrease in risk that would be achieved.]

- There are now several books on acceptable risk.
- A television series was titled "Acceptable Risk".
- At http://definedterm.com/acceptable_risk you can find a paper that lists at least six definitions of acceptable risk.
- There are a dozen more definitions at: https://www.bing.com/search?q= acceptable+risk+definition&form=EDGSPH&mkt=en-us&httpsmsn=1&refig =3df3f92163a54547a279b17d40a1690f&PC=HCTS&sp=3&ghc=1&qs=AS& pq=acceptable+risk+&sk=LS1AS1&sc=8-16&cvid=3df3f92163a54547a279 b17d40a1690f&cc=US&setlang=en-US (n.d.).

WITH RESPECT TO THE FOREGOING CITATIONS

Some commonly used definitions of risk allow top management to decide what the tolerable risk level is to be for an organization. Be careful of such definitions. Worker fatality data indicates that in some organizations, perhaps only a few, the fatality rate considered tolerable is quite high.

Statistics as shown in the following appeared in a report issued by the Center for Safety & Health Sustainability in August 2017 and titled "The Need for Standardized Sustainability Reporting Practices: Issues Relating to Corporate Disclosure of Information on Occupational Health & Safety Performance"

KEY FINDINGS (p. 4)
The number of reporters providing information on fatalities increased (from 38 reporters to 50).

- 12 reported more than one work-related death.
- 4 reported 10 or more fatalities.
- 2 reported more than 20 work-related fatalities (20 and 27).
- 1 reported a total of 63 deaths over a 3-year period.
- No organization specifically mentioned fatalities related to occupational diseases.

It is difficult to image that the deaths of 63 workers over three years—an annual average of 21 fatalities—are considered tolerable by management. That is a big number. Acceptable risks should be as low as reasonably practicable and tolerable in the setting being considered.

SUMMARY TO THIS POINT

As the cited references illustrate, the concept of acceptable risk has been broadly adopted internationally, and the term acceptable risk is becoming normal in use. Those who assert that they are safety professionals and are still reluctant to adopt the concept of acceptable risk, would do well to recognize that they have an obligation to be current with respect to the state-of-the-art and reconsider their views.

THE NATURE AND SOURCE OF RISK

Risk is expressed as an estimate of the probability of a hazard-related incident or exposure occurring and the severity of harm or damage that could result. All risks with which safety professionals deal derive from hazards. There are no exceptions.

518 ACHIEVING ACCEPTABLE RISK LEVELS: THE OPERATIONAL GOAL

A hazard is defined as the potential for harm. Hazards include all aspects of technology and activity that produce risk. Hazards include the characteristics of things (equipment, dusts, chemicals, etc.) and the actions or inactions of people.

The probability aspect of risk is defined as the likelihood of an incident or exposure occurring that could result in harm or damage—for a selected unit of time, events, population, items, or activity being considered.

The severity aspect of risk is defined as the degree of harm or damage that could reasonably result from a hazard-related incident or exposure.

Comparable statements and definitions appear in much of the current literature on risk and acceptable risk. One resource has been chosen for citation here because of its broad implications. The following excerpts are taken from the *Framework for Environmental Health Risk Management* issued by The Presidential/Congressional Commission on Risk Assessment and Risk Management (1997).

What is "risk"

Risk is defined as the probability that a substance or situation will produce harm under specified conditions. Risk is a combination of two factors:

- The **probability** that an adverse event will occur;
- The **consequences** of the adverse event.

Risk encompasses impacts on public health and on the environment, and arises from **exposure** and **hazard**. Risk does not exist if exposure to a harmful substance or situation does not or will not occur. Hazard is determined by whether a particular substance or situation has the potential to cause harmful effects. Risk ... is the probability of a specific outcome, generally adverse, given a particular set of conditions.

Residual risk ... is the health risk remaining after risk reduction actions are implemented, such as risks associated with sources of air pollution that remain after implementation of maximum achievable control technology.

A ZERO RISK LEVEL IS NOT ATTAINABLE

It has long been recognized by researchers, authors, and practitioners that a zero risk level is not attainable. If a facility exists or an activity proceeds, it is not possible to realistically conceive of a situation in which there is no probability of an adverse incident or exposure occurring.

William W. Lowrance was one of the most significant and influential authors on the concept of acceptable risk. Lowrance writes in his previously cited book *Of Acceptable Risk: Science and the Determination of Safety* that:

> Nothing can be absolutely free of risk. One can't think of anything that isn't, under some circumstances, able to cause harm. Because nothing can be absolutely free of risk, nothing can be said to be absolutely safe. There are degrees of risk, and consequently there are degrees of safety. (p. 8)

Similar comments are made in ISO/IEC Guide 51: *Safety Aspects—Guidelines for their inclusion in standards*, which was also previously cited. Under the caption "The Concept of Safety" (Section 5), this appears.

There can be no absolute safety: some risk will remain, defined in this Guide as residual risk. Therefore, a product, process or service can only be relatively safe. Safety is achieved by reducing risk to a tolerable level, defined in this Guide as tolerable risk.

In the real world, attaining a zero-risk level is not possible, whether in the design or redesign processes or in facility operations. That said, after risk avoidance, elimination, reduction, or control measures are taken, the residual risk must be acceptable, as judged by the decision-makers.

Also, it is necessary to recognize that inherent risks that are acceptable and tolerable in some occupations would not be tolerable in others. For example, some work conditions considered tolerable in deep sea fishing (e.g., a pitching and rolling work floor, the ship's deck) would not be tolerable elsewhere.

In other and opposite situations—such as for certain chemical or radiation exposures that function at higher than commonly accepted exposure levels—the residual risk will be judged as unacceptable and operations at those levels would not be permitted.

Nevertheless, society accepts continuation of certain operations in which the occupational and environmental risks are high. That is demonstrated by the fatality rate data published annually by the U.S. Bureau of Labor Statistics in its National Census of Fatal Occupational Injuries (2018). For the Bureau, the fatality rate is the rate per 100,000 full-time equivalent workers. For all private industries in 2017, the national fatality rate was 3.5.

Although fatality rates among all employment categories are highest for the occupations shown in Chart 25.1, the public has not demanded discontinuation of the operations in which they occur. Inherent risks in the high-hazard categories are considered "tolerable" in their settings in relation to the benefit attained.

Occupation	Fatality Rate
Fishers and related fishing workers	98.8
Logging workers	84.3
Aircraft pilots and flight engineers	48.6
Roofers	45.2
Refuse and recyclable materials collectors	35.0
Structural iron and steel workers	33.4
Drivers/sales workers and truck drivers	26.8
Farmers, ranchers, and other agricultural managers	24.0
First-line supervisors of landscaping, lawn service, and grounds keeping workers	20.1
Electrical power-line installers and repairers	18.7

Chart 25.1 Civilian Occupations with High Fatal Work Injury Rates, 2017 (BLS)

OPPOSITION TO IMPOSED RISKS

Literature is abundant on the resistance people have to being exposed to risks they believe are imposed on them. For some, their aversion to adopting the acceptable risk concept derives from their view that imposed risks are objectionable and are to be rebelled against. Conversely, they accept the significant risks of activities in which they choose to engage (e.g., skiing, pot smoking, bicycle riding, driving an automobile).

This idea needs exploration, which commences with a statement that can withstand a test of good logic. Richard A. Stephans (2004) says in *System Safety for the 21st Century* that:

> The safety of an operation is determined long before the people, procedures, and plant and hardware come together at the worksite to perform a given task. (p. 13)

Start from the beginning in the process of creating a new facility as the creditability of Stephan's statement is validated. Consider, first, a site survey for ecological considerations, soil testing, and then move into the design, construction and fitting out of a facility.

Thousands of safety-related decisions are made in the design processes that result in an imposed level of risk. Usually, those decisions meet (or exceed) applicable safety-related codes and standards with respect to issues such as the contour of exterior grounds, sidewalks and parking lots; building foundations; facility layout and configuration; floor materials; roof supports; process selection and design; determination of the work methods; aisle spacing; traffic flow; hardware; equipment; tooling; materials to be used; energy choices and controls; lighting, heating, and ventilation; fire protection; and environmental concerns.

Designers and engineers make decisions on the foregoing during the original design processes. Those decisions establish what the designers implicitly believe to be acceptable risk levels. Thus, the occupational and environmental risk levels have been largely imposed before a facility begins operations.

Indeed, if persons employed in such settings conclude that the imposed risks are not acceptable, communication systems should be in place to allow them to express their views and to have them resolved.

MINIMUM RISK AS AN INADEQUATE SUBSTITUTE FOR ACCEPTABLE RISK

Invariably, those who oppose the use of the term acceptable risk offer substitutions. One frequent suggestion is that designers and operators should achieve minimum risk levels, or minimize the risks? That sounds good, until application of the terms is explored. Minimum means—the least amount or the lowest amount.

CONSIDERATIONS IN DEFINING ACCEPTABLE RISK

Minimization means—to reduce something to the lowest possible amount or degree. Note particularly, "least" and "lowest possible."

Assume that the threshold limit value (TLV) for a chemical is 4 parts per million. For 10 million dollars, a system can be designed, built and installed that will operate at 2 parts per million—and be the best in the world. But, for an additional 100 million dollars, a 1 part per million exposure level can be achieved. Increase the investment to 200 million dollars and the result is an exposure level of 0.1 parts per million.

At 2 parts per million, the exposure level is acceptable, but not the minimum because a lower exposure level can be achieved. At that level, the exposure level is not the least or the lowest possible.

Requiring that systems be designed and operated to minimum risk levels—that risks be minimized—is impractical because the investments necessary to do so may be so high that the cost of the product required to recoup the investment and make a reasonable profit would not be competitive in the market place.

CONSIDERATIONS IN DEFINING ACCEPTABLE RISK

If the residual risk for a task or operation cannot be zero, for what risk level does one strive? It is the norm that resources are always limited. There is never enough money to address every hazard identified. Such being the case, safety professionals have a responsibility to give counsel so that the greatest good to society, to employees, to employers, and to the product users is attained through applying available resources practicably and economically to obtain acceptable risk levels.

Determining whether a risk is acceptable requires consideration of many variables. An additional excerpt from ISO/IEC Guide 51, Section 5, helps in understanding the concept of designing and operating for risk levels as low as reasonably practicable.

> Tolerable risk [acceptable risk] is determined by the search for an optimal balance between the ideal of absolute safety and the demands to be met by a product, process or service, and factors such as benefit to the user, suitability for purpose, cost effectiveness, and conventions of the society concerned.

Understanding cost effectiveness has become a more important element in risk acceptance decision-making. That brings the discussion to ALARA and ALARP.

ALARA and ALARP are commonly used acronyms in risk assessment and risk reduction literature. ALARA stands for as low as reasonably achievable. ALARP is short for as low as reasonably practicable.

The use of the ALARA concept as a guideline originated in the atomic energy field. This is taken from the U.S. Nuclear Regulation Commission website.

> As defined in Title 10, Section 20.1003, of the *Code of Federal Regulations* (10 CFR 20.1003), ALARA is an acronym for "as low as (is) reasonably achievable," which means making every reasonable effort to maintain

exposures to ionizing radiation as far below the dose limits as practical, consistent with the purpose for which the licensed activity is undertaken, taking into account the state of technology, the economics of improvements in relation to state of technology, the economics of improvements in relation to benefits to the public health and safety, and other societal and socioeconomic considerations, and in relation to utilization of nuclear energy and licensed materials in the public interest. (*Reviewed/Updated, March 21, 2019*)

Implications in the foregoing that executives are to "make every reasonable effort to maintain exposures to ionizing radiation as far below the dose limits as practical" provides conceptual guidance in striving to achieve acceptable risk levels in all classes of operations.

ALARP seems to be an adaptation from ALARA. But ALARP has become the more frequently used term for operations other than atomic and it appears more often in the literature.

ALARP is defined as that level of risk which can be further lowered only by an increment in resource expenditure that is disproportionate in relation to the resulting decrement of risk.

Concepts embodied in the terms ALARA and ALARP apply to the design of products, facilities, equipment, work systems and methods, and environment controls.

In the real world of decision-making, benefits represented by the amount of risk reduction to be obtained and the costs to achieve those reductions become important factors. Trade-offs are frequent and necessary. An appropriate goal is to have the residual risk be as low as reasonably achievable. Paraphrasing the terms contained in the definition of ALARA previously gives help in explaining the process:

1. Reasonable efforts are to be made to identify, evaluate, and eliminate or control hazards so that the risks deriving from those hazards are acceptable.
2. In the design and redesign processes for operating systems and for the work methods, risk levels for injuries and illnesses to employees and the public, and property and environmental damage, are to be as far below what would be achieved by applying current standards and guidelines as is economically practicable.
3. For items 1 and 2, decision-makers are to take into consideration;
   - The purpose of the undertaking.
   - The state of the technology.
   - The costs of improvements in relation to benefits to be obtained.
   - Whether the expenditures for risk reduction in a given situation could be applied elsewhere with greater benefit.

Since resources are always limited, spending an inordinate amount of money to reduce the risk only a little through costly engineering and redesign is inappropriate, particularly if that money could be better spent otherwise. That premise can be demonstrated through an example that uses a risk assessment matrix as a part of the decision-making.

THE ALARP PRINCIPLE

ALARP promotes a management review, the intent of which is to achieve acceptable risk levels. Practical, economic risk trade-offs are frequent and necessary in the benefit/cost deliberations that take place when determining whether the costs to reduce risks further can be justified "by the resulting decrement in risk."

Several depictions of the ALARP concept begin with an inverted triangle, the purpose being to indicate that the risk is greater at the top and much less at the bottom. Figure 25.1 shows the concept combined with the elements in the risk assessment matrix. ALARP "at a glance," (n.d.) issued by the Health and Safety Executive in the United Kingdom, was a resource for this figure.

RISK ASSESSMENT MATRICES

A Risk Assessment Matrix that assigns numbers to risk levels at the beginning serves well for the purposes of this chapter and for the following demonstration showing the application of the ALARP principle. This matrix is recommended because of experience with it. Many matrices use verbal categories to establish risk levels—such as Low, Moderate, Serious, and High—as does this one.

But, in operations in which this matrix was used at the shop floor level, employees said that it was easier to relate risk categories to each other, such as a Moderate risk level to a Serious risk level, by first establishing in their minds the relationship between numbers—such as 6 to 12.

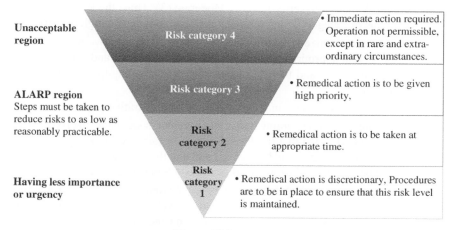

Figure 25.1 ALARP

524 ACHIEVING ACCEPTABLE RISK LEVELS: THE OPERATIONAL GOAL

An example follows illustrating how a team used the matrix within a risk assessment process and applied the ALARP concept to make a decision about acceptable risk. At the end, the team took the opportunity to suggest to management that, rather than spend money on additional risk reduction through major physical process changes, they would prefer that the money be spent to build a wellness center.

1. A chemical operation was built 15 years ago. While engineering modifications have been made in the system over the years, management is aware that its operations are not at the current state-of-the-art.
2. A risk assessment team was convened to consider the chemically related risks in a particular process in the overall system.
3. In the deliberations, the group refers to its established Hierarchy of Controls, which follows.
   a. Risk avoidance.
   b. Eliminate or reduce risks in a redesign process.
   c. Reduce risks by substituting less hazardous methods or materials.
   d. Incorporate safety devices.
   e. Provide warning systems.
   f. Apply administrative controls (work methods, training, work scheduling, etc.).
   g. Provide personal protective equipment.
4. The group first considers and holds open the possibility of completely redesigning and replacing the process. Substitution of materials or methods is considered and it is determined that action on such opportunities have already been taken. Safety devices and warning systems have been updated and are considered state-of-the-art. Maintenance is considered superior.
5. Occurrence probability for a chemically related illness is judged to be Moderate (3) and the Severity Level is Moderate (3). Thus, the Risk Score is 9, which is in Category 3 and remedial action is to be given high priority.
6. The team recognized that to reduce the risk further, improvements must be made so that appropriate training is given and repeated, and standard operating procedures and the use of personal protective equipment are rigidly enforced.
7. Management agrees to fund the necessary administrative improvements.
8. Assuming that these administrative improvements are made, the risk assessment group decided that the Probability of occurrence of an illness from a chemical exposure would be Low (2) and that the Severity of harm expected would be Low (2). Thus, the Risk Score is 4—in Category 1.
9. Re-engineering and replacing the process would reduce the Probability level to Very Low (1) and the Severity level to Very Low (1), thereby achieving a Risk Score of 1, which also is in Category 1. The estimated cost of completely redesigning and replacing the process—$1,500,000—was

considered disproportionate with respect to the amount of risk reduction to be obtained.
10. The risk assessment group took the opportunity to say to management that they would prefer having money spent on a wellness center.

DEFINING ACCEPTABLE RISK

Risk acceptance is a function of many factors and varies considerably across industries, e.g., mining vs. medical devices vs. farming. Even at locations of a single global company, the acceptable risk levels can vary. The culture dominant in a company and the culture of a country in which a facility is domiciled play an important role in risk acceptability as has been experienced by colleagues working in global companies.

Training, experience, and resources can also influence acceptable risk levels. Risk acceptability is also time dependent, in that what is acceptable today may not be acceptable tomorrow, next year, or the next decade.

A sound and workable definition of acceptable risk must encompass hazards, risks, probability, severity, and economic considerations. This author believes that the definition of acceptable risk included in the previously mentioned standard *ANSI/ASSE* Z590.3-2011 (R2016) represents the most practical usage of the term developed in recent several years. It is repeated here.

> **Acceptable Risk:** That risk for which the probability of an incident or exposure occurring and the severity of harm or damage that may result are as low as reasonably practicable (ALARP) in the setting being considered.

All of the foregoing having been said, this author states that, after applying all of the steps in a Hierarchy of Controls, the test for an acceptable risk level should be having the risk fall in the "Low Risk" Category shown in Figure 25.2 but not higher than the "Moderate Risk" Category, temporarily. There could be exceptions, such as for a rescue mission.

One of the most interesting definitions of acceptable risk found on the Internet that is compatible with the definition of ALARP is placed there by Encyclopedia.com. This definition, as follows, was accessed on June 21, 2019 at https://www.encyclopedia.com/education/encyclopedias-almanacs-transcripts-and-maps/acceptable-risk

Encyclopedia.com—Acceptable Risk

> The term "acceptable risk" describes the likelihood of an event whose probability of occurrence is small, whose consequences are so slight, or whose benefits (perceived or real) are so great, that individuals or groups in society are willing to take or be subjected to the risk that the event might occur. The concept of acceptable risk evolved partly from the realization that absolute

526   ACHIEVING ACCEPTABLE RISK LEVELS: THE OPERATIONAL GOAL

safety is generally an unachievable goal, and that even very low exposures to certain toxic substances may confer some level of risk. The notion of virtual safety corresponding to an acceptable level of risk emerged as a risk management objective in cases where such exposures could not be completely or cost-effectively eliminated.

Acceptable risk, then, must have a probability that is "small" and "whose consequences are slight". That would place acceptable risk in the low-risk category as in Figure 25.2.

Risk assessment matrix

Severity Levels and Values	Occurrence Probabilities and Values				
	Unlikely (1)	Seldom (2)	Occasional (3)	Likely (4)	Frequent (5)
Catastrophic (5)	5	10	15	20	25
Critical (4)	4	8	12	16	20
Marginal (3)	3	6	9	12	15
Negligible (2)	2	4	6	8	10
Insignificant (1)	1	2	3	4	5

Numbers were intuitively derived. They are qualitative, not quantitative. They have meaning only in relation to each other.

Incident or Exposure Severity Descriptions

Catastrophic: One or more fatalities, total system loss and major business down time, environmental release with lasting impact on others with respect to health, property damage, or business interruption.
Critical: Disabling injury or illness, major property damage and business down time, environmental release with temporary impact on others with respect to health, property damage, or business interruption.
Margina   Medical treatment or restricted work, minor subsystem loss or property damage, environmental release triggering external reporting requirements.
Negligible: First aid or minor medical treatment only, non serious equipment or facility damage, environmental release requiring routine cleanup without reporting.
Insignificant: Inconsequential with respect to injuries or illnesses, system loss or down time, or environmental release.

Incident or Exposure Probability Descriptions

Unlikely: Improbable, unrealistically perceivable.
Seldom: Could occur but hardly ever.
Occasional: Could occur intermittently.
Likely: Probably will occur several times.
Frequent: Likely to occur repeatedly.

Risk Levels

Combining the Severity and Occurrence Probability values yields a risk score in the matrix. The risks and the action levels are categorized below.

Risk Categories, Scoring, and Action Levels

Category	Risk Score	Action Level
Low risk	to 5	Remedial action discretionary.
Moderate risk	6 to 9	Remedial action to be taken at appropriate time.
Serious risk	1 to 14	Remedial action to be given high priority.
High risk	15 or greater	Immediate action necessary. Operation not permissible except in an unusual circumstance, closely monitored and limited exception with approval of the person having authority to accept the risk.

Figure 25.2   Risk Assessment Matrix

THE STATE OF THE ART IN RISK ASSESSMENT

Safety professionals must understand that risk assessment is as much an art as science and that subjective judgments—educated, to be sure—are made on incident or exposure probability and the severity of outcome to arrive at a risk category. Also, it must be recognized that economically applicable risk assessment methodologies have not been developed to resolve all risk situations.

As an example, this author was asked "How would you assess the cumulative risk in an operation in which there was an unacceptable noise level and toluene was used in the process?" It was hoped that search into relative resource material, such as EPA's (May 2003) *Framework for Cumulative Risk Assessment* would provide an answer. That inquiry was not successful. EPA is cautionary about cumulative risk assessment methods. Their paper says:

> It should be acknowledged by all practitioners of cumulative risk assessment that in the current state of the science there will be limitations in methods and data available. (p. 31)
>
> Finding a common metric for dissimilar risks is not an analytical process, because some judgments should be made as to how to link two or more separate scales of risks. These judgments often involve subjective values, and because of this, it is a deliberative process. (p. 55)
>
> Calculating individual stressor risks and then combining them largely presents the same challenges as combination toxicology but also adds some statistical stumbling blocks. (p. 66)

Where multiple, diverse hazards exist, the practical approach is to treat each hazard independently, with the intent of achieving acceptable risk levels for all. In the noise and toluene example, the hazards are indeed independent. In complex situations, or when competing solutions to complex systems must be evaluated, the assistance of specialists with knowledge of more sophisticated risk assessment methodologies such as Hazard and Operability Analysis (HAZOP) or Fault Tree Analysis (FTA) may be required. However, for most applications, this author does not recommend that diverse risks be summed through what could be a questionable methodology.

CONCLUSION

Risk acceptance is the deliberate decision to assume a risk. This author postulates that an assumed risk is an acceptable risk if the probability of a hazard-related incident or exposure occurring and the severity of harm or damage that may result are as low as reasonably practicable, and tolerable in a given situation. In an ideal world, all personnel who are impacted would be involved in or be informed of risk acceptance decisions.

Use of the term acceptable risk has arrived. It has become a norm. In organizations with advanced safety management systems, that idea—achieving practicable and acceptable risk levels throughout all operations—is a cultural value. It is suggested that safety professionals adopt the concept of attaining acceptable risk levels as a goal to be embedded in every risk avoidance, elimination, reduction, or control action proposed. To achieve that goal, safety professionals must educate others on the beneficial effects of applying the concept.

Also, safety professionals must have the capability to work through the greatly differing views people can have about risk levels, incident and exposure probabilities, and severity potential. Workers views about risks should be considered for their value. With respect to environmental risks, community views must be taken into consideration as well.

In arriving at acceptable risk levels where the hazard/risk scenarios are complex and risk assessments are to be made, it is best to gather a team of experienced personnel for their contributions and for their buy-in to the conclusions.

REFERENCES

Additional Resources and Definitions on Acceptable Risk. Accessed June 3, 2019 (n.d.).

ALARP "at a glance." London, UK: Health and Safety Executive. Available at http://www.hse.gov.uk/risk/theory/alarpglance.htm. Accessed June 3, 2019 (n.d.).

ANSI B11.0-2010. *Safety of Machinery—General Safety Requirements and Risk Assessments*. POB 690905 Houston, TX 77269, USA: Secretariat and Accredited Standards Developer: B11 Standards, Inc.

ANSI/ASSE Z244.1-2003 (R2009). American National Standard for *Control of Hazardous Energy: Lockout/Tagout and Alternative Methods*. Park Ridge, IL: American Society of Safety Professionals.

ANSI/ASSE Z590.3-2011 (R2016). *Prevention Through Design: Guidelines for Addressing Occupational Hazards and Risks in Design and Redesign Processes*. Des Plaines, IL: American Society of Safety Engineers, 2016.

ANSI/PMMI B155.1-2011. American National Standard for *Safety Requirements for Packaging Machinery and Packaging-Related Converting Machinery*. Arlington, VA: Packaging Machinery Manufacturers Institute.

AS/NZS 4360: 2004. *Risk Management Standard*. Strathfield, NSW, Australia: Standards Association of Australia.

BS OHSAS 18001:2007. *Occupational Health and Safety Management Systems—Requirements*. London, UK: British Standards Institution (BSI).

Framework for Cumulative Risk Assessment. EPA/630/P-02/001F. Washington, DC: Risk Assessment Forum, U.S. Environmental Protection Agency, May 2003.

Framework for Environmental Health Risk Management. Washington, DC: The Presidential/Congressional Commission on Risk Assessment and Risk Management, 1997.

Hunter, P.R. and L. Fewtrell, "Water quality: guidelines, standards and health." In *Water Quality: Guidelines, Standards and Health*, The World Health Organization (WHO), Lorna Fewtrell and Jamie Bartram, eds. London, UK: IWA Publishing, 2001.

ISO/IEC Guide 51:1999(E), *Safety Aspects—Guidelines for their Inclusion in Standards*. Geneva, Switzerland: International Organization for Standardization (ISO), 1999. (IEC stands for the International Electrotechnical Commission.)

IUD v. API, 448 U.S. at 655. Supreme Court Decision, Benzene, 1980. Available at https://supreme.justia.com/cases/federal/us/448/607/. Accessed June 3, 2019.

Lowrance, W.F. *Of Acceptable Risk: Science and the Determination of Safety*. Los Altos, CA: William Kaufman, Inc., 1976.

Machine Safety: Prevention of Mechanical Hazards. Quebec, Canada: The Institute for Research for Safety and Security at Work and The Commission for Safety and Security at Work in Quebec, 2009. Available at https://www.irsst.qc.ca/en/publications-tools/publication/i/100361/n/machine-safety-prevention-of-mechanical-hazards-fixed-guards-and-safety-distances-rg-597. Accessed June 3, 2019.

National Census of Fatal Occupational Injuries in 2017. Washington, DC: Bureau of Labor Statistics, US Department of Labor. USDL 08-1182: Released December 18, 2018. Available at https://www.bls.gov/news.release/pdf/cfoi.pdf. Accessed June 3, 2019.

OSHA's Voluntary Protection Program (VPP). Section C. in CSP 03-01-002—TED 8.4 Voluntary Protection Programs (VPP): Policies and Procedures, 2003. Available at https://www.osha.gov/enforcement/directives/csp-03-01-002#chapter3. Accessed June 3, 2019.

Risk Management Definitions. Washington, DC: The U.S. Department of Transportation, Pipeline and Hazardous Materials Safety Administration (PHMSA), 2005. (DOT, Office of Hazardous Materials Safety.) Available at https://definedterm.com/acceptable_risk. Accessed June 3, 2019.

Sci-Tech Dictionary, 2009. Available at https://www.answers.com/search?q=Acceptable+Risk. Accessed June 3, 2019.

Stephans, R.A. *System Safety for the 21st Century*. Hoboken, NJ: John Wiley & Sons, 2004.

"The Need for Standardized Sustainability Reporting Practices: Issues Relating to Corporate Disclosure of Information on Occupational Health & Safety Performance." Brussels, Belgium: Center for Safety & Health Sustainability, August 2017. Available at http://centershs.org/assets/docs/NeedForSustainabilityReporting-Final-August.pdf. Accessed June 3, 2019.

UN ISDR. *Terminology: Basic Terms of Disaster Risk Reduction*. United Nations, NY, 2009. Available at https://www.unisdr.org/we/inform/publications/7817. Accessed June 3, 2019.

FURTHER READING

ALARA—U.S. Nuclear Regulatory Commission. Available at https://www.nrc.gov/reading-rm/basic-ref/glossary/alara.html. Accessed June 21, 2019 (n.d.).

ANSI/ASSP Z10.0-2019. *Occupational Health and Safety Management Systems*. Park Ridge, IL: American Society of Safety Professionals, 2019.

ANSI/ASSP/ISO 45001-2018. *Occupational Health and Safety Management Systems—Requirements with Guidance for Use*. Park Ridge, IL: American Society of Safety Professionals, 2018.

INDEX

Acceptable Level of Risk (ALOR), 25, 301, 302
Acceptable risk, 181, 292
  ALARP, 523
  ALARA, 521-522
  defining, 88, 511, 525
  fatality data, 517
  fundamental, 512
  guidelines, 514
  levels, 12–13
  minimum risk, 520–521
  nature and source of risk, 517–518
  opposition, imposed risks, 520
  progression, 513–516
  risk assessment matrices, 523–525
  state of risk assessment, 525
  zero risk level, 518–519
Accident Evolution and Barrier Function (AEB) Model, 448
Accident model, 439
Accident Trajectory, 483, 485
Active errors, 442
American National Standards Institute (ANSI), 489

American Society of Safety Engineers, 489
An Engineer's View of Human Error, 137, 159
ANSI/ASSE Z590.3
  application, 291
  definitions, 291
  design safety reviews, 293
  hazard analysis and risk assessment process, 293–294
  hierarchy of controls, 294–295
  purpose statement, 290
  relations with suppliers, 292–293
  risk assessment techniques, 294
  roles and responsibilities, 292
  scope, 290
ANSI/ASSP/ISO 45001-2018
  Annex A, 46–47
  annexes, bibliographies, and guidance manuals, 44
  compatibility and harmonization, 19
  emergency preparedness and response, 38

Advanced Safety Management: Focusing on Z10.0, 45001, and Serious Injury Prevention, Third Edition. Fred A. Manuele.
© 2020 John Wiley & Sons, Inc. Published 2020 by John Wiley & Sons, Inc.

532 INDEX

ANSI/ASSP/ISO 45001-2018 *(contd.)*
  forewords and introductions for, 22–23
  goal of, 16
  hazard analysis and risk assessment, 170–171
  hierarchy of controls, 6, 33–34, 238–239
  history, development, and consensus, 18–19
  improvement in, 43
  internal audits, 40–41
  management leadership, 2
  management of change, 34–38, 308
  management review, 41–42
  normative references, 24
  operation in, 31–32
  organizational culture, 2
  organization context, 25–26
  outsourcing, 38
  overview, 1–2
  PDCA concept, 19–20
  performance evaluation, 38–42
  planning process, 27–29
  safety design reviews, 251–252
  scope, 24
  support in, 29–31
  terms and definitions in, 25
  themes, 21
  training, 46
  worker participation, 2
  and Z10.0, 16–18
ANSI/ASSP Z10.0-2019
  annexes, bibliographies, and guidance manuals, 44
  applicable life cycle phases, 36
  audits, 40–41
  compatibility and harmonization, 19
  contractors, 37
  corrective action, 41, 390–391
    audits, 390
    feedback and organizational learning, 393
    incident investigation, 389–390
    management review in 45001, 391–393
    monitoring, measurement and assessment, 388
  definitions, 25
  design review and management of change, 34–38, 45
  emergency preparedness, 38
  employee participation, 26–27
  evaluation and corrective action, 38–42
  forewords and introductions for, 22–23
  hazard analysis and risk assessment, 170–171
  hierarchy of controls, 6, 33–34, 238–239
  history, development, and consensus, 18–19
  implementation and operation, 31–32
  incident investigation, 39–40, 419–420
  and ISO 45001, 16–18
  leadership and worker participation, 26–27
  management leadership, 2, 26–27
  management of change system, 308–309
  management review, 42–43
  occupational health, 37, 46
  organizational culture, 2
  organization context, 25–26
  overview, 1–2
  PDCA concept, 19–20
  planning process, 27–29
  process verification, 36
  procurement requirements, 36
  references, 24
  revisions, 15–16
  safety design reviews, 252–253
  scope, purpose, and application, 24
  support in, 29–31
  themes, 21
  training requirements, 45
  worker participation, 2
Arnoldy, Frank, 77, 78
As low as reasonably practicable (ALARP), 25, 47, 88, 210, 292, 502, 516, 521–523
Audit requirements, 9
  auditor competency, 404–405
  evaluations, auditors, 404
  exit interview, 403–404
  guidelines, audit system, 406

INDEX

hazardous situations, 403
  safety culture, 402–403
  sector-specific hazards and risks, 405
  self-built audit system, 406
Audits, corrective action, 390

Bach, Jonathan A., 284–290
Barriers
  causation models, 444–446
  serious injury and fatality prevention, 145–146
  Swiss Cheese model, 484
Barriers and Accident Prevention, 145, 439, 442, 444, 467, 484
Basic Guide To System Safety, 507
Benner, Ludwig, 430
The Blame Machine: Why Human Error Causes Accidents, 52, 141, 163, 165, 424
Bow-Tie Risk Assessment Methodology, 448
Brander, Roy, 258
British Standards Institute (BSI), 16
Browning, R.L., 505, 507
Bureau of Labor Statistics (BLS), 134

Causation models
  barriers and controls, 444–446
  categories of
    epidemiological accident models, 442–443
    sequential accident models, 442
    systems accident models, 443
  Heinrichean premise, 439–442
  incident investigation, 437
  principles, 438
  resources on, 446–447
  safety professionals, 439
Center for Chemical Process Safety, 440
Change agents. *See also* Culture change agent
  characteristics of, 70–71
  definition, 3, 70, 79
Chapanis, Alphonse, 160
Clause 10, management review
  in 45001, 395–398
  continual improvement, 397–398
  incident, nonconformity, and corrective action, 396
  in Z10.0, 398–399

Clements, Pat L., 183, 497
Collaborative thinking, 110
Complexity theory, 106
Complex systems, 105, 106
Composite management system, MOC, 333–336
Continual improvement process, 397–398
Controlling hazards, mathematical evaluations, 222–236
Corrective action
  audits, 390
  feedback and organizational learning, 393
  incident investigation, 389–390
  management review in 45001, 391–393
  monitoring, measurement and assessment, 388
  in Z10.0, 390–391
Culture change
  basic guide for achieving, 81–82
  initiatives, failures of, 79–80
Culture change agents, 3, 75–79

Data collection system design, 110
Decision-making process, 257–258
Dekker, Sidney, 74, 95, 106, 112, 115, 116, 143, 164–166, 439, 441–442
The Deming Management Method, 93
Deming, W. Edwards, 93, 94, 137
Department of Defense Standard Practice for System Safety, 178
Design and purchasing guidelines
  circuit installation, 369
  electrical-design and construction, 364–366
  environmental impact/hazard evaluation, 361
  equipment and fixtures, 356
  ergonomics, 358–360
  industrial hygiene, 358
  machine and process control, 360–361
  maintenance considerations, 369–370
  mechanical-design and construction, 362–364
  procurement process, 354
  purpose, 356

Design and purchasing guidelines (*contd.*)
  remote interlocks, 372
  requirements, 357
  wire colors, 370–371
  wiring practices, 367–369
Designing For Safety (DFS), 284
"Drift" concept, 74–75
Drift into Failure, 74, 106

Emery, Fred, 115
Epidemiological accident models, 442–443
Ergonomics
  design and purchasing guidelines, 358–360
  human error avoidance, 158
  procurement process, 354
  safety design reviews, 260
Ericson, Clifton A. II., 497, 506
Evaluation and corrective action, 8–9. *See* also Corrective action

Failure Modes and Effects Analyses (FMEAs), 186–187, 190, 202
  detection ranking criteria, 206
  form, 221
  occurrence ranking criteria, 206
  ranking criteria, 205–208
  severity ranking criteria, 205
Failures, culture change initiatives, 79–80
Fault tree analysis (FTA), 187–188
Ferry, T., 93
The Field Guide to Understanding Human Error, 95, 112, 115, 143, 166, 439, 441–442
The Fifth Discipline: The Art and Science of the Learning Organization, 104
Five why problem-solving technique, 10, 448
  applications of, 457–462
  culture change necessary, 457
  observations, 462–463
  opposition, 456–457
  procedures for use, 455–456
  significant incidents, 454–455
  training, 456

Frequency, 208
Frequency of Exposures (FE), 211, 214

Galloway, Shawn M., 79
Grimaldi, John V., 506
Grose, Vernon L., 202
Guidelines for Management of Change for Process Safety, 315
Guidelines for Preventing Human Error in Process Safety, 95, 137, 162, 423, 430, 440, 481

Haddon, William, 145, 444–445, 484
Hammer, Willie, 137, 161
Handbook of System and Product Safety, 137, 161
Hazard analysis and risk assessment
  acceptable risk, 181
  additional resources, 189–190
  codes, 193
  definition, 170–171, 183
  documentation, 176
  exposure frequency and duration, 174
  failure modes, 173–174
  initial risk, 174
  management decision levels, 181–183
  matrix, 172–173
  occurrence probability, 174
  parameters, 173
  preliminary hazard analysis, 183–184
  probability and severity, 176–178
  reassess the risk, 176
  residual risk, 175
  risk, 171
  risk acceptance decision-making, 176
  risk assessment matrices, 178–181, 194
  risk coding system, 184–185
  risk prioritization, 175
  risk reduction and control methods, 175
  severity, 174
  unrealistic expectations, 189
  What–If Analysis, 185–189
Hazard and Operability (HAZOP) Analysis, 187
Hazard identification
  process, 31–32

and risk assessment techniques, 499–501
Hazards, 214, 292
  definition, 87, 197
  prevent, 200
Hazards analysis and risk assessment, 6
Heath, E.D., 93
Heinrichean causation model
  The Field Guide to Understanding Human Error, 441–442
  Guidelines for Preventing Human Error in Process Safety, 440
  Human Error, 440–441
  safety management, techniques of, 441
  safety practitioners, 439–440
  The Ten Basic Principles of Safety, 441
Heinrich, H.W., 142, 439
Heinrich's principles
  emphasis on frequency, 142–143
  88-10-2 ratios, 142
Hendrick, Kingsley, 430
Hierarchies of controls, 6, 292
  administrative controls, 242
  45001 and Z10.0, 238–239
  ANSI/ASSE Z590.3, 294–295
  application of, 243
  deciding and taking action, 249
  elimination, 239–240
  engineering controls, 241
  European Communities, Council, 244–245
  goals, 246
  hazardous methods, materials, 240–241
  measuring for effectiveness, 249
  MIL-STD-882E-2012, 245–246
  National Safety Council, 243–244
  personal protective equipment, 242–243
  practice of safety, 91–92
  problem identification and analysis, 248
  problem-solving technique, 246–248
  system safety, 501–503
  variations, 243
  warning, 241–242
Hollcroft, Bruce, 149

Hollnagel, Erik, 109, 145, 439, 442, 444, 467, 484
Human error avoidance, 5–6
  CCPS, 162–163
  Chapanis, Alphonse, 160
  concept of drift, 165–166
  defining, 159–160
  Dekker, Sidney, 164
  ergonomics, 158
  Hammer, Willie, 161
  Johnson, W., 163–164
  Norman, Donald A., 162
  organizational transition, 158–159
  Reason, James, 161–162
  safety professionals, 157–158
  Strater, Oliver, 164
  Whittingham, R.B., 163
Human Errors, 94–95, 143–145, 159, 161–162, 440–441, 482, 485, 486

Incident causation models, 10, 112, 426
  barriers and controls, 444–446
  categories of
    epidemiological accident models, 442–443
    sequential accident models, 442
    systems accident models, 443
  Heinrichean, 439–442
  incident investigation, 437
  principles, 438
  resources on, 446–447
  safety professionals, 439
Incident investigations, 10, 112
  and analysis, 152–153
  author's research, 420–421
  causal factors, 433–435
  corrective actions, 389–390, 433–435
  cultural implications, 423–425
  problem solving technique, 428–429
  research result, 421
  resources, 429–431
  safety practitioners, 437
  self-evaluation, culture, 426–427
  supervisors, 422, 425
  worker unsafe acts, 421
  in Z10.0, 419–420
Industrial Accident Prevention, 142, 439
Initial risk, 174

Innovations, serious injury and fatality
    prevention, 4–5
  barriers, 145–146
  change/pre-job planning, 152
  convincing management, OSHA, 141
  culture in place, 140
  employee injuries, 133–134
  fatality history, 134–135
  fatality prevention, 138–140
  Heinrich's principles, 142–143
  incident investigation and analysis,
    152–153
  organization's culture, 140–141
  prevention through design, 150–151
  risk assessments, 147–150
  safety engineers, 146–147
  safety professionals, 135–137
  serious injuries and fatalities, 137–138
  unsafe acts, 143–145
In *Risk Assessment: A Practical Guide
    for Assessing Operational Risks*,
    448

"J" formula worksheet, 234–236
Job hazard analysis, 197
  important of, 197
  value of, 198
John, Ciaran, 104
Johnson, W.G., 466
Johnson, William, 137, 163–164
Juran, Joseph M., 94

Kase, D.W., 100
Kelvin, Lord, 202
Kletz, Trevor, 137, 159
Kotter, John P., 78
Krause, Thomas, 125, 310

Latent errors, 443
Leadership and training, 93–94
Leadership in Energy & Environmental
    Design (LEED), 287
Leading and lagging indicators, 100
Leveson, Nancy, 105
Likelihood of Occurrence (LO), 211
*The Loss Rate Concept In Safety
    Engineering*, 505, 507
Lowrance, William W., 513, 518
Lyon, Bruce, 149

Macro thinking, 109
Main, Bruce W., 150, 190, 255, 256, 297
Management leadership, 72–73
Management leadership and employee
    participation
  hazard/risk problem solving, 53
  OHSMS, 62–64
  results/outcomes, 51
  safety and health professionals, 51
  safety director, 65–66
Management Leadership and Worker
    Participation, 397
Management of change (MOC) systems,
    7–8, 125–126
  in 45001, 308
  activities, 314
  alpha corporation, 324–325
  application, 311
  beta company, 325–328
  composite management system,
    333–336
  conglomerate, 321
  documentation, 317
  educational purpose, 321
  employee signatures, 320
  epsilon corporation, 338–341
  food company, 321
  gamma corporation, 329–332
  high risk multiproduct manufacturing
    operation, 320
  implementation, 315–316
  international multi-operational entity,
    322
  operation producing mechanical
    components, 318–319
  OSHA, 312, 333–336
  paper guidance, 320
  policy and procedures, 317, 320
  prescreening questionnaire, 320
  process, 312–313
  purpose of, 311
  request form, 314
  responsibility levels, 313–314
  risk assessments, 316
  significance, training, 316–317
  specialty construction contractor, 319
  studies, 310–311
  in Z10.0, 308–309
  zeta corporation, 343–348

Management Oversight and Risk Tree
(MORT), 11, 71
  barriers, 444
  barriers and controls, 467–468
  causation models, 446–447
  controls and barriers, 473–474
  educational resource, 466
  hazard analysis and risk assessment, 188
  incident investigation, 429–430
  losses and assumed risks, 469
  main branches of, 468–469
  management systems, deficiencies, 469–473
  revision of, 466–467
  Specific & Management Oversights & Omissions, 469
  unusual aspects, 465–466
  work and processes, 474–478
Management review, 9
  in 45001, 395–398
  continual improvement, 397–398
  corrective action in 45001, 391–393
  incident, nonconformity, and corrective action, 396
  in Z10.0, 398–399
Management systems, 88–89
Management system standard *vs.* specification standard, 21
Managing the Risks of Organizational Accidents, 124, 140, 161, 165, 431, 481, 483, 488
Meadows, Donella H., 104, 105, 107
Micro thinking, 110
Mogford, J., 56
Moriarty, Brian, 497, 507
MORT Safety Assurance Systems, 137, 466

National Institute for Occupational Safety and Health (NIOSH), 255, 489, 497
  education activities, 289
  encouraging research, 289
  policy activities, 290
  practice activities, 290
  prevention through design, 282–283
National Safety Council, 203–205, 243–244, 254–255

Naughtin, Pat, 78
Negative safety culture, 56–57, 74, 86
NFPA Life Safety Code 101, 34
Norman, Donald A., 162

Occupational Health and Safety Management System (OHSMS), 62–64, 387
Occupational Safety and Health (OSH), 353, 391
Occupational Safety and Health Administration (OSHA)
  audit system, 406
  composite management system, 333–336
  convincing management, 141
  Voluntary Protection Program (VPP), 312
OHSMS Issues, 27, 71, 87
"On the Escape of Tigers: An Ecologic Note," 445
On The Practice of Safety, 429, 489
Operational risk management, 91, 188
  incident investigation, 437
  socio-technical model for (*see* Socio-technical operational risk management systems)
Organizational Accidents Revisited, 441, 488
Organization's culture, 86
  significance of, 110, 117, 118
  and worker participation, 64–65 (*see also* Management leadership and employee participation)
Out of the Crisis, 93, 94
Overarching, 70

PDCA. *See* Plan-Do-Check-Act (PDCA)
Perception surveys, 58
Personal protective equipment (PPE), 303
Petersen, Dan, 138, 441
Plan-Do-Check-Act (PDCA), 19–20
  corrective action, 393
  management review, 399–400
Positive safety culture, 86
  characteristics, 55–56, 119
  definition, 53–54, 72
  management decision-making, 55

Povov, Georgi, 149
Practice of safety
   defining, 90–91
   effective leadership and training, 93–94
   hierarchy of controls, 91–92
   human errors, 94–95
   incident causation, 97–98
   performance measures, 99–100
   prevention through design, 95–96
   safety audits, 100
   setting priorities and utilizing resources effectively, 96–97
Pre-job planning, 125–126
Pre-Job Planning and Safety Analysis system, 318, 319
Preliminary hazard analysis (PHA), 183–184
Pre-Project & Pre-Task Safety and Health Planning, 319
Prevention through design, 7, 95–96, 121–123
   safety design reviews, 253
Prevention through design (PtD)
   ANSI/ASSE Z590.3, 290–295
   Bach, Jonathan A., 284–290
   characteristics, 297–298
   document history, 304
   elements, 302
   goals, 295–296
   implementation, 301–302
   NIOSH, 284–290
   organizational benefits, 301
   overview, 281–284
   patience, 297
   purpose & goal, 301
   risk assessment, 302–303
   safety and protection, 298–300
   safety professionals, 296
   scope, 301
   serious injury and fatality prevention, 150–151
   stakeholders, 304
   supplier agreements, 304
   supporting application of, 280–281
   sustainable control, 304
Probability, 88, 214, 292
Probability and severity, 176–178
Problem solving technique

hierarchies of control, 246–248
incident investigation, 428–429
Procurement process, 8
   45001 and Z10.0, 350
   culture change, 351
   design and purchasing guidelines, 354
   ergonomics, 354
   health and safety, 352–353
   pre-work necessary, 352
   safety-related specifications, 350–351
   significance of, 351–352
Program Evaluation Profile (PEP), 406
The Psychology of Everyday Things, 162

Rasmussen, Jens, 74, 165
Reason, James, 11–12, 124, 140, 143, 159, 161–162, 165, 422, 431, 440–441, 481–490
Reason's Swiss Cheese Model, 441, 448
Reductionism, 106
Residual risk, 88, 175, 292
Risk, 214, 292, 518
   acceptable, 181
   acceptance decision-making, 176
   coding system, 184–185
   defined, 88, 171
   initial, 174
   minimum, 520–521
   prioritization, 175
   reassess, 176
   reduction and control methods, 175
   score formula, 214–215
   two factors, 171
   zero risk level, 518–519
Risk Assessment:Challenges and Opportunities, 297
Risk assessment code (RAC), 204, 205
Risk assessment matrix, 172–173, 178–181
   acceptable risk, 523–525
   numerical gradings, 180
   system safety, 501
Risk assessments, 89–90, 120–121, 181
   process, 32
   transitions in, 202
   value and trending, 147–150
Risk Assessment Tools for Safety Professionals, 467, 484